Atlas on the Human Testis

Davor Ježek
Editor

Atlas on the Human Testis

Normal Morphology and Pathology

Springer

Editor
Davor Ježek, M.D., Ph.D.
Department of Histology and Embryology
University of Zagreb
Zagreb
Croatia

ISBN 978-1-4471-6965-9 ISBN 978-1-4471-2763-5 (eBook)
DOI 10.1007/978-1-4471-2763-5
Springer London Heidelberg New York Dordrecht

Softcover re-print of the Hardcover 1st edition 2013

Printed on acid-free paper

Springer is part of Springer Science+Business Media (www.springer.com)

To my wife, Vesna, and my sons, Bartol and Tadej, who have filled my life with love, joy, and happiness.
To my parents, who inspired me to study medicine.

Davor Ježek

Preface

In modern reproductive medicine, testicular biopsy is used for both diagnosing and treating the most difficult cases of male infertility, such as azoospermia. The correct interpretation of histologic findings related to the morphology of the testicular biopsy sample is crucial for therapeutic and in vitro fertilization procedures. The main method of treating obstructive and nonobstructive azoospermia is (micro) surgical and is aimed at successfully isolating spermatozoa from the biopsy specimen and microinjecting them into the oocyte. Therefore, a good knowledge of the normal development, morphology, and pathology of the human testis is the basis for managing infertility. Furthermore, early detection of carcinoma in situ and more advanced pathologic changes in the testicular parenchyma based on evaluation of the testicular biopsy sample leads to high cure rates of testicular neoplasms.

The aim of this atlas is to familiarize experts such as clinical embryologists, gynecologists, andrologists, urologists, pathologists, and other physicians/reproductive biology scientists with the normal morphology and pathology of the testis. Briefly, the atlas presents (a) methods of testicular tissue sampling; (b) various histologic procedures for tissue processing and freezing–thawing; (c) normal development of the early, fetal, and postnatal testis; (d) normal spermatogenesis and morphology of the interstitial tissue; (e) damage of spermatogenesis, cryptorchidism, Klinefelter's syndrome, testicular dysgenesis syndrome, and detection of carcinoma in situ; (f) testicular neoplasms such as seminomas and nonseminomas; (g) other related pathology; and (h) new approaches to diagnosing male infertility, such as DNA microarray analysis and advanced microscopic techniques.

This atlas might guide experts in the reproductive medicine field toward the correct interpretation of histologic findings and standardization of testicular biopsy analysis/procedures in their laboratories. The book presents images from paraffin and semithin sections of the human testis; such sections are used routinely to assess testicular tissue morphology.

Many people sacrificed their time and energy to contribute to this book. Therefore, I am deeply grateful to the authors and coauthors of all chapters. Without the valuable help of my young assistants, Andreja Vukasović and Viviana Kozina, and their schematic representations, it would be impossible for me to finalize some chapters and submit them to the publisher. I am indebted to the Springer crew, especially our developmental editor, Georgette Forgione, for the continuous support and encouragement. Finally, without the understanding of my wife and family during the many nights I spent amending pictures and writing text and legends, this book would not have become a reality.

Davor Ježek

Contents

Contributors

Kristian Almstrup, M.Sc., Ph.D. University Department of Growth and Reproduction, Rigshospitalet, Copenhagen, Denmark

Ljerka Banek, M.D., Ph.D. Department of Histology and Embryology, School of Medicine, University of Zagreb, Zagreb, Croatia

Tomislav Banek, M.D., Ph.D. Department of Nuclear Medicine, Dubrava Clinical Hospital, School of Medicine, University of Zagreb, Zagreb, Croatia

Martin Bergmann, Ph.D. Institute of Veterinary Anatomy, Histology, and Embryology, Justus Liebig University, Giessen, Germany

Floriana Bulić-Jakuš, M.D., Ph.D. Department of Biology, School of Medicine, University of Zagreb, Zagreb, Croatia

Ivan Damjanov, M.D., Ph.D. Department of Pathology, University of Kansas School of Medicine, Kansas City, KS, USA

Pilar González-Peramato, M.D., Ph.D. Department of Pathology, University Hospital La Paz, Madrid, Spain

Department of Pathology, University Autonoma of Madrid, Madrid, Spain

Đurđica Grbeša, M.D., Ph.D. Department of Histology and Embryology, School of Medicine, University of Zagreb, Zagreb, Croatia

Marta Himelreich Department of Histology and Embryology, School of Medicine, University of Zagreb, Zagreb, Croatia

Davor Ježek, M.D., Ph.D. Department of Histology and Embryology, School of Medicine, University of Zagreb, Zagreb, Croatia

Gordana Jurić-Lekić, M.D., Ph.D. Department of Histology and Embryology, School of Medicine, University of Zagreb, Zagreb, Croatia

Marina Kos, M.D., Ph.D. Clinical Department of Pathology "Ljudevit Jurak", Clinical Hospital Center "Sestre milosrdnice", Zagreb, Croatia

Viviana Kozina, MEd Department of Histology and Embryology, School of Medicine, University of Zagreb, Zagreb, Croatia

Ivan Krhen, M.D., Ph.D. Clinic for Urology, Clinical Hospital Centre "Zagreb", School of Medicine, University of Zagreb, Zagreb, Croatia

Marta Lekić, University of Zagreb, Zagreb, Croatia

Tanja Leniček, M.D. Clinical Department of Pathology "Ljudevit Jurak", Clinical Hospital Center "Sestre milosrdnice", Zagreb, Croatia

Iva Marolt, DMD Department of Histology and Embryology, School of Medicine, University of Zagreb, Zagreb, Croatia

Gregor Mikuz, M.D., Ph.D. Institute of Pathology, Medical University Innsbruck, Innsbruck, Austria

John E. Nielsen, M.Sc. University Department of Growth and Reproduction, Rigshospitalet, Copenhagen, Denmark

Manuel Nistal, M.D., Ph.D. Department of Pathology, University Hospital La Paz, Madrid, Spain

Department of Anatomy, Histology, and Neuroscience, University Autonoma of Madrid, Madrid, Spain

Ewa Rajpert-De Meyts, M.D., DMSc University Department of Growth and Reproduction, Rigshospitalet, Copenhagen, Denmark

Peter N. Schlegel, M.D. Department of Urology, Weill Cornell Medical College of Cornell University, New York, NY, USA

Wolfgang Schulze, M.D. Department of Andrology, University Hospital Hamburg-Eppendorf, Hamburg, Germany

Hans-Christian Schuppe, M.D. Andrology Unit, Department of Urology, Pediatric Urology, and Andrology, Justus Liebig University, Giessen, Germany

Niels E. Skakkebæk, M.D., DMSc University Department of Growth and Reproduction, Rigshospitalet, Copenhagen, Denmark

Andrej-Nikolai Spiess, Ph.D. Department of Andrology, University Hospital Hamburg-Eppendorf, Hamburg, Germany

Andreja Vukasović, M.D. Department of Histology and Embryology, School of Medicine, University of Zagreb, Zagreb, Croatia

Nikica Zaninović, MS, Ph.D. Department of Reproductive Medicine, Weill Cornell Medical College of Cornell University, New York, NY, USA

1 "Open" Biopsy of the Testis and Testicular Sperm Extraction

Ivan Krhen, Andreja Vukasović, and Davor Ježek

Modern reproductive medicine offers new possibilities for treating infertility in men, including, among other approaches, the freezing of gametes or parts of the male or female sex gonad. In fertility practice, clinical indications for cryopreservation include storage of spermatozoa, testicular and ovarian tissues, and early embryos. Apart from freezing spermatozoa, the introduction of intracytoplasmic sperm injection into the oocyte (ICSI) in the late 1980s also created a need to freeze the testicular parenchyma. In the ICSI procedure, an oocyte is fertilized by injecting a single spermatozoon. Men with a low spermatozoa count, low spermatozoa motility, or a lack of healthy spermatozoa must provide a sample with a few viable male gametes for ICSI. Several techniques for spermatozoa retrieval from the epididymis and testis have been described and may successfully be combined with ICSI, including microsurgical epididymal sperm aspiration (MESA), percutaneous epididymal sperm aspiration (PESA), testicular spermatozoa aspiration (TESA), and testicular spermatozoa extraction (TESE). TESE has been used to treat cases of obstructive and nonobstructive azoospermia. Thus, frozen–thawed pieces of testicular tissue are usually used for TESE in the case of severe male infertility, including in men with azoospermia and testicular tumors. If the tumor is not too large and the surrounding testicular parenchyma is relatively preserved, one may obtain multiple testicular biopsy samples from that area and use them for TESE/ICSI.

This chapter briefly describes "open" testicular biopsy as a method for obtaining tissue for cryopreservation, as well as the assessment of the histologic structure of the testicular parenchyma, including detection of carcinoma in situ. This chapter also illustrates the handling of testicular biopsy samples in cases of limited testicular tumors. Details regarding the histologic processing of the testicular biopsy specimen and freezing and thawing are presented in Chap. 3.

I. Krhen, M.D., Ph.D.
Clinic for Urology, Clinical Hospital Centre "Zagreb", School of Medicine, University of Zagreb, Zagreb, Croatia

A. Vukasović, M.D. • D. Ježek, M.D., Ph.D. (✉)
Department of Histology and Embryology, School of Medicine, University of Zagreb, Zagreb, Croatia
e-mail: davor@mef.hr, davorjezek@yahoo.com

D. Ježek (ed.), *Atlas on the Human Testis*,
DOI 10.1007/978-1-4471-2763-5_1, © Springer-Verlag London 2013

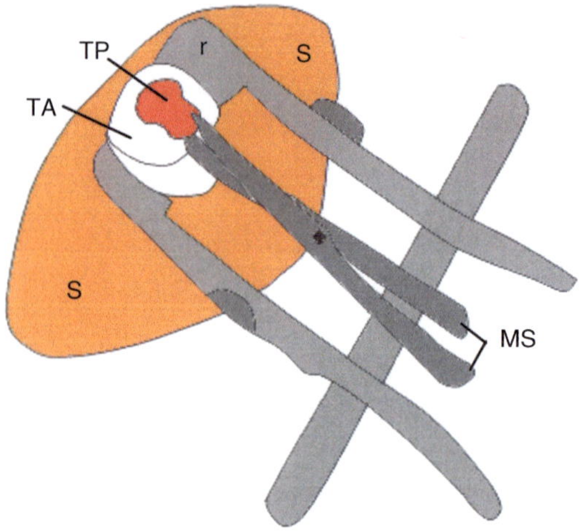

Fig. 1.1 "Open" biopsy of the testis. The operation (in the vast majority of cases) is performed under local or spinal anesthesia. After exposure of the tunica albuginea (*TA*), a small incision of approximately 3–5 mm is made. Following the incision, the testicular tissue/parenchyma (*TP*) spontaneously protrudes and may be removed with microscissors (*MS*). Because the testicular parenchyma is well vascularized, microscissors are used to minimize bleeding. Thus, several pieces of tissue may be removed using a single- or multiple-site approach (i.e., several incisions and biopsies at one point or different points of the parenchyma). Special attention should be paid to the position of the epididymis, which should be kept fixed and out of the incision plane. *s* skin of the scrotal sac, *r* retractor

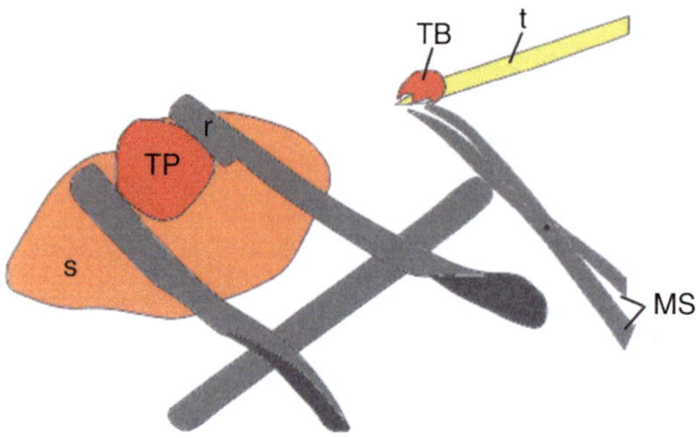

Fig. 1.2 After excision, the testicular biopsy sample (*TB*) is immediately immersed in a transport/freezing medium or fixative. To prevent any tissue squeezing or trauma, a sterilized toothpick (*t*) or needle may be used to transfer the tissue to the proper medium. Testicular parenchyma, in particular, adheres very well to the wooden surface of a toothpick. Thus, tweezers or similar devices are avoided. *TP* testis parenchyma, *s* skin of the scrotal sac, *r* retractor, *MS* microscissors

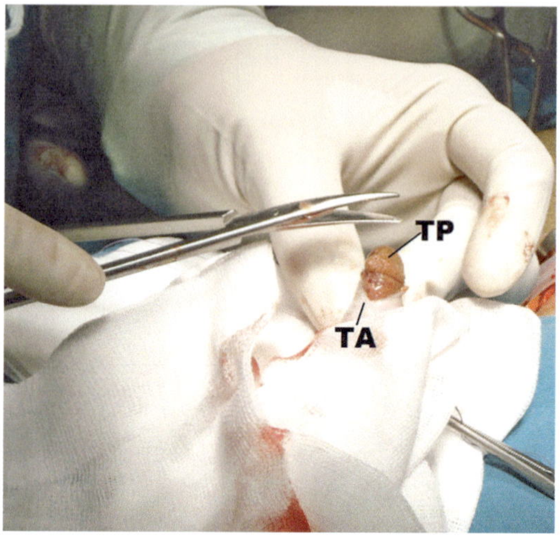

Fig. 1.3 Performing the testicular biopsy. After incision of the tunica albuginea (*TA*), the testis parenchyma (*TP*) is exposed and a small piece (the size of a rice grain, approximately 2–3 mm^3) is removed with scissors

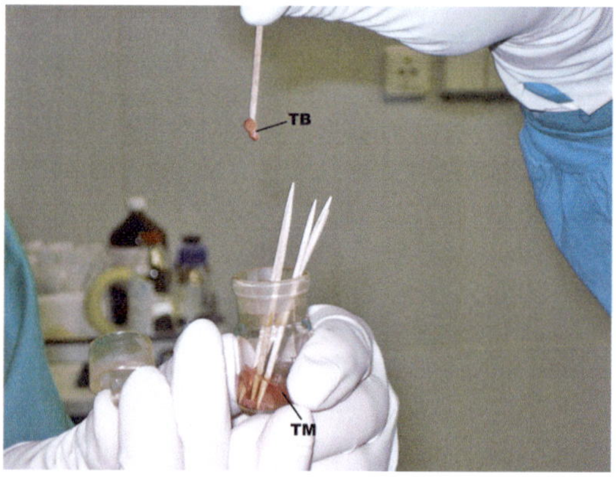

Fig. 1.4 Within the operational theater, the biopsy specimen (*TB*) is attached to a sterile toothpick or needle and it is immediately transferred to the transport medium (*TM*), freezing medium (for cryopreservation), or fixative (i.e., Bouin's fluid) for histologic processing

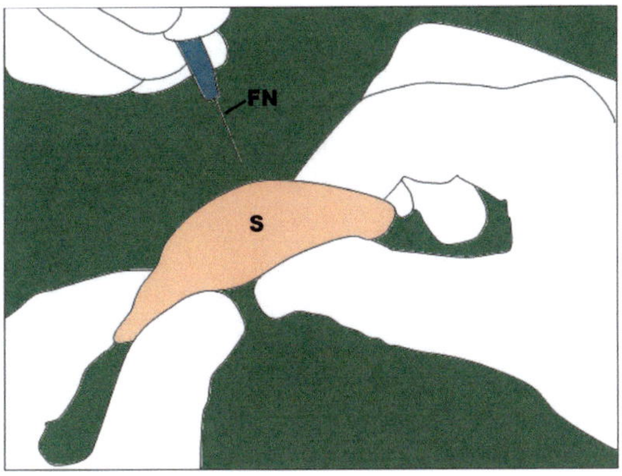

Fig. 1.5 Spermatozoa also may be retrieved by fine-needle (*FN*) aspiration, known as testicular sperm aspiration. The principle of TESA is shown here. An assistant holds the testis and the epididymis in a fixed manner, thus enabling the testis to be punctured without affecting the epididymis. Multiple aspirations (or mini-biopsies) may be done during "testicular mapping," the systematic assessment of various regions of the testis for the presence of spermatozoa. TESA is recommended in cases of obstructive azoospermia. *S* skin of the scrotal sac

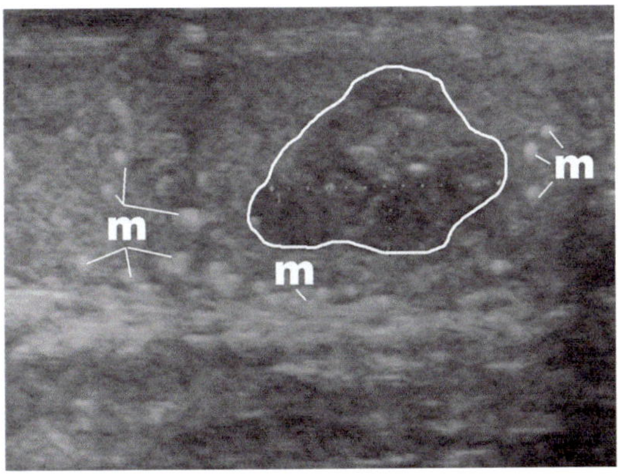

Fig. 1.6 Tumor of the right testis: an ultrasound finding. Both testes have pronounced microlithiasis (*m*), and a limited neoplasm (*encircled area*) may be seen near the lower pole of the testis. The contralateral testis was affected during childhood as the result of a viral infection (mumps). The part of the testicular parenchyma distal to the tumor (which seems unaffected) may be used for cryopreservation of spermatozoa, if present

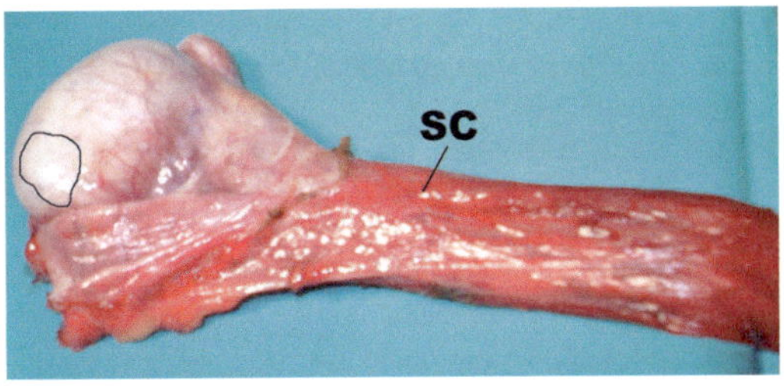

Fig. 1.7 Orchiectomy of the testis illustrated in the previous figure. The *encircled area* denotes the limits of the tumor, based on macroscopic inspection. *SC* spermatic cord

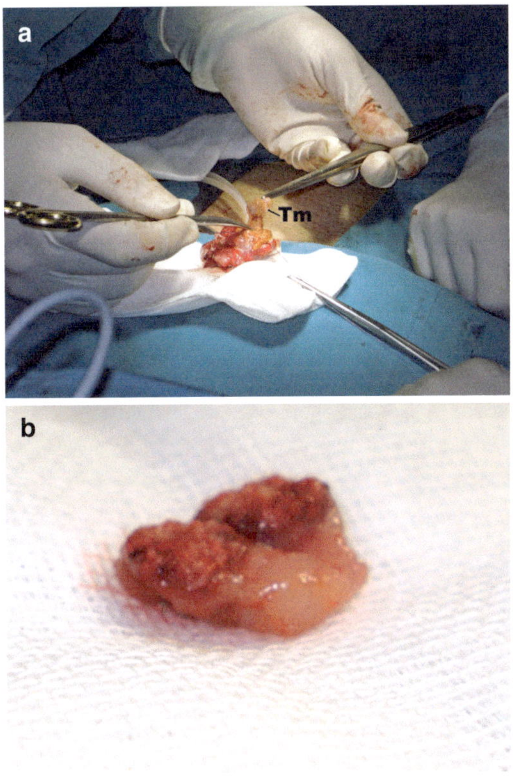

Fig. 1.8 Extirpation of the testicular neoplasm illustrated in Fig. 1.7. (**a**) The tumor tissue (*Tm*) occupies a space of approximately 15×10 mm and is surrounded by the rest of the testicular parenchyma. (**b**) The tumor is incompletely divided into halves and shows "rough" edges

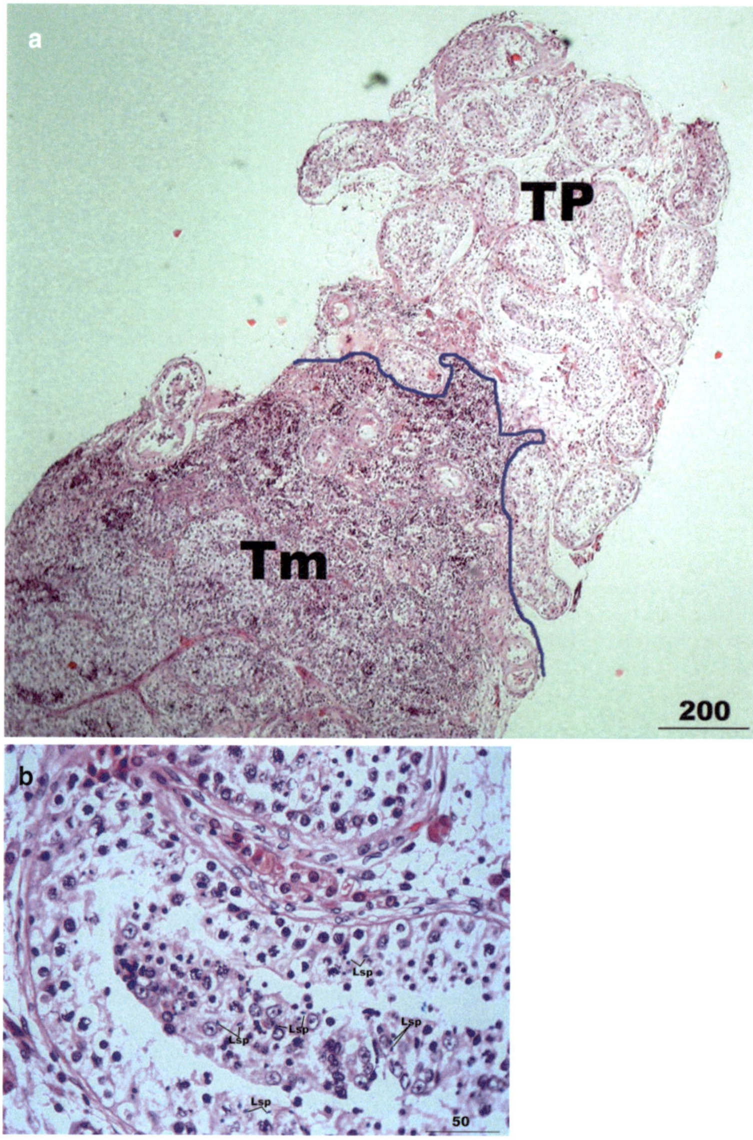

Fig. 1.9 Histologic structure of the testicular tumor and surrounding tissue illustrated in Fig. 1.8. (**a**) Tumor tissue (*Tm*; seminoma) is separated from a relatively preserved testicular parenchyma (*TP*) by a *blue line* (hematoxylin and eosin stain, ×40; *scale bar*, 200 μm). (**b**) Part of the testicular parenchyma next to the tumor. In some areas of seminiferous epithelium, late spermatids (*Lsp*) can be recorded. These cells may be used for TESE (hematoxylin and eosin stain, ×400; *scale bar*, 50 μm)

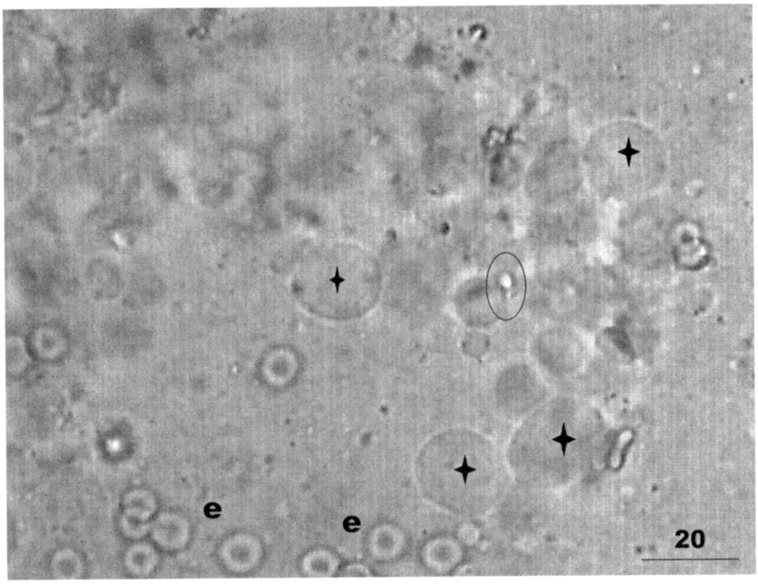

Fig. 1.10 Testicular biopsy sample from a patient with nonobstructive azoospermia after freezing–thawing and enzymatic digestion. The specimen contains rare spermatozoa (*circled area*), some spermatogenic cells (*crosslets*), and erythrocytes (*e*) (wet preparation, ×600)

Recommended Literature

Banz-Jansen C, Nikorowitsch C, Drechsler T, Felberbaum R, Schmutzler A, Diedrich K, Finas D. Incidence of testicular malignancies and correlation to risk factors in a TESE population of subfertile men. Arch Gynecol Obstet. 2012;285(1):247–53.

Dancet EA, Spiessens C, Blocquiaux L, Sermeus W, Vanderschueren D, D'Hooghe TM. Testicular biopsy before ART: the patients' perspective on the quality of care. Hum Reprod. 2010;25(12):3072–82.

Ishikawa T. Surgical recovery of sperm in non-obstructive azoospermia. Asian J Androl. 2012;14(1):109–15.

Ježek D, Knuth UA, Schulze W. Successful testicular spermatozoa extraction (TESE) in spite of high serum FSH and azoospermia: correlation between testicular morphology, TESE results, semen analysis and serum hormone values in 103 infertile men. Hum Reprod. 1998;13(5):1230–4.

Ježek D, Knezević N, Kalanj-Bognar S, Vukelić Z, Krhen I. From testicular biopsy to human embryo. Verh Dtsch Ges Pathol. 2004;88:136–43.

McLachlan RI, Rajpert-De Meyts E, Hoei-Hansen CE, de Kretser DM, Skakkebaek NE. Histological evaluation of the human testis – approaches to optimizing the clinical value of the assessment: mini review. Hum Reprod. 2007;22(1):2–16.

Nowroozi MR, Ahmadi H, Ayati M, Jamshidian H, Sirous A. Testicular fine-needle aspiration versus testicular open biopsy: comparable sperm retrieval rate in selected patients. Indian J Urol. 2012;28(1):37–42.

Salihu HM, Aliyu MH. Sperm retrieval in infertile males: comparison between testicular sperm extraction and testicular sperm aspiration techniques. Wien Klin Wochenschr. 2003;115(11):370–9.

Schulze W, Knuth UA, Ježek D, Benson DM, Fischer R, Naether OGJ, et al. Intratesticular sperm extraction: basis for successful treatment of infertility in men with ejaculatory azoospermia. Adv Exp Med Biol. 1997;424:81–8.

Schulze W, Thoms F, Knuth UA. Testicular sperm extraction: comprehensive analysis with simultaneously performed histology in 1418 biopsies from 766 subfertile men. Hum Reprod. 1999;14 Suppl 1:82–96.

Shah R. Surgical sperm retrieval: techniques and their indications. Indian J Urol. 2011;27(1):102–9.

Tüttelmann F, Werny F, Cooper TG, Kliesch S, Simoni M, Nieschlag E. Clinical experience with azoospermia: aetiology and chances for spermatozoa detection upon biopsy. Int J Androl. 2011;34(4):291–8.

2 MicroTESE and Embryo Development

Nikica Zaninović and Peter N. Schlegel

Assisted reproductive technologies have dramatically advanced the opportunities for treating the most challenging cases of male infertility. In no case has this opportunity been clearer than for men with the most severe defect in spermatogenesis: nonobstructive azoospermia (NOA). In men with NOA, sperm production is so defective that no sperm reach the ejaculate. The morphology of these sperm routinely is abnormal, and our experience is that many in vitro fertilization (IVF) laboratories consider sperm obtained from these men to be unusable for assisted reproductive techniques. Sperm retrieval requires invasive techniques to identify limited foci of sperm production within the testis. Clearly, the appearance of sperm from these men is abnormal, but few data have been published on the effect of using these sperm on IVF outcomes, including embryo quality. In this chapter, we describe the outcomes of intracytoplasmic sperm injection (ICSI) using testicular sperm from men with NOA, comparing the results with those obtained with testicular sperm from men with obstructive azoospermia (controlling for testicular source) and from men with sperm in the ejaculate using donor oocytes (controlling for oocytic factors). We also present data on the use of fresh and frozen testicular sperm, as well as sperm retrieved on the day of versus the day before oocyte retrieval. The results show that use of testicular sperm from men with NOA has a substantial effect on fertilization rates (72.6 % vs. 53.3 %; $P<0.05$, compared with obstructed testicular sperm) but with similar embryo development (66.2 % day 3 "good" embryos [grade 1–2] vs. 53.8 % with obstructed testicular sperm). Clinical pregnancy rates were statistically better (52.8 % vs. 41.4 %) using testicular sperm from men with obstruction versus those with NOA.

N. Zaninović, MS, Ph.D. (✉) • P.N. Schlegel, M.D.
Department of Reproductive Medicine and Department of Urology, Cornell Reproductive Medicine Institute, Weill Cornell Medical College
New York-Presbyterian Hospital,
530E 70st, M901, New York, NY 10021, USA
e-mail: nizanin@med.cornell.edu

Sperm Retrieval

Testicular sperm may be retrieved with a variety of techniques, including fine-needle aspiration, biopsy, or multiple biopsies (testicular sperm extraction [TESE]). We have developed a refined technique of microdissection TESE (microTESE) that allows a detailed search of these dysfunctional testes to identify the very rare sites of sperm production that are present within them. This approach limits the amount of tissue that must be removed from the testes and appears to increase the sperm retrieval rate by 50 % over that obtained with multiple testis biopsies, while limiting the effects on the testicle. Because less tissue is delivered to the embryology laboratory with a higher yield of sperm, the work required for tissue processing and sperm identification may be limited. The microsurgical

D. Ježek (ed.), *Atlas on the Human Testis*,
DOI 10.1007/978-1-4471-2763-5_2, © Springer-Verlag London 2013

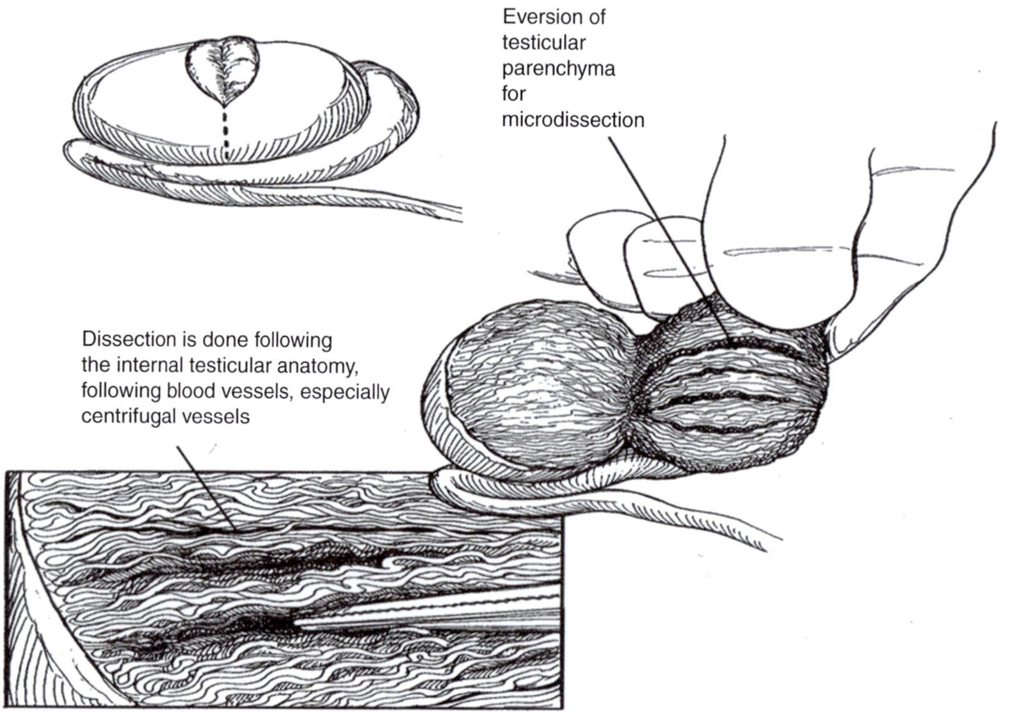

Fig. 2.1 Testicular tissue exposure for microTESE (*Courtesy of* © the Brady Urology Foundation, 2012)

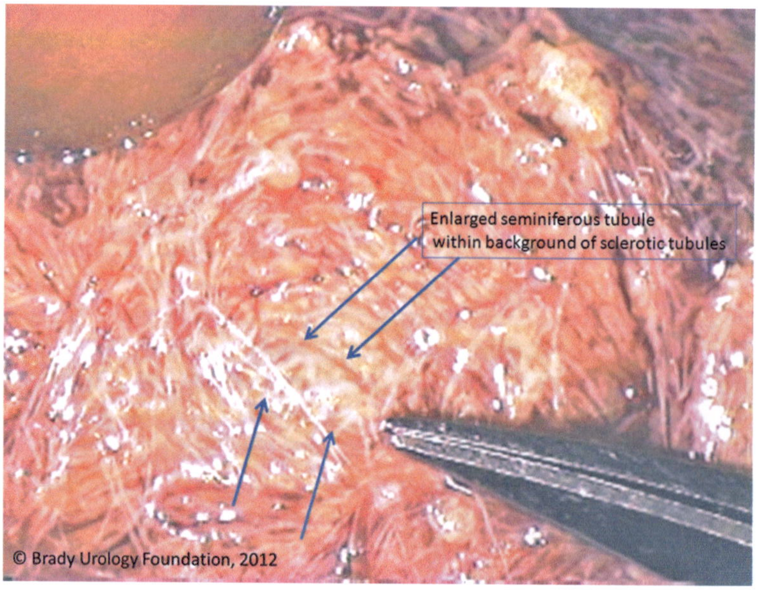

Fig. 2.2 Intraoperative demonstration of enlarged seminiferous tubules in a patient with NOA during a microTESE procedure. In this case, as well as in Fig. 2.3, the entire extent of the enlarged tubule is removed, if possible without vascular compromise (*Courtesy of* © the Brady Urology Foundation, 2012)

search for sites of sperm production involves the wide opening and exposure of testicular tissue, following the natural vascular and tissue planes of the testicle (Fig. 2.1). Individual seminiferous tubules most likely to contain sperm can be identified with an operating microscope at ×10 to ×15 magnification, allowing selection of the widest and most opaque tubules identifiable (Figs. 2.2 and 2.3).

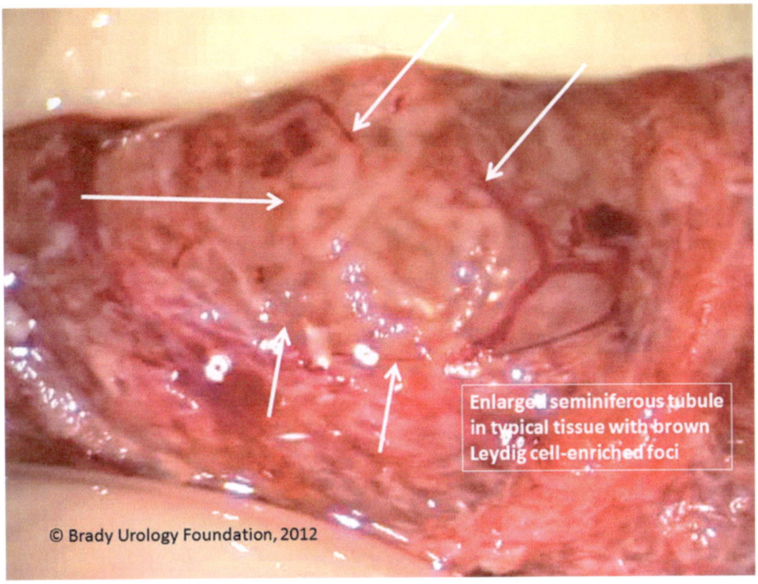

Fig. 2.3 Intraoperative demonstration of enlarged seminiferous tubules in a patient with NOA during a microTESE procedure. In this case, as well as in Fig. 2.2, the entire extent of the enlarged tubule is removed, if possible without vascular compromise (*Courtesy of* © the Brady Urology Foundation, 2012)

We have attempted sperm retrieval on the day of, as well as the day before, oocyte retrieval. In both cases, a final semen specimen is examined using an extended sperm preparation technique, allowing us to identify sperm and avoid surgery in 5–10 % of men previously diagnosed as having NOA. Sperm retrieval on the day before oocyte retrieval allows greater flexibility to process tissue and identify sperm during the nearly 24 h before oocyte retrieval is done. The effect of such an approach on IVF outcomes is described.

With microTESE, we have been able to retrieve sperm from 57 % of men with NOA at Weill Cornell Medical Center, despite most patients having had prior failed attempts at treatment elsewhere (and despite the exclusion of 5–10 % of men thought to have NOA but actually having cryptozoospermia).

Obstructive Versus Nonobstructive Azoospermia

Testicular sperm are obtained primarily from male patients with either obstructive azoospermia (OBST; congenital or acquired obstructions) or NOA azoospermia. The comparison of ICSI efficiency between OBST and NOA has limitations, as sperm from various surgical sources (epididymal vs. testicular) are compared. However, when we compared these two groups at Cornell (Table 2.1), we observed an overall lower fertilization rate in the NOA versus the OBST group. This is expected, as epididymal and TESE sperm samples from men with obstruction generally have more sperm and greater motility, which may reflect greater fertilizing ability. Other centers also have shown this trend, with fertilization rates of 70 % for OBST and 50 % for NOA. Interestingly, if motile epididymal or testicular sperm are injected, no differences in fertilization or pregnancy rates are observed in patients with obstructive azoospermia. Furthermore, embryos created from TESE sperm have apparently higher developmental potential than epididymal sperm in obstructed cases. At Cornell, higher fertilization rates in OBST also are reflected in higher pregnancy and implantation rates and higherpercentages of frozen embryos (blastocyst stage), with no differences in miscarriage rates (Table 2.1). Similar results were obtained elsewhere in a meta-analysis that revealed no difference in implantation and miscarriage rates. Regarding embryo quality, Vernaeve et al. (2003) observed no differences in the average percentage of good-quality embryos between OBST and NOA (58 % vs. 52 %) or among types of NOA. Weill Cornell data for 2010 also show no differences in

Table 2.1 Obstructive versus nonobstructive azoospermia (1995–2010)

Parameter	OBST (TESE, frozen TESE, EPI, frozen EPI)	NOA (TESE, frozen TESE)	*P* value (Fisher or X^2)
Cases, *n*	638	801	–
Avg. age, years	34.6	32.6	–
Oocytes, *n* (avg.)	7,743 (12.1)	9,984 (12.5)	–
Mature/injected, *n* (%)	5,995 (77)	7,824 (78)	–
Fertilizations, *n* (%)	4,353 (72.6)	4,174 (53.3)	<0.05
Day 3 embryos transferred, *n* (avg. grade)	559 (2.4)	686 (2.0)	–
Day 5 embryos transferred, *n*	39	27	–
Embryos transferred, *n* (avg.)	1,796 (3)	1,805 (2.5)	–
Frozen embryos, *n*	804	404	–
Freezing rate, %	18.5	9.7	<0.05
Clinical pregnancies, *n* (%)	316 (52.8)	295 (41.4)	<0.05
Babies, *n* (% implantations)	451 (25)	390 (21.6)	<0.05
Losses and biochemical pregnancies, *n* (%)	98 (16.4)	113 (15.8)	NS
Good embryos (grade 1–2) on day 3, *n* (%)	13 (53.8)	68 (66.2)	–
Cycles, *n*	6	33	–
Embryos grade ≥ 2BB on day 5 (2010 only), *n* (%)	73 (78.1)	53 (71.7)	–
Cycles, *n*	39	27	–

Avg. average, *EPI* epididymal sperm, *NS* not significant

the percentage of good-quality embryos transferred on day 3 versus day 5 (Table 2.1).

Fresh Versus Frozen TESE

In cases in which spermatozoa are found with TESE, surplus sperm and testicular tissue may be cryopreserved for future use and possibly represent the only option for preserving fertility in severe cases of NOA. In comparing cases using fresh TESE (OBST and NOA) with those using frozen TESE, we found no difference in fertilization rates (Table 2.2); however, pregnancy and implantation rates, as well as the percentage of frozen embryos, were higher in the fresh versus frozen sperm patients. These data reflect sperm obtained from both OBST and NOA cases, as well as all female age groups. It is not immediately clear why frozen sperm have poorer outcomes, as even the percentage of good-quality embryos did not differ between groups. Nevertheless, it is possible that sperm DNA damage due to freezing and thawing might influence postimplantation embryo development. Of note, many, but not all, patients had a first attempt at ICSI using freshly retrieved testicular spermatozoa. In contrast, our observations show that miscarriage rates did not increase in cases using frozen TESE samples (Table 2.2). These results are consistent with other reports comparing first-attempt IVF–ICSI cases. The same occurrences were observed in comparing fresh versus frozen TESE OBST cases, with significantly higher fertilization and pregnancy rates using motile versus nonmotile spermatozoa.

Day-of Versus Day-Before TESE

The timing of the microTESE sperm procedure is important if freshly retrieved spermatozoa are to be used, to synchronize with oocyte retrieval and optimal patient (female and male) treatment. If microTESE is done the same day as oocyte retrieval (day of), pressure is generated on the patient, surgeon, and laboratory staff to find enough sperm for ICSI. Time constraints in completing the male surgery (which may take between 2 and 4 h or longer) may influence the outcome, as the best chance of sperm retrieval is within the first 2–4 h of operative time. In addition, at Weill Cornell, in analyzing

Table 2.2 Fresh versus frozen testicular sperm extraction (1995–2010)

Parameter	Fresh TESE	Frozen TESE	*P* value (Fisher or X^2)
Cases, *n* (avg.)	643 (12.9)	202 (10.6)	–
Avg. age, years	32.0	35.3	–
Oocytes, *n*	8,315	2,149	–
Mature/injected, *n* (%)	6,526 (78)	1,679 (78)	–
Fertilizations, *n* (%)	3,557 (54.5)	876 (52.1)	NS (0.08)
Day 3 embryos transferred, *n* (avg. grade)	557 (2.1)	165 (1.9)	–
Day 5 embryos transferred, *n*	23	7	–
Embryos transferred, *n* (avg.)	1,459 (2.6)	455 (2.7)	–
Frozen embryos, *n*	379	63	–
Freezing rate, %	10.7	7.2	<0.05
Clinical pregnancies, *n* (%)	256 (44.1)	57 (33.1)	<0.05
Babies, *n* (% implantations)	337 (23)	77 (17)	<0.05
Losses and biochemical pregnancies, *n* (%)	91 (16)	27 (15.7)	NS
Good embryos (grade 1–2) at day 3 transfer, *n* (%)	49 (62.2)	22 (68.2)	–
Cases, *n*	6	11	–
Embryos grade ≥ 2BB on day 5 (2010 only), *n* (%)	45 (71.1)	17 (88.2)	–
Cases, *n*	23	9	–

sperm retrieval success and the time spent finding the sperm postoperatively (extensive laboratory search), only 7 % of the time was sperm found after extensive searches in the laboratory when spermatozoa were not identified in the operating room (OR). Experienced embryologists are essential to identify the rare sperm identified after tissue digestion. Another crucial factor in TESE–ICSI outcome is the optimal window of sperm injection. Oocytes must be injected within 4–6 h after retrieval for optimal outcome; difficult cases of sperm retrieval with microTESE and subsequent tissue processing on the day of oocyte retrieval might cause that window to be missed.

Sperm retrieval on the day before oocyte retrieval appears to be the more optimal approach (and is our preferred strategy for microTESE cases). This approach gives us the flexibility to examine TESE samples adequately. We then have time to cancel oocyte retrieval in cases of no sperm, to use donor sperm as a backup, or to obtain the patient's consent for oocyte cryopreservation. To evaluate the efficiency of performing TESE on the day of or the day before oocyte retrieval, we compared the combined data for the last 7 years in OBST and NOA patients (Table 2.3). The fertilization rate is decreased slightly in the day-before TESE group (51.7 % vs. 55.2 %); on the contrary, pregnancy and implantation rates were no different, and no major differences in embryo quality were found (Table 2.3).

There are ongoing questions whether sperm retrieved from NOA patients with microTESE have a lower fertilization capacity than ejaculated sperm and what role of the male gamete is in embryo development and implantation. To shed light on these questions, we compared data from shared oocyte donors in cases in which one recipient used ejaculated sperm and the other used TESE sperm using the same oocyte cohort, removing the female factor (Table 2.4). Although the sample size was small, these preliminary results show fertilization rates slightly lower using TESE sperm (63 % vs. 79 %). Pregnancy and implantation rates did not differ, resulting in overall good embryo quality on days 3 and 5.

To our knowledge, there are no studies adequately addressing embryo quality using testicular sperm from men with NOA. These studies would be very difficult to perform because of the differences among patient populations and male and female factors. In general, viable sperm from

Table 2.3 Day-of versus day-before testicular sperm extraction (2003–2010)

Parameter	Day-of TESE	Day-before TESE	*P* value (Fisher or X^2)
Cases, *n*	149	265	–
Avg. age, years	32.2	30.9	–
Oocytes, *n* (avg.)	1,888 (12.7)	3,497 (13.2)	–
Mature/injected, *n* (%)	1,446 (77)	2,791 (79.8)	–
Fertilizations, *n* (%)	799 (55.2)	1,444 (51.7)	<0.05
Day 3 embryos transferred, *n* (avg. grade)	126 (1.9)	219 (2.3)	–
Day 5 embryos transferred, *n*	4	18	–
Embryos transferred, *n* (avg.)	305 (2.3)	515 (2.2)	–
Frozen embryos, *n*	53	186	–
Freezing rate, %	6.7	12.9	<0.05
Clinical pregnancies, *n* (%)	56 (43)	106 (45)	NS
Babies, *n* (% implantations)	72 (23.6)	138 (26.7)	NS
Losses and biochemical pregnancies, *n* (%)	16 (12.3)	38 (16)	NS (0.07)
Good embryos (grade 1–2) on day 3, *n* (%)	17 (76.5)	29 (58.6)	–
Cases, *n*	8	14	–
Embryos grade ≥ 2BB on day 5 (2010 only), *n* (%)	8 (50)	36 (15)	–
Cases, *n*	4	18	–

Table 2.4 Ejaculated versus testicular-extracted sperm in split donor oocyte cycles[a]

Parameter	Ejaculated sperm	TESE/frozen TESE	*P* value (Fisher or X^2)
Cases, *n*	6	6	–
Avg. age, years			
Recipient	41.0	44.3	–
Donor	27.8	27.8	–
Oocytes, *n*	64	62	–
Mature/injected, *n* (%)	53 (83)	54 (87)	–
Fertilizations, *n* (%)	42 (79.2)	34 (63)	NS (0.08)
Day 3 embryos transferred, *n*	3	4	–
Day 5 embryos transferred, *n*	3	2	–
Embryos transferred, *n* (avg.)	14 (2.3)	13 (2.2)	–
Frozen embryos, *n*	11	11	
Freezing rate, %	26.2	32.4	NS
Clinical pregnancies, *n* (%)	4 (67)	4 (67)	NS
Babies, *n* (% implantations)	7 (50)	5 (38.5)	NS
Losses and biochemical pregnancies, *n*	2	1	–
Good embryos (grade 1–2) on day 3, *n* (%)	3 (100)	4 (100)	–
Embryos grade ≥ 2BB on day 5, *n* (%)	3 (100)	2 (100)	–

[a]Six donors split among 12 recipients

TESE (OBST and NOA) have a similar, if not the same, fertilization and embryo developmental capacity as ejaculated sperm. More large case studies are needed with selection based on male history, infertility, and sperm characteristics, with careful analysis of embryo development.

Preparation of Testicular Tissue (TESE)

At Weill Cornell, microTESE generally is limited to NOA cases in which low sperm numbers are anticipated. The main goal is to retrieve an adequate amount of sperm, preferably motile, that may be used for ICSI. In the OR, the urologist performing the TESE procedure rinses blood from the tissue piece before mincing and mechanically processing the testicular sample suspension sequentially through a 24-gauge angiocatheter to confirm adequate disruption of the testicular tissue. The tissue then is passed to the IVF andrologist for evaluation. A detailed description of the TESE technique performed by Peter N. Schlegel at Weill Cornell may be found in the article by Ostad et al. (1998). After the microtubules are minced and broken up, the sample is placed into either a 1.5-mL microcentrifuge tube or a 5-mL round-bottom culture tube, depending on the volume of each tissue sample. The IVF andrologist in the OR places a small aliquot (5–7 μL) of the liquid portion of the sample on a disposable counting chamber slide and examines it under phase-contrast microscopy at ×200 magnification. The surgeon is given information on whether sperm are present, the morphologic characteristics of the sperm identified, and the presence or absence of germ cells observed. Additional samples are taken using microdissection as necessary until adequate sperm are observed, or until all areas of tissue that can be safely accessed are examined without positive results.

Sperm Observed in the OR

After the number of matured oocytes for injection has been determined, a decision may be made regarding how much of the sample should be processed to retrieve adequate sperm for ICSI. Depending on the sperm concentration and the total volume of the sample, samples may be either combined or processed separately. The tissue suspension from the sample tube(s) is mixed gently, and the liquid portion from the sample tube(s) is removed carefully. The tissue particles may be combined with the remaining tissue for cryopreservation. The liquid portion is washed with sperm wash medium and centrifuged for 5 min at 1,800 × *g*. The supernatants are removed and pellets resuspended in a small amount of medium. The processed sample is ready for ICSI. When very rare sperm are seen in perhaps one tube, the use of enzymatic digestion can maximize the number of sperm recovered.

No Sperm Observed in the OR: Enzymatic Digestion

Enzymatic Digestion of Testicular Tissue (Collagenase/DNAse Treatment)

Pretreating with an erythrocyte-lysing buffer for 5 min at 37 °C may be beneficial before proceeding with digestion. Prepared collagenase/DNAse is stored frozen in small cryovials, each containing 1 mL of prepared enzyme solution. The enzyme solution is thawed at 30 °C and added at a 1:1 (v:v) ratio to the TESE tissue suspension. The tissue is incubated at 37 °C for 1 h (agitating the tubes every 10–15 min to facilitate digestion). After incubation is complete, centrifugation is performed at 50 × *g* for 5 min to pellet the undigested tissue and separate the supernatant. Enzymes are removed by adding an equal amount of sperm wash medium. To remove the supernatant, centrifugation is carried out at 1,800 × *g* for 5 min (×2). Resuspended pellets are mixed thoroughly, and a small aliquot of the sample is examined for the presence of sperm. Microdroplets are made for sperm searching and ICSI by mixing the final digested sample. The total volume of the final suspension is critical to the success of the search throughout the microdroplets. If the suspension is too concentrated, the samples will be too dense and difficult to search. We prefer to see a monolayer of cells/debris with a little space between each of the components. For frozen TESE samples, the same tissue digestion may be used successfully with cryoprotectants completely removed before enzymatic digestion.

Pentoxifylline Treatment

Pentoxifylline is a phosphodiesterase inhibitor shown to increase motility and enhance the acrosome reaction (Terriou et al. 2000). Prolonged exposure of sperm to pentoxifylline may be damaging and must be avoided. One milliliter of the sample is diluted in sperm wash medium, 20 μL of 180-mM pentoxifylline solution (final concentration = 3.53 mM) is added, and the sample is incubated for 20 min at 37 °C. The sample is washed twice by centrifugation/resuspension (1,800 × *g* for 5 min). Alternatively, pentoxifylline may be added directly to the microdroplet in cases of extremely low sperm concentrations (TESE). Two microliters of the originally prepared pentoxifylline solution (180 mM) is added to 1 mL of sperm wash medium. One microliter of this diluted pentoxifylline solution is then added to a 5-μL microdroplet containing the tissue/sperm suspension. Motility should be enhanced after approximately 20 min of treatment. Sperm selected from the microdroplet are washed in 7 % polyvinylpyrrolidone before injection (ICSI).

TESE Performed the Day Before Oocyte Harvest

When TESE is performed the day before oocyte retrieval and no sperm are found, one has the option of cancelling oocyte retrieval, freezing oocytes, or proceeding immediately with donor backup once oocytes are collected. Once it is determined that at least a few sperm are present, the digested sample may be resuspended in sperm wash medium, sealed tightly, and placed in the dark at room temperature (20–25 °C) until the next day.

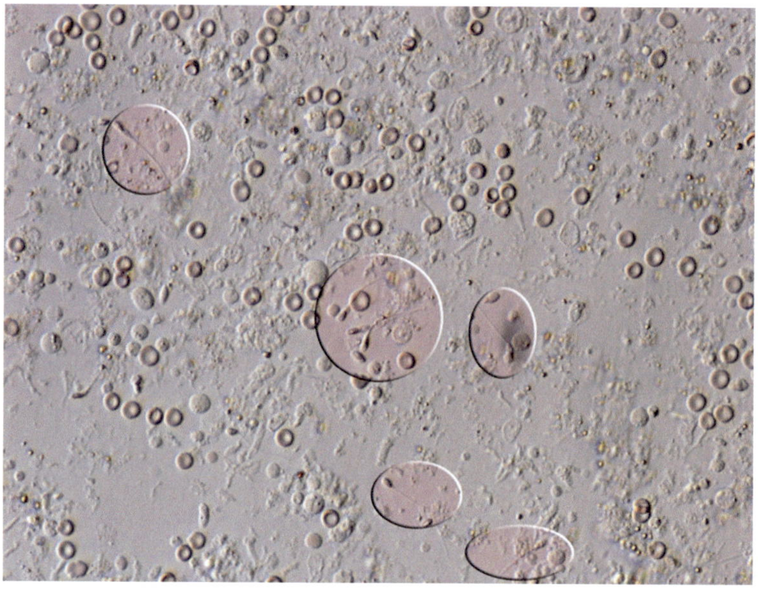

Fig. 2.4 Testicular biopsy sample after tissue digestion. Various spermatozoa forms are seen, featuring mature and immature (cytoplasmic neck droplet) sperm cells (*circled areas*). Some spermatozoa have a neck defect and others a small acrosome (×400)

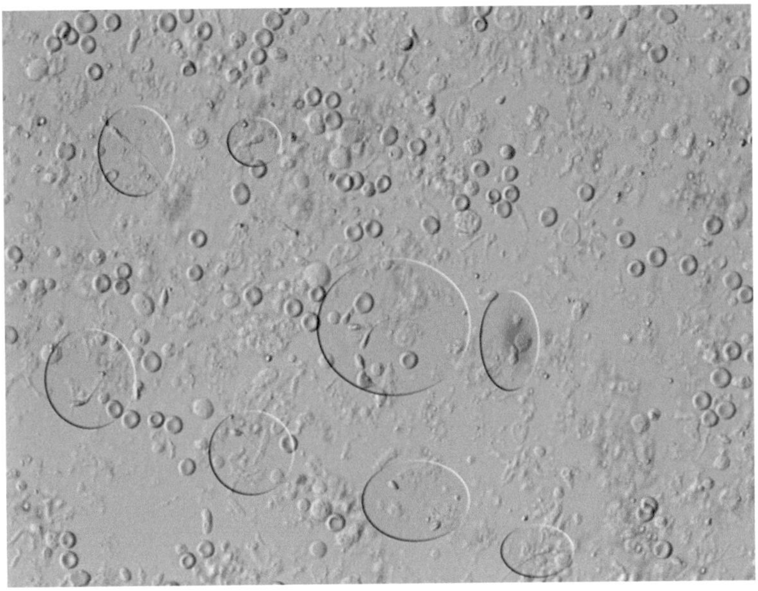

Fig. 2.5 Another TESE sample. Most of the spermatozoa have a neck defect and cytoplasmic droplet (×400)

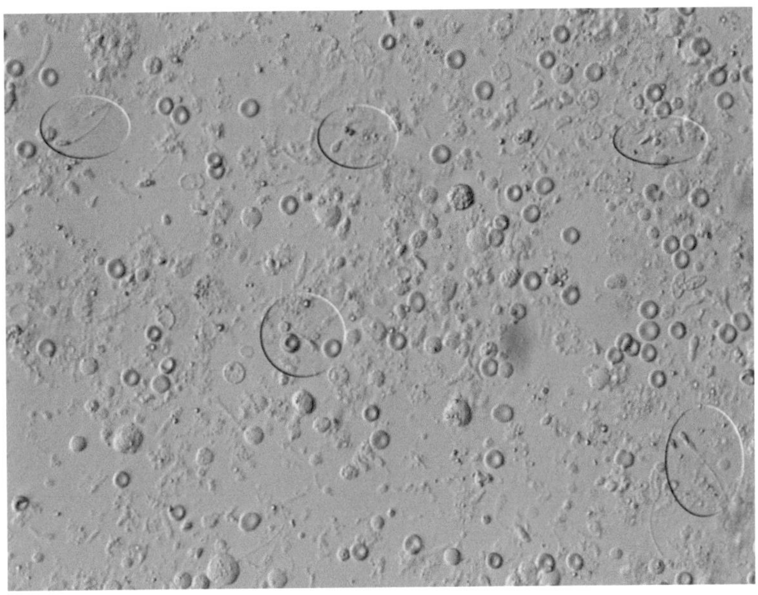

Fig. 2.6 TESE sample after digestion. More mature forms of spermatozoa are seen. With the addition of pentoxifylline, some of them show motility (×400)

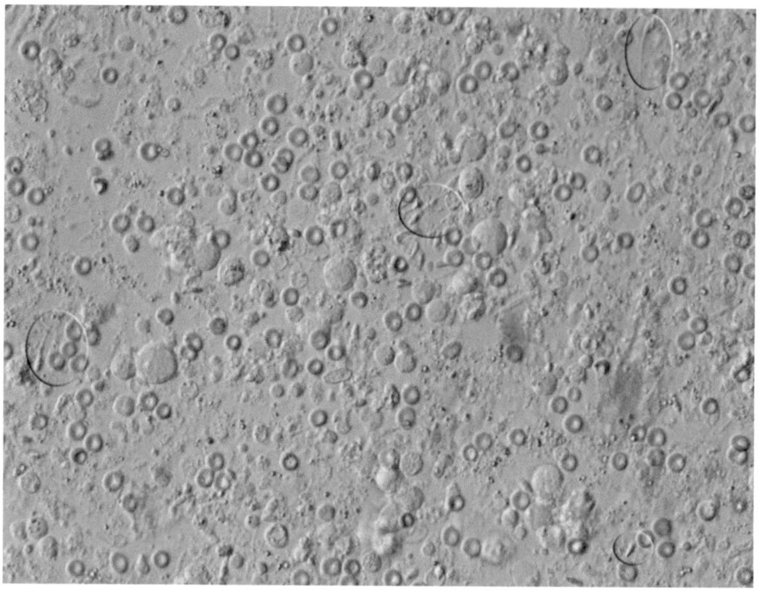

Fig. 2.7 TESE sample with abundant cells and tissue. A few spermatozoa may be seen, most with amorphous heads or having only a head (*bottom right*; ×400)

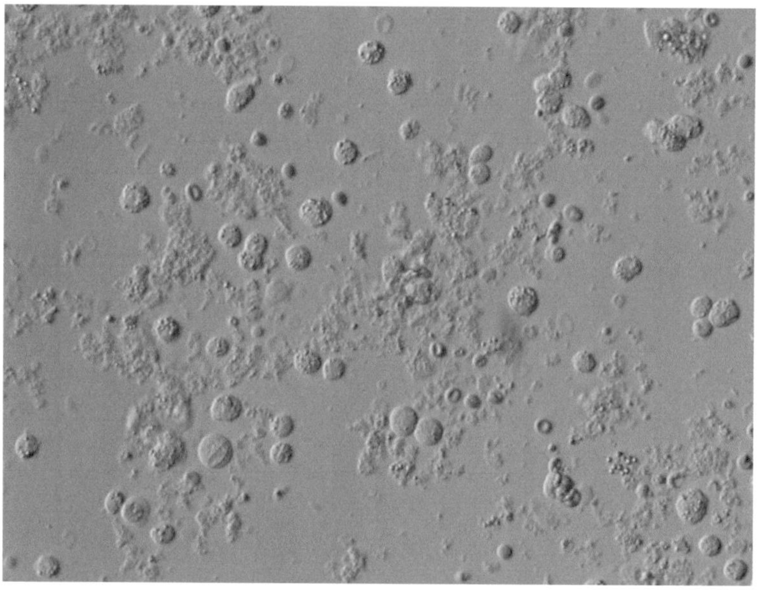

Fig. 2.8 TESE tissue after digestion. Large Sertoli cells are seen with some connective tissue. No germ cells or spermatozoa could be recorded (×400)

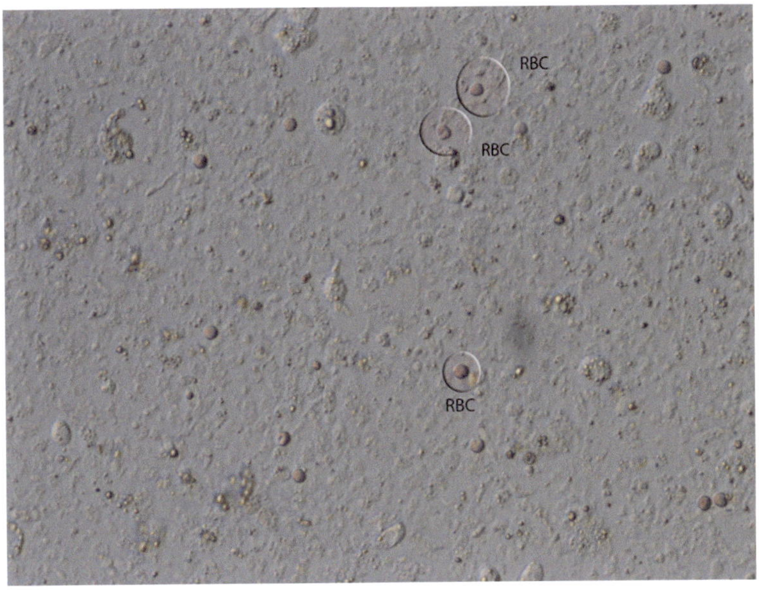

Fig. 2.9 Testicular tissue after digestion. The tissue is very depleted with no germ cells or spermatozoa and few red blood cells (*RBCs*; ×400)

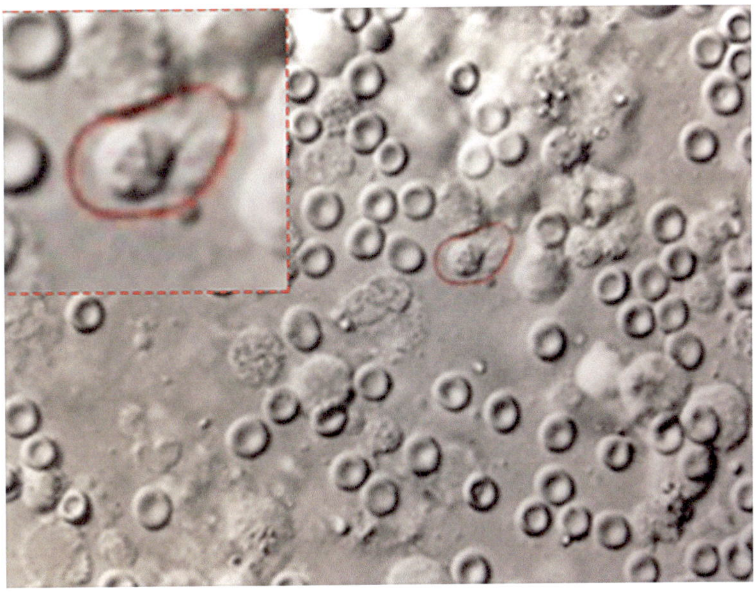

Fig. 2.10 Postdigestion TESE tissue in which only early elongated (immature) spermatids (*red circle*) were found. These cells may be injected with the patient's consent (×400)

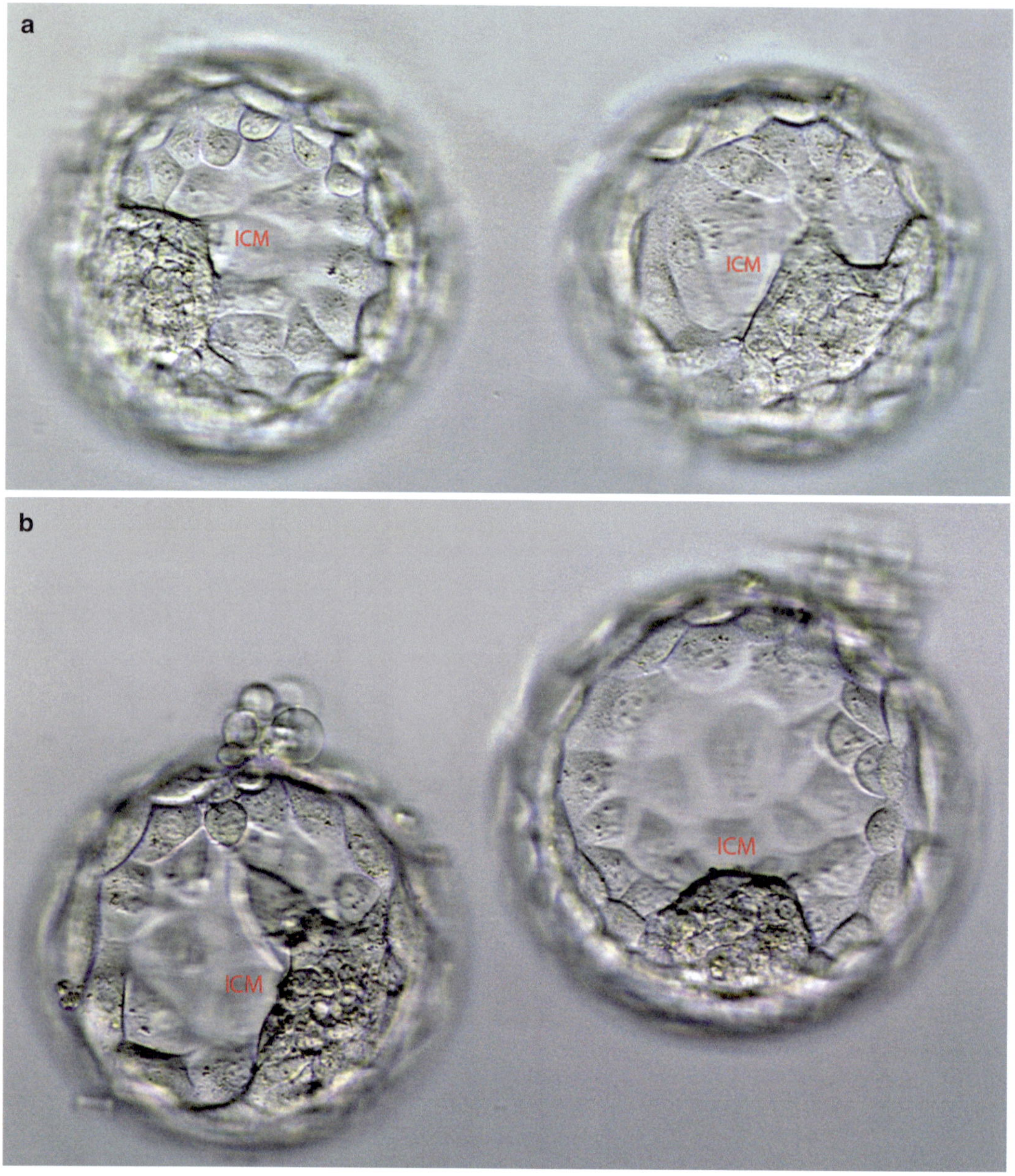

Fig. 2.11 Embryos at the time of transfer (day 5) in donor oocyte recipients (sharing the same egg donor). One patient received ejaculated spermatozoa (**a**), the other TESE spermatozoa (**b**). Both images show a high grade of blastocysts. The patient in *panel A* (ejaculated spermatozoa) had a miscarriage, whereas the patient in *panel B* (TESE) had two babies. Inner cell mass (*ICM*) is indicated in both panels (×200)

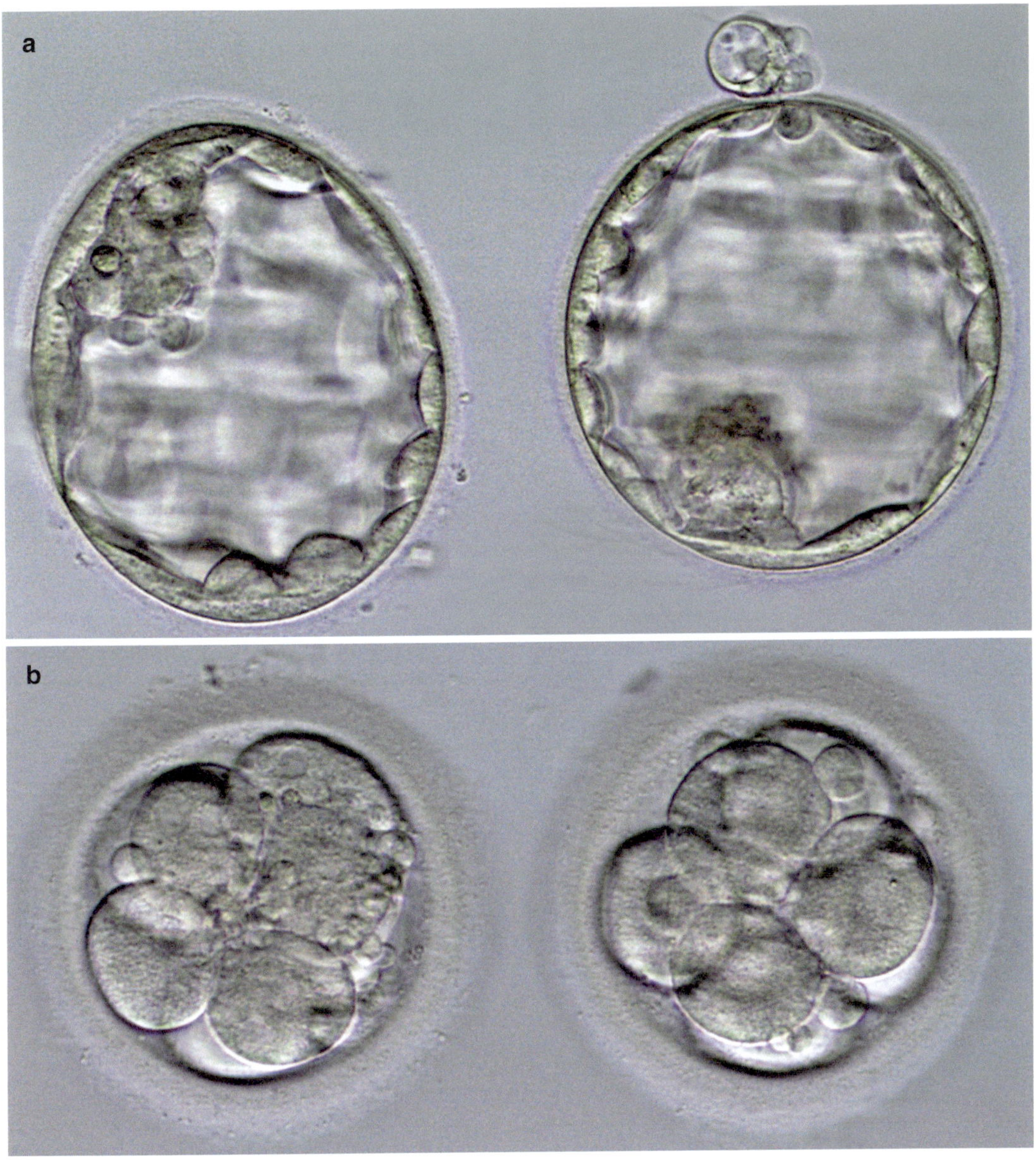

Fig. 2.12 Embryos at the time of transfer in shared donor oocyte recipients. (**a**) This patient used ejaculated sperm, transfer on day 5. High-quality blastocysts resulted in delivery of one baby. (**b**) This patient used TESE sperm, transfer on day 3. Although the embryos were of good quality in this case no pregnancy resulted (×200)

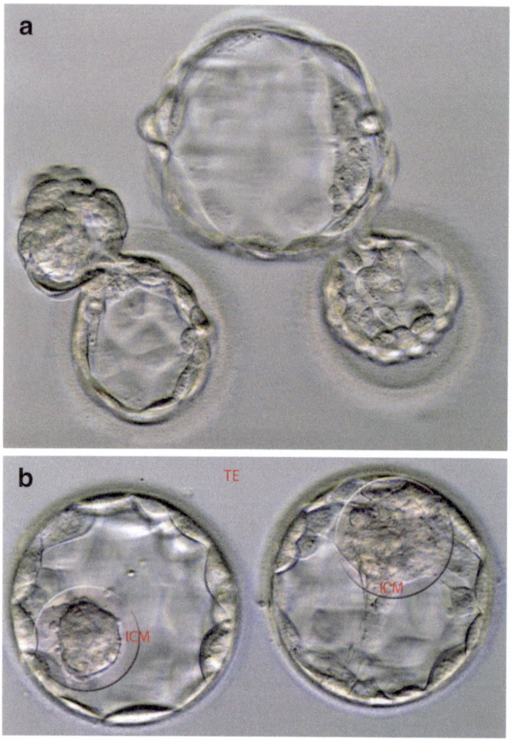

Fig. 2.13 Embryos at the time of transfer on day 5 in shared donor oocyte recipients. (**a**) In this recipient using ejaculated spermatozoa, preimplantation genetic diagnostic testing resulted in the birth of two babies. (**b**) This image from recipient using TESE sperm recipient shows high-grade blastocysts with prominent ICMs (*circle*), which resulted in the birth of one baby (×200). *TE* trophectoderm

Recommended Literature

Aoki VW, Wilcox AL, Thorp C, Hamilton BD, Carrell DT. Improved in vitro fertilization embryo quality and pregnancy rates with intracytoplasmic sperm injection of sperm from fresh testicular biopsy samples vs. frozen biopsy samples. Fertil Steril. 2004;82(6):1532–5.

Crabbe E, Verheyen G, Silber S, Tournaye H, Van de Velde H, Goossens A, et al. Enzymatic digestion of testicular tissue may rescue the intracytoplasmic sperm injection cycle in some patients with non-obstructive azoospermia. Hum Reprod. 1998;13(10):2791–6.

Dozortsev D, Neme R, Diamond MP, Abdelmassih S, Abdelmassih V, Oliveira F, et al. Embryos generated using testicular spermatozoa have higher developmental potential than those obtained using epididymal spermatozoa in men with obstructive azoospermia. Fertil Steril. 2006;86(3):606–11.

Hammadeh ME, Askari AS, Georg T, Rosenbaum P, Schmidt W. Effect of freeze-thawing procedure on chromatin stability, morphological alteration and membrane integrity of human spermatozoa in fertile and subfertile men. Int J Androl. 1999;22(3):155–62.

Hu Y, Maxson WS, Hoffman DI, Ory SJ, Licht MR, Eager S. Clinical application of intracytoplasmic sperm injection using in vitro cultured testicular spermatozoa obtained the day before egg retrieval. Fertil Steril. 1999;72(4):666–9.

Kanto S, Sugawara J, Masuda H, Sasano H, Arai Y, Kyono K. Fresh motile testicular sperm retrieved from nonobstructive azoospermic patients has the same potential to achieve fertilization and pregnancy via ICSI as sperm retrieved from obstructive azoospermic patients. Fertil Steril. 2008;90(5):2010.e5–7.

Levran D, Ginath S, Farhi J, Nahum H, Glezerman M, Weissman A. Timing of testicular sperm retrieval procedures and in vitro fertilization-intracytoplasmic sperm injection outcome. Fertil Steril. 2001;76(2):380–3.

Nagy ZP, Verheyen G, Tournaye H, Devroey P, Van Steirteghem AC. An improved treatment procedure for testicular biopsy specimens offers more efficient sperm recovery: case series. Fertil Steril. 1997;68(2):376–9.

Nicopoullos JD, Gilling-Smith C, Almeida PA, Norman-Taylor J, Grace I, Ramsay JW. Use of surgical sperm retrieval in azoospermic men: a meta-analysis. Fertil Steril. 2004;82(3):691–701.

Ostad M, Liotta D, Ye Z, Schlegel PN. Testicular sperm extraction for nonobstructive azoospermia: results of a multibiopsy approach with optimized tissue dispersion. Urology. 1998;52(4):692–6.

Palermo GD, Schlegel PN, Hariprashad JJ, Ergun B, Mielnik A, Zaninovic N, et al. Fertilization and pregnancy outcome with intracytoplasmic sperm injection for azoospermic men. Hum Reprod. 1999;14(3):741–8.

Park YS, Lee SH, Song SJ, Jun JH, Koong MK, Seo JT. Influence of motility on the outcome of in vitro fertilization/intracytoplasmic sperm injection with fresh vs. frozen testicular sperm from men with obstructive azoospermia. Fertil Steril. 2003;80(3):526–30.

Ramasamy R, Fisher ES, Ricci JA, Leung RA, Schlegel PN. Duration of microdissection testicular sperm extraction procedures: relationship to sperm retrieval success. J Urol. 2011a;185(4):1394–7.

Ramasamy R, Reifsnyder JE, Bryson C, Zaninovic N, Liotta D, Cook CA, et al. Role of tissue digestion and extensive sperm search after microdissection testicular sperm extraction. Fertil Steril. 2011b;96(2):299–302.

Schlegel PN. Testicular sperm extraction: microdissection improves sperm yield with minimal tissue excision. Hum Reprod. 1999;14(1):131–5.

Schlegel PN. Causes of azoospermia and their management. Reprod Fertil Dev. 2004;16(5):561–72.

Terriou P, Hans E, Giorgetti C, Spach JL, Salzmann J, Urrutia V, et al. Pentoxifylline initiates motility in spontaneously immotile epididymal and testicular spermatozoa and allows normal fertilization, pregnancy, and birth after intracytoplasmic sperm injection. J Assist Reprod Gen. 2000;17(4):194–9.

Vernaeve V, Tournaye H, Osmanagaoglu K, Verheyen G, Van Steirteghem A, Devroey P. Intracytoplasmic sperm injection with testicular spermatozoa is less successful in men with nonobstructive azoospermia than in men with obstructive azoospermia. Fertil Steril. 2003;79(3):529–33.

3 Histologic Procedures and Testicular Biopsy Freezing

Andreja Vukasović, Viviana Kozina, and Davor Ježek

The purpose of this chapter is to describe the biopsy handling and histologic processing of testicular tissue. This chapter is intended to serve as a practical guide to performing routine procedures in the histopathologic examination and cryopreservation of the testicular biopsy specimen. Some basic staining and step-by-step laboratory procedures are described.

Paraffin Embedding, Paraffin Sections, and Hematoxylin-Eosin Staining

When an "open" biopsy of the testis is performed, several pieces of tissue are taken using a single- or multiple-site approach from different parts of the male gonad. All pieces are transferred immediately to a transport medium for freezing and histologic analysis. Each piece of testicular tissue is divided into two parts. One part is immersed in 0.5 mL of sperm-freezing medium (e.g., MediCult [Origio, Måløv, Denmark]) and, after programmed freezing, is stored in a tissue bank; the other part is histologically processed (Fig. 3.1). Histologic examination, a substantial step in processing the testicular biopsy specimen, starts with fixation. In our laboratory, we use Bouin's fluid for paraffin-embedded tissue and buffered 5.5 % glutaraldehyde for resin-embedded samples. The tissue in fixative is rinsed several times and dehydrated. Dehydration is followed by clearing and infiltration with an embedding medium. After hardening, the paraffin block is sectioned with a microtome. Ribbons of sections are transferred to the surface of a warm bath to flatten them; then they are placed on microscope slides and left on a hotplate for thorough drying. Hematoxylin (or hemalaun)-eosin, a basic staining method, is performed routinely. After deparaffinization in xylene or its substitutes, the slides are rehydrated briefly and then stained with hematoxylin. After thorough rinsing and washing in water, eosin is applied. Ethanol is used for rinsing; then we continue with dehydration, clearing, and covering.

A. Vukasović, M.D. (✉) • V. Kozina, MEd
D. Ježek, M.D., Ph.D.
Department of Histology and Embryology, School of Medicine, University of Zagreb, Šalata 3, 10000 Zagreb, Croatia
e-mail: andreja_vukasovic@yahoo.com

D. Ježek (ed.), *Atlas on the Human Testis*,
DOI 10.1007/978-1-4471-2763-5_3, © Springer-Verlag London 2013

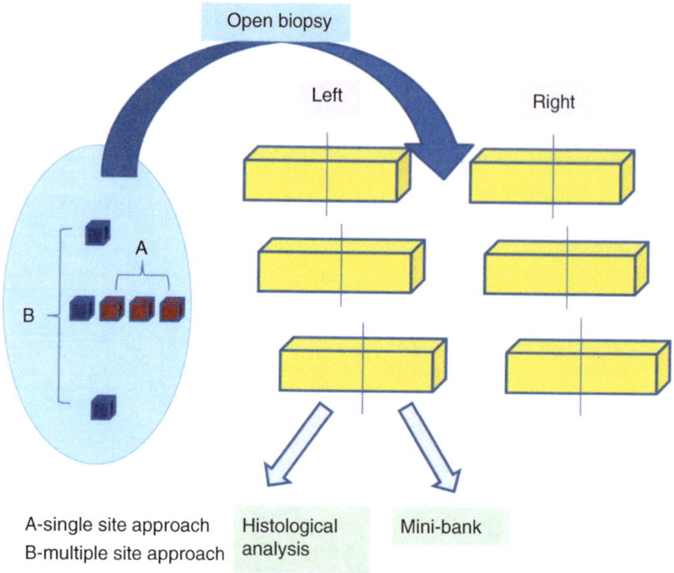

Fig. 3.1 Bilateral testicular biopsy. Typically, three small pieces of tissue are taken from the left and three from the right testicle. Each piece is divided further into two parts. One part is subjected to histologic analysis; the other is frozen for potential TESE/ ICSI. The frozen parts form an individual mini-bank of biopsies from a particular patient. The same procedure is applied to the left and right testes

Fig. 3.2 A small piece of testicular tissue is immersed in a fixative (in this case, Bouin's fluid). The fixed small piece of testis parenchyma (*arrow*) has an average size of 1–3 mm^3. The bilateral testicular biopsy sample consists of six small pieces, three from each testicle

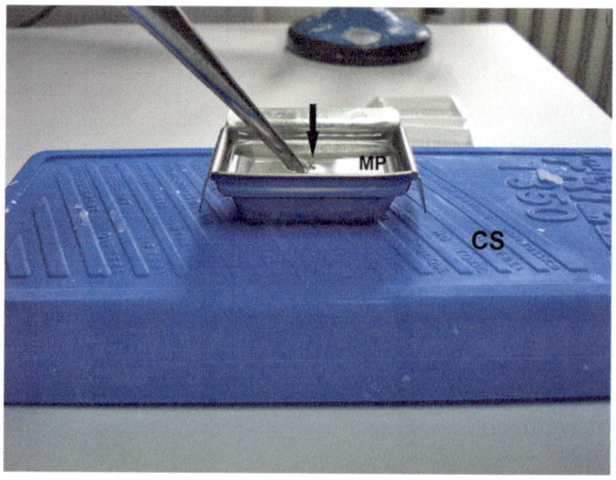

Fig. 3.3 The embedding mold is filled with molten paraffin (*MP*), and the testicular tissue is inserted in it (*arrow*). To harden the paraffin, the mold is placed on a cold surface (*CS*). A hot instrument (in this case, tweezers) is applied along the upper surface to release air from the paraffin

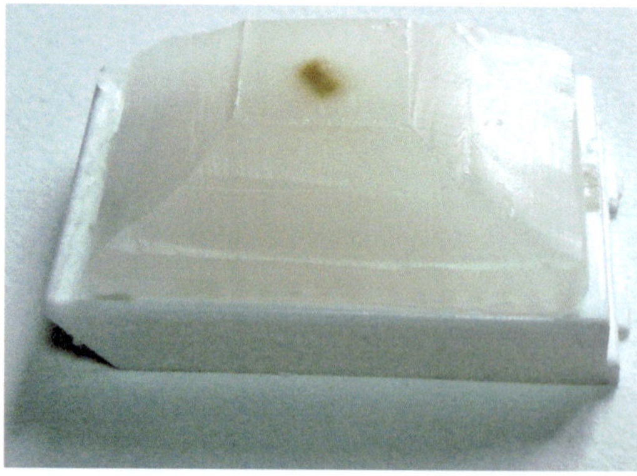

Fig. 3.4 A block of paraffin-embedded tissue, trimmed and ready for cutting

Fig. 3.5 Ribbons (*S*) of 5-μm-thick sections are cut by a rotary microtome. The *arrow* points to the tissue block. *B* blade, *BH* tissue block holder

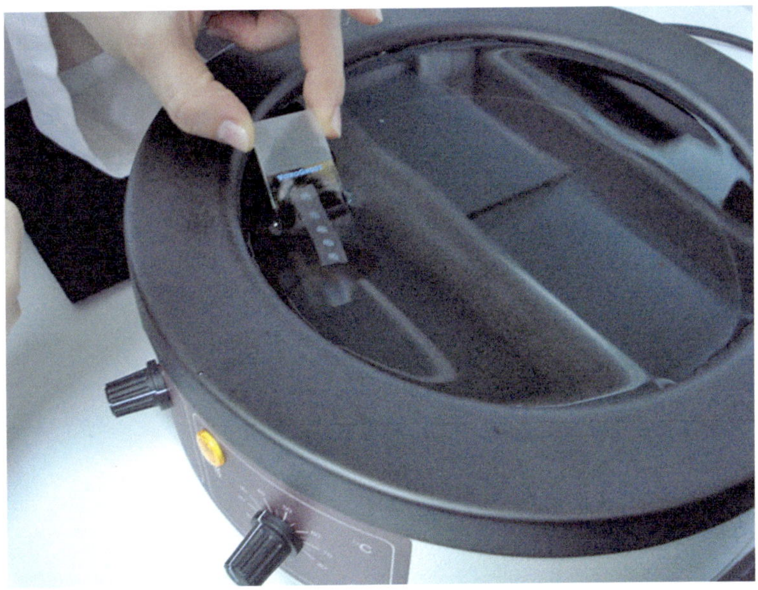

Fig. 3.6 Sections are carefully picked up from the water bath (37 °C) carefully and placed on slides. For better adherence, the slides are silane coated. It is very important not to leave sections floating on a warm water surface for too long because this may soften the paraffin and the section might be lost

Fig. 3.7 The slides are kept on a hotplate (37 °C) for several hours to dry

Fig. 3.8 A staining set prepared for routine hematoxylin-eosin staining in a histology laboratory. The glass jars filled with the necessary liquids are now ready to use

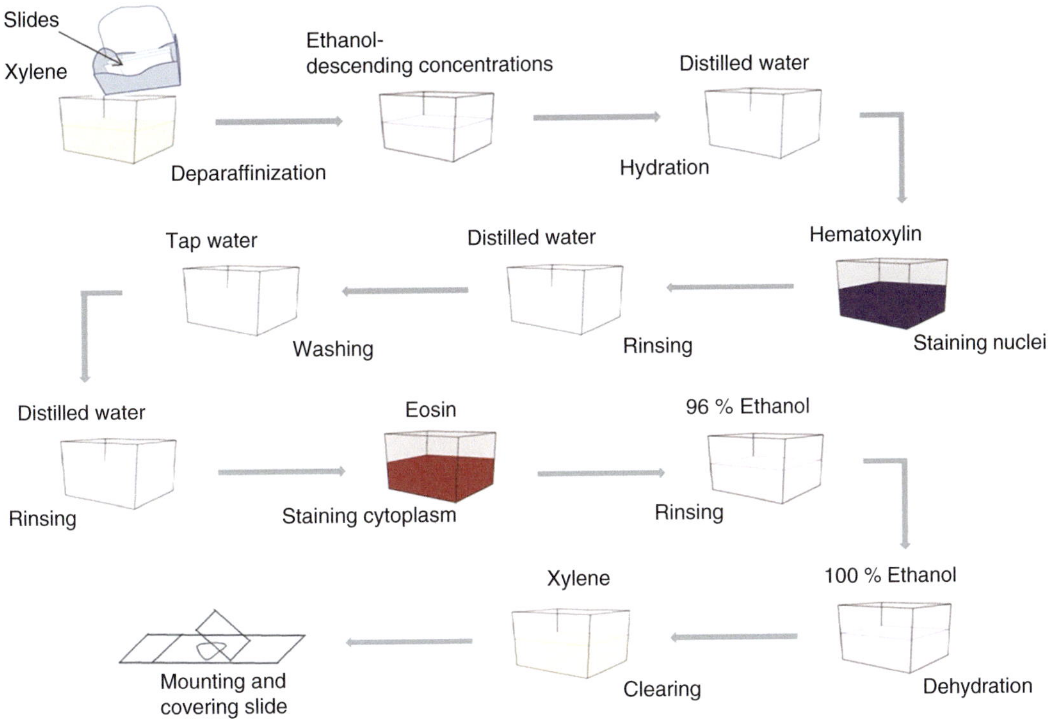

Fig. 3.9 Hematoxylin-eosin staining, step-by-step

Immunohistochemistry

Infertile men have a much higher frequency of carcinoma in situ. Therefore, the bioptic material should be checked routinely for malignant changes. Tissue sections 3–5 μm thick are cut from paraffin-embedded tissue blocks, as described earlier, and placed on object silane-coated slides. The slides containing tissue sections are incubated for 20 min at 60 °C. The sections are then deparaffinized and incubated for 3–5 min in 10 mmol/L of citrate buffer (pH 6.0) in a microwave oven at 800 and 400 W for antigen retrieval. Next, the tissue slides are washed with phosphate-buffered saline (PBS; pH 7.2), and the endogenous peroxidase activity (optional; some new kits do not require this step) is blocked by a 5-min treatment with H_2O_2. The slides are then washed with PBS buffer and incubated in a humid chamber for 30 min with primary antibody (i.e., placental alkaline phosphatase) at room temperature. After washing in PBS, the secondary biotinylated antibody (or the appropriate kit with EnVision complex [Dako, Hamburg, Germany]) is added for a 30-min incubation. The slides are then washed with PBS buffer and treated with streptavidin-horseradish peroxidase for 30 min. The tissue sections are washed once more in PBS; then chromogen is added for 5 min. The slides are washed in distilled water, stained with hematoxylin for 1 min, washed with water, dehydrated with alcohol (96 %), cleared with xylene, and mechanically covered. It is important to note that the choice of immunohistochemical (IHC) procedure depends on the antibodies used and the instructions provided by the manufacturer of the antibodies/kit. Thus, one should check the procedure carefully and include slides for the appropriate positive (e.g., histologic sections of placenta or testicular tumor) and negative control (e.g., omission of the primary antibody).

Fig. 3.10 A batch of slides ready for immunohistochemical staining

Fig. 3.11 For antigen retrieval, citrate buffer is usually applied

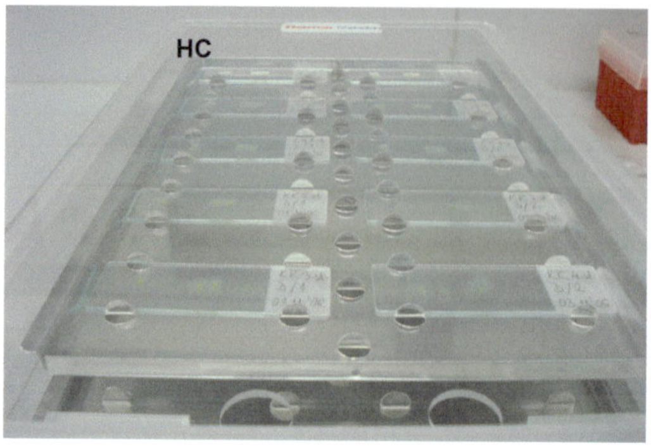

Fig. 3.12 Incubation with primary and secondary antibodies is carried out in a humid chamber (*HC*)

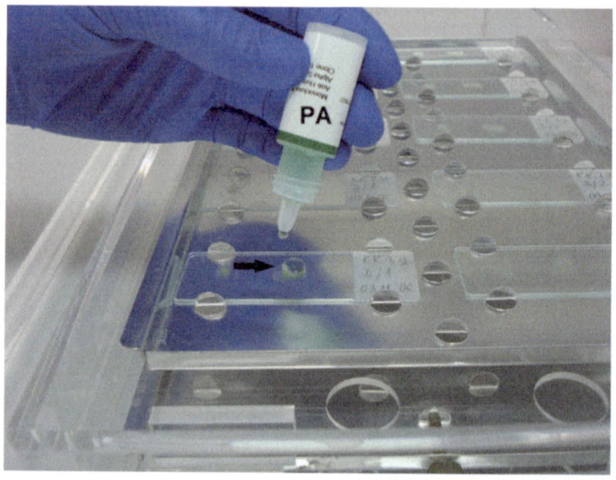

Fig. 3.13 A primary antibody (*PA*) is applied to the sections. The primary antibody should be added carefully to cover the whole section (*arrow*)

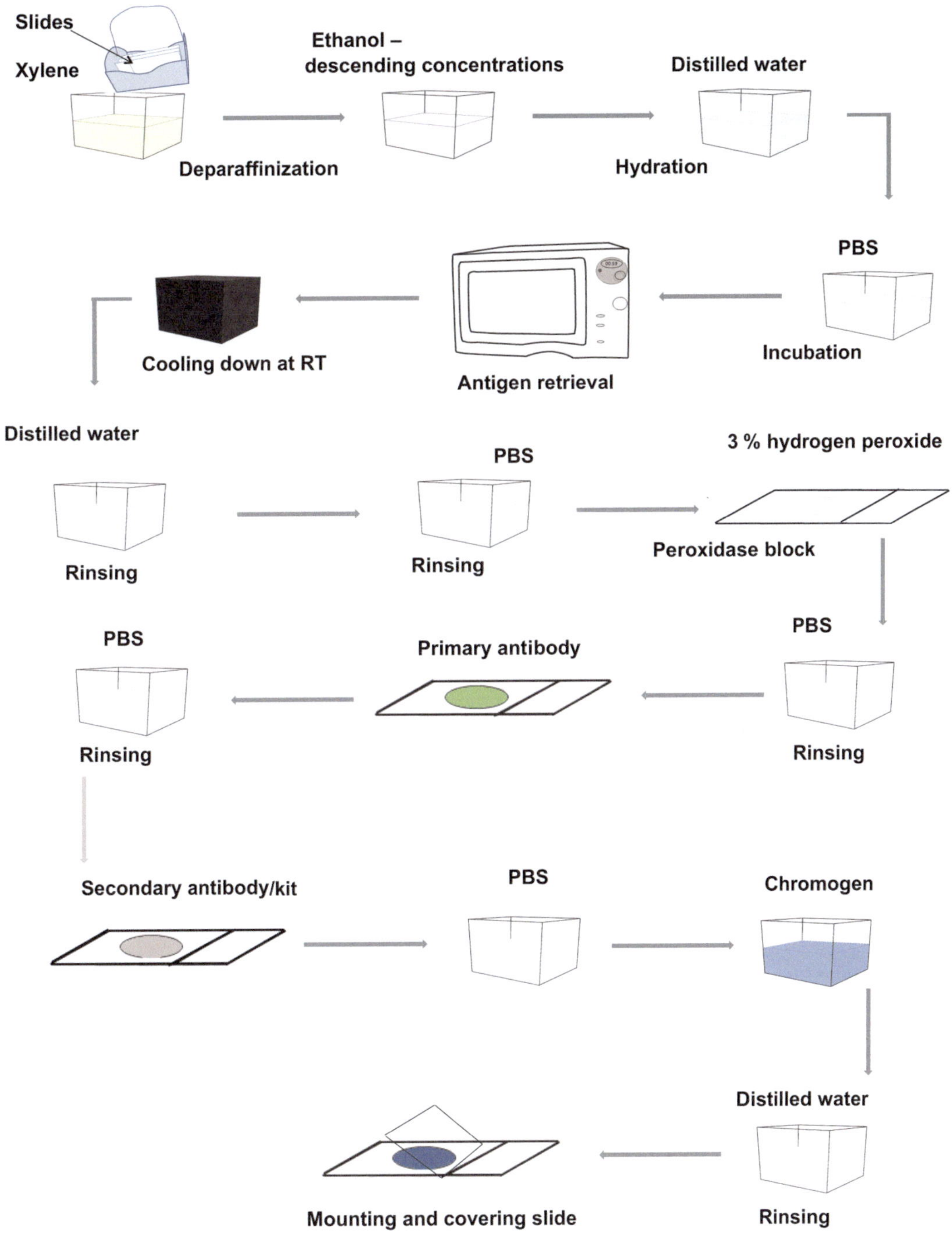

Fig. 3.14 Immunohistochemical staining. The IHC procedure should be adjusted carefully according to different primary and secondary antibodies or kits. The appropriate positive and negative controls/slides should be used for each IHC staining

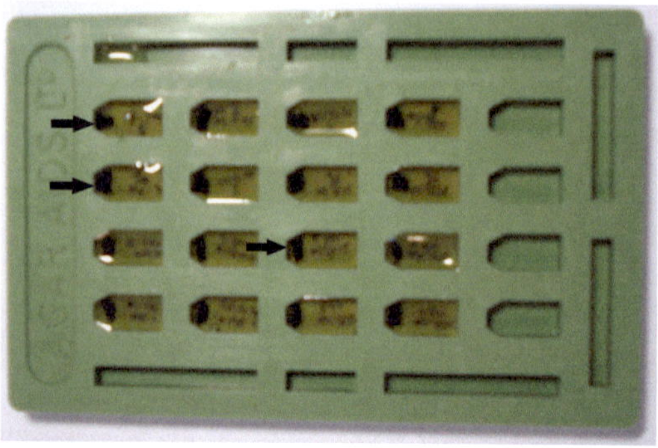

Fig. 3.15 A flat embedding mold. A liquid resin mixture (with a consistency similar to that of honey) is poured into the molds, and the tissue samples are placed at the tips (*arrows*). The tissue samples are black because of osmium tetroxide postfixation. The molds are kept at 63 °C in the incubator for 72 h for resin hardening

Plastic Embedding and Semithin Sections

Because of their small thickness (<1 μm), semithin sections offer excellent insight into testis morphology. For example, based on semithin section analysis, one can distinguish between residual bodies and spermatids or spermatozoa heads. In addition, semithin sections may reveal cancer in situ cells. Semithin sections (section thickness = 0.9 μm) from tissue blocks embedded in Durcupan (Sigma-Aldrich, St. Louis, MO) or other plastic embedding material are cut with an ultramicrotome and stained with 1 % toluidine blue. The sections are analyzed with a bright field microscope. In some cases, ultrathin sections (section thickness = 70 nm) must be made, contrasted with lead citrate and uranyl acetate, and examined under a transmission electron microscope. Histologic evaluation of the testicular biopsy is performed in the standard fashion.

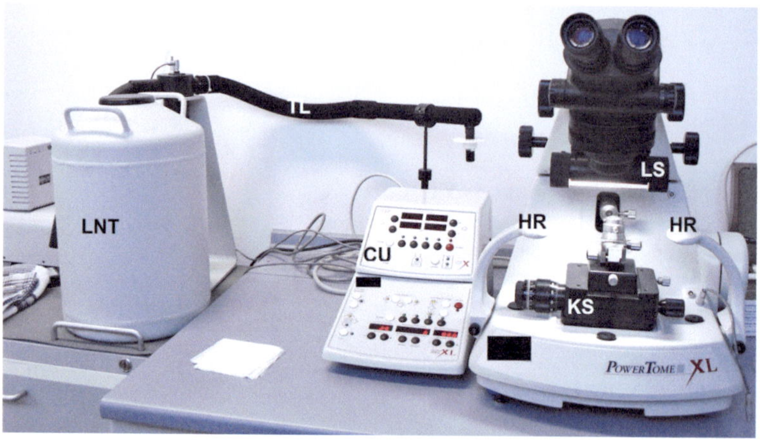

Fig. 3.16 Semithin and ultrathin sections are cut with an ultramicrotome. If needed for ultrathin cryosections, a cryochamber is added to the microtome and connected to a liquid nitrogen supply. *CU* control unit, *HR* hand rest, *KS* micrometer-adjustable knife stage, *LNT* liquid nitrogen tank, *LS* light source, *TL* liquid nitrogen transfer line

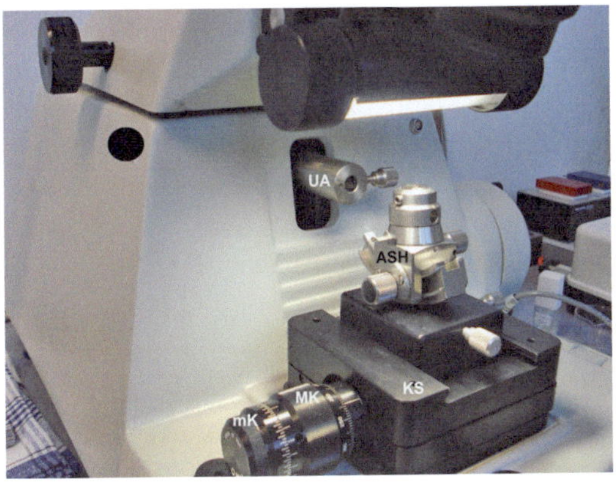

Fig. 3.17 A tissue block is placed in the holder and manually trimmed with a blade using an eyepiece magnifier. *ASH* arc segment holder, *KS* micrometer-adjustable knife stage, *MK* macro-knob, *mK* micro-knob, *UA* ultramicrotome arm

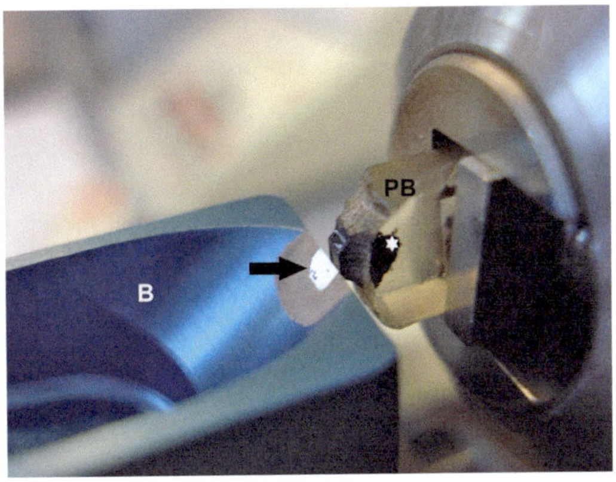

Fig. 3.19 A plastic block (*PB*) with the specimen tissue embedded at its tip (*star*) is cut with the diamond knife (*arrow*). *B* boat of the diamond knife

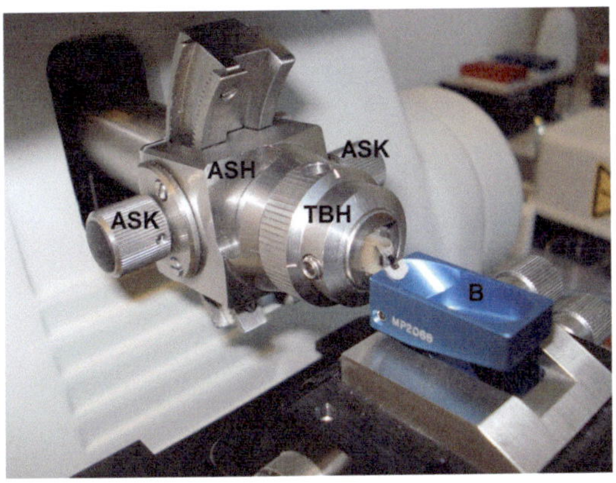

Fig. 3.18 A tissue block is placed in the holder so that two-thirds of the block is inside the holder and one-third protrudes from it. When the block is trimmed, a tissue block holder (*TBH*) is placed in the arc segment holder (*ASH*) and positioned on the ultramicrotome arm. *ASK* arc segment knob, *B* boat of the diamond knife

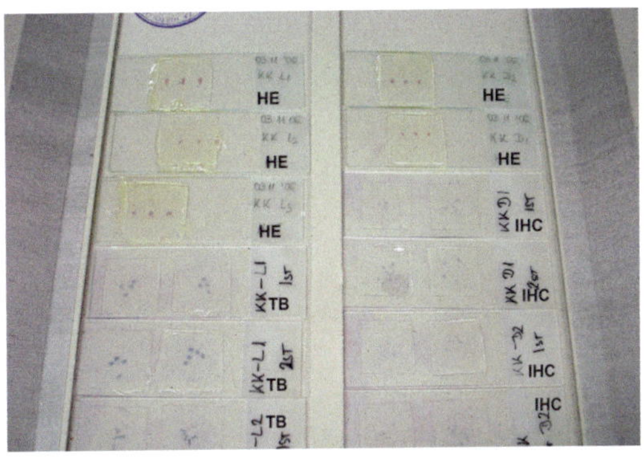

Fig. 3.20 A selection of slides after complete histologic analysis of a testicular biopsy specimen. *HE* hematoxylin-eosin staining, *IHC* immunohistochemical staining with placental alkaline phosphatase for detection of carcinoma in situ, *TB* semithin sections stained with toluidine blue

Tissue Freezing, Testicular Biopsy Mini-bank, and Testicular Sperm Extraction

Tissue in the sperm-freezing medium is subjected to programmed freezing using a freezing device (in our case, Nicool 100 [Air Liquide, Marne-la-Vallée, France]). During the first 5 min, the tissue is cooled to −70 °C, then exponentially to −120 °C during the next 55 min. Finally, the tissue samples are stored in the liquid nitrogen until a TESE/ICSI (testicular sperm extraction/intracytoplasmic sperm injection into the oocyte) attempt.

For the planned TESE/ICSI attempt, the TESE is performed as follows. Frozen samples are retrieved from the liquid nitrogen and immediately transferred to a water bath (37 °C). The sperm-freezing medium (in our case, MediCult) is replaced with 1 mL of pre-warmed (37 °C) sperm-preparation medium (MediCult) and incubated for 1 h. This is followed by the addition of 1 mL of the pre-warmed sperm-preparation medium with 0.8 mg of collagenase (type IA, Sigma-Aldrich) and 0.2 μg of trypsin inhibitor (Sigma-Aldrich). After 2 h of incubation, the samples are macerated with sterile needles and centrifuged for 10 min at 800 × *g* and 37 °C. After the supernatant is removed, the pellet is checked for the presence of spermatozoa. If the aforementioned procedure for testicular biopsy is followed (see Fig. 3.1), the TESE procedure may be expected to be successful in 63 % of patients with nonobstructive azoospermia, whereas in cases of obstructive azoospermia the success rate is 100 %.

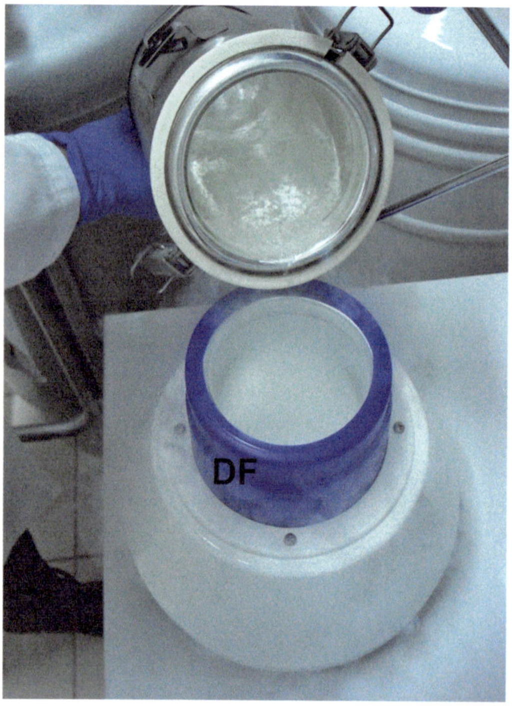

Fig. 3.21 Liquid nitrogen is poured into the Dewar flask (*DF*) of a freezing device

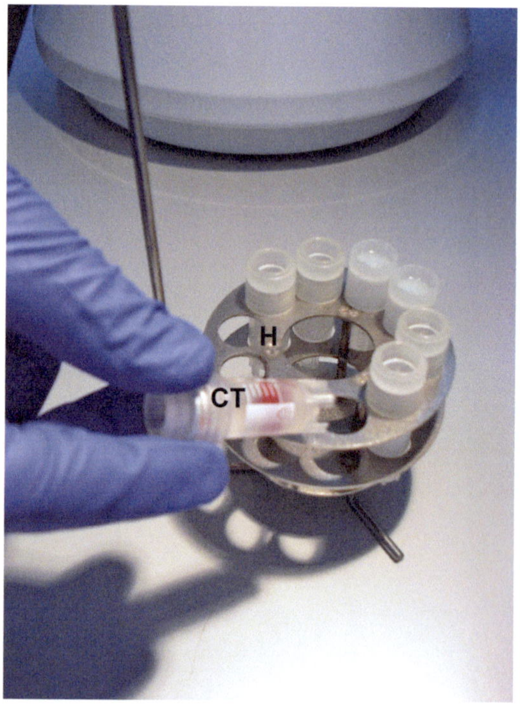

Fig. 3.22 Cryotubes (*CT*) containing cryomedium with testicular biopsy tissue are placed in the vertical holder (*H*) of a freezing device

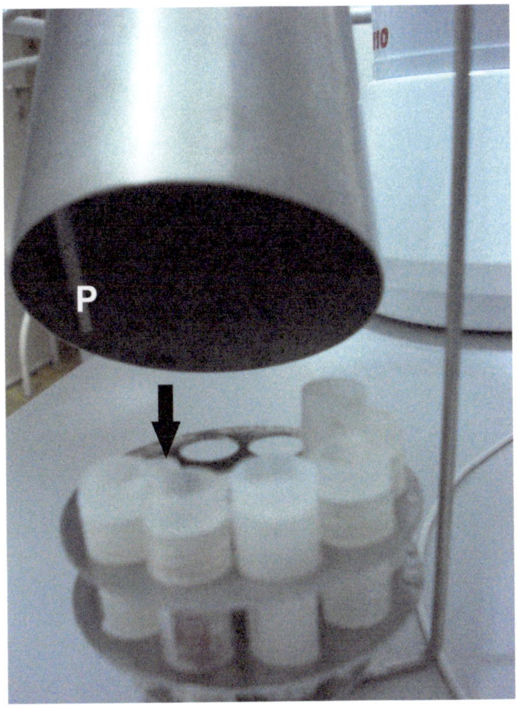

Fig. 3.23 One cryotube (*arrow*) contains cryomedium only; a probe (*P*) connected to a digital temperature indicator will be inserted in that tube

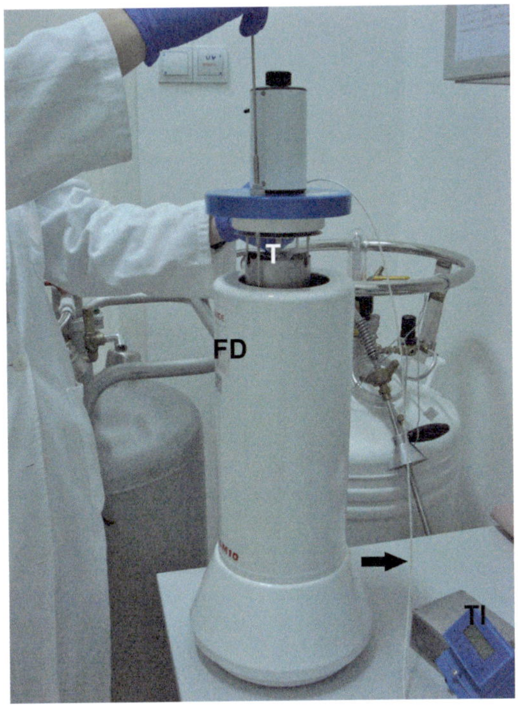

Fig. 3.24 A holder with cryotubes is secured vertically inside the freezing device (*FD*). The specimens are kept above the liquid nitrogen. A turbine (*T*) at the top of the device cools the specimens through gaseous convection. For programmed freezing and temperature monitoring, the probe is connected (*arrow*) to the digital temperature indicator (*TI*)

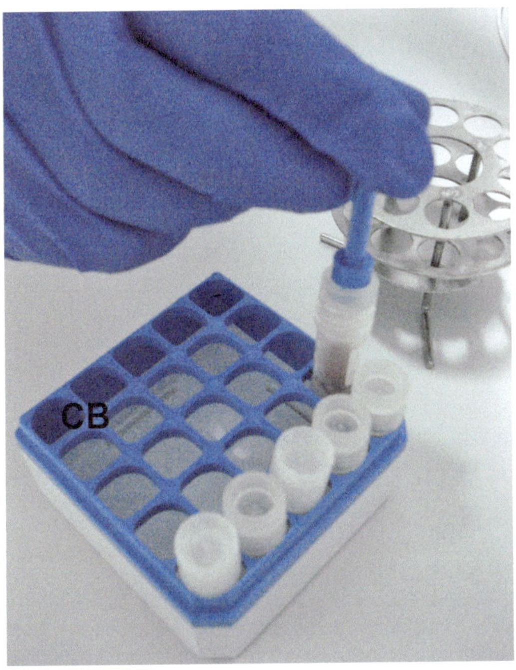

Fig. 3.25 After programmed freezing, the cryotubes are placed inside a cryobox (*CB*)

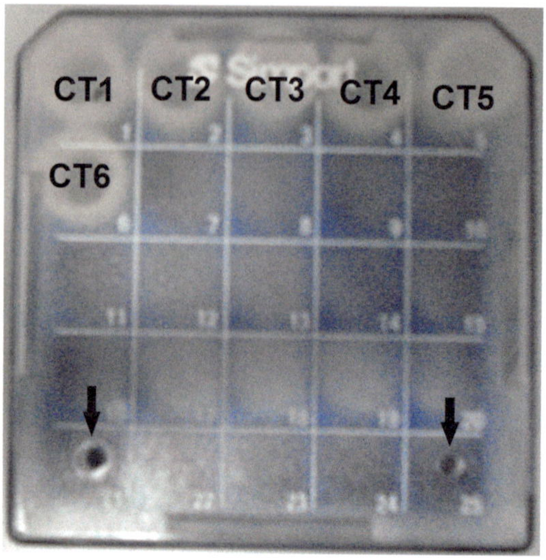

Fig. 3.26 A mini-bank of testicular biopsy samples from the same patient is formed and positioned in the cryobox. The corresponding cryotubes (*CT*) occupy strictly defined positions (1–6). To enhance the rapid inflow of liquid nitrogen into the box, the lid contains small holes (*arrow*). Note that each cryobox has the capacity to hold 25 cryotubes

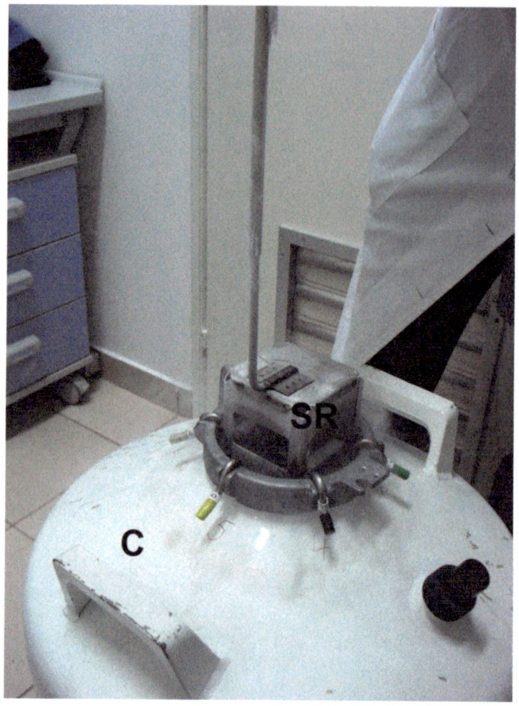

Fig. 3.27 This cryocontainer (*C*) has the capacity to hold six vertically arranged box-type storage racks (*SR*). Five cryoboxes may be placed vertically in each rack. Each cryobox with frozen samples is placed in a defined position within the rack and plunged into the liquid nitrogen. The samples are stored at −196 °C until an attempt at TESE/ICSI

Recommended Literature

Anger JT, Gilbert BR, Goldstein M. Cryopreservation of sperm: indications, methods and results. J Urol. 2003; 170(4 Pt 1):1079–84.

Beliveau ME, Turek PJ. The value of testicular 'mapping' in men with non-obstructive azoospermia. Asian J Androl. 2011;13:225–30.

Bergmann M, Behre HM, Nieschlag E. Serum FSH and testicular morphology in male infertility. Clin Endocrinol (Oxf). 1994;40:133–6.

Bunge RG, Sherman JK. Fertilizing capacity of frozen human spermatozoa. Nature. 1953;172(4382):767–8.

Capanna E. Lazzaro Spallanzani: at the roots of modern biology. J Exp Zool. 1999;285:178–96.

Chan PT, Schlegel PN. Diagnostic and therapeutic testis biopsy. Curr Urol Rep. 2000;1:266–72.

Diemer T, Hauptmann A, Weidner W. Treatment of azoospermia: surgical sperm retrieval (MESA, TESE, micro-TESE) [in German]. Urologe A. 2011;50:38–46.

Jezek D, Knuth UA, Schulze W. Successful testicular sperm extraction (TESE) in spite of high serum follicle stimulating hormone and azoospermia: correlation between testicular morphology, TESE results, semen analysis and serum hormone values in 103 infertile men. Hum Reprod. 1998;13:1230–4.

Jezek D, Knezevic N, Kalanj-Bognar S, Vukelic Z, Krhen I. From testicular biopsy to human embryo. Verh Dtsch Ges Pathol. 2004;88:136–43.

McLachlan RI, Rajpert-De Meyts E, Hoei-Hansen CE, de Kretser DM, Skakkebaek NE. Histological evaluation of the human testis – approaches to optimizing the clinical value of the assessment: mini review. Hum Reprod. 2007;22:2–16.

Navarro J, Risco R, Toschi M, Schattman G. Gene therapy and intracytoplasmatic sperm injection (ICSI) – a review. Placenta. 2008;29(Suppl B):193–9.

Palermo GD, Cohen J, Alikani M, Adler A, Rosenwaks Z. Development and implementation of intracytoplasmic sperm injection (ICSI). Reprod Fertil Dev. 1995;7: 211–7.

Perloff WH, Steinberger E, Sherman JK. Conception with human spermatozoa frozen by nitrogen vapor technique. Fertil Steril. 1964;15:501–4.

Polge C, Smith AU, Parkes AS. Revival of spermatozoa after vitrification and dehydration at low temperatures. Nature. 1949;164(4172):666.

Prabakaran SA, Agarwal A, Sundaram A, Thomas Jr AJ, Sikka S. Cryosurvival of testicular spermatozoa from obstructive azoospermic patients: the Cleveland Clinic Experience. Fertil Steril. 2006;86:1789–91.

Schlegel PN. Causes of azoospermia and their management. Reprod Fertil Dev. 2004;16:561–72.

Schoor RA. Prostatitis and male infertility: evidence and links. Curr Urol Rep. 2002;3:324–9.

Schoor RA, Elhanbly S, Niederberger CS, Ross LS. The role of testicular biopsy in the modern management of male infertility. J Urol. 2002;167:197–200.

Schulze W, Knuth UA, Jezek D, Benson DM, Fischer R, Naether OG, et al. Intratesticular sperm extraction. Basis for successful treatment of infertility in men with ejaculatory azoospermia. Adv Exp Med Biol. 1997;424:81–8.

Schulze W, Thoms F, Knuth UA. Testicular sperm extraction: comprehensive analysis with simultaneously performed histology in 1418 biopsies from 766 subfertile men. Hum Reprod. 1999;14 Suppl 1:82–96.

Sherman JK. Improved methods of preservation of human spermatozoa by freezing and freeze-drying. Fertil Steril. 1963;14:49–64.

Sherman JK, Bunge RG. Effect of glycerol and freezing on some staining reactions of human spermatozoa. Proc Soc Exp Biol Med. 1953;84:179–80.

Tanrikut C, Goldstein M. Obstructive azoospermia: a microsurgical success story. Semin Reprod Med. 2009;27: 159–64.

Tunc L, Kirac M, Gurocak S, Yucel A, Kupeli B, Alkibay T, Bozkirli I. Can serum Inhibin B and FSH levels, testicular histology and volume predict the outcome of testicular sperm extraction in patients with non-obstructive azoospermia? Int Urol Nephrol. 2006;38: 629–35.

van Casteren NJ, de Jong J, Stoop H, Steyerberg EW, de Bekker-Grob EW, Dohle GR, et al. Evaluation of testicular biopsies for carcinoma in situ: immunohistochemistry is mandatory. Int J Androl. 2009;32:666–74.

Early Development of the Human Testis

4

Gordana Jurić-Lekić, Marta Himelreich, Marta Lekić, Đurđica Grbeša, and Floriana Bulić-Jakuš

The gonads (testes and ovaries) are derived from three sources: the mesothelium (mesodermal epithelium) lining the posterior abdominal wall, underlying mesenchyme (embryonic connective tissue), and primordial germ cells.

The initial stages of gonadal development occur during the fifth week, when a thickened area of mesothelium develops on the medial side of the mesonephros. Proliferation of this epithelium and the underlying mesenchyme produces a bulge on the medial side of the mesonephros, which is called the *genital* or *gonadal ridge* [1]. At first, independently from the genetic sex, the gonadal primordium is colonized by the primordial germ cells [2]. They originate from the epiblast, migrate through the primitive streak, and, by the third week, reside among endodermal cells in the wall of the yolk sac close to the allantois. They migrate along the dorsal mesentery of the hindgut during the fourth week, arrive at the primitive gonad at the beginning of the fifth week, and invade the genital ridges in the sixth week [3]. Fingerlike epithelial (gonadal) cords grow into the underlying mesenchyme. The indifferent gonad now consists of an external cortex and an internal medulla [1]. If the germ cells fail to reach the ridges, the gonads do not develop. Hence, the primordial germ cells have an inductive influence on development into ovaries or testes [3].

As the embryo grows, the gonadal ridge gradually becomes pinched off from the mesonephros. However, some cells of mesonephric origin join the gonadal ridge. Furthermore, the gonadal ridge remains connected to the remnant of mesonephros by a fold of peritoneum, the mesorchium, or the mesovarium [4].

During the seventh week, sexual distinction in the gonadal ridge becomes perceptible. The differentiation of the gonadal ridge into a testis is a rapid phenomenon, in contrast to the slow and late development of the ovary. Seminiferous tubules originate from the mesonephros, the "medulla," whereas ovarian tissue originates from the secondary sex cords formed from the germinal epithelium, the "cortex."

Inside the seminiferous tubules, the germ cells are large. They divide actively but do not enter meiosis. The Sertoli cells are smaller than the germ cells, tend to surround them and prepare the future seminiferous tubules. A basal membrane is formed, isolating the tubules from the surrounding mesenchymal tissue. The individualization of tubules and the synthesis of anti-Müllerian hormone precede Leydig cell differentiation. Between the eighth and ninth weeks, the Leydig

G. Jurić-Lekić, M.D., Ph.D. (✉) • M. Himelreich
M. Lekić • Đ. Grbeša, M.D., Ph.D.
Department of Histology and Embryology, School of Medicine, University of Zagreb, Zagreb, Croatia
e-mail: gjuric@mef.hr

F. Bulić-Jakuš, M.D., Ph.D.
Department of Biology, School of Medicine, University of Zagreb, Zagreb, Croatia

D. Ježek (ed.), *Atlas on the Human Testis*,
DOI 10.1007/978-1-4471-2763-5_4, © Springer-Verlag London 2013

cells differentiate from interstitial tissue and then spread progressively in the intertubular spaces between the 14th and 18th weeks. They secrete testosterone from the eighth week of development [2].

The following microphotographs, which show human testicular development up to the tenth week, are from the image collection of human embryos and fetuses from the Archive of the Department of Histology and Embryology, School of Medicine, University of Zagreb. This collection consists of histologic slides with serial sections of celloidin-embedded material, routinely processed and stained.

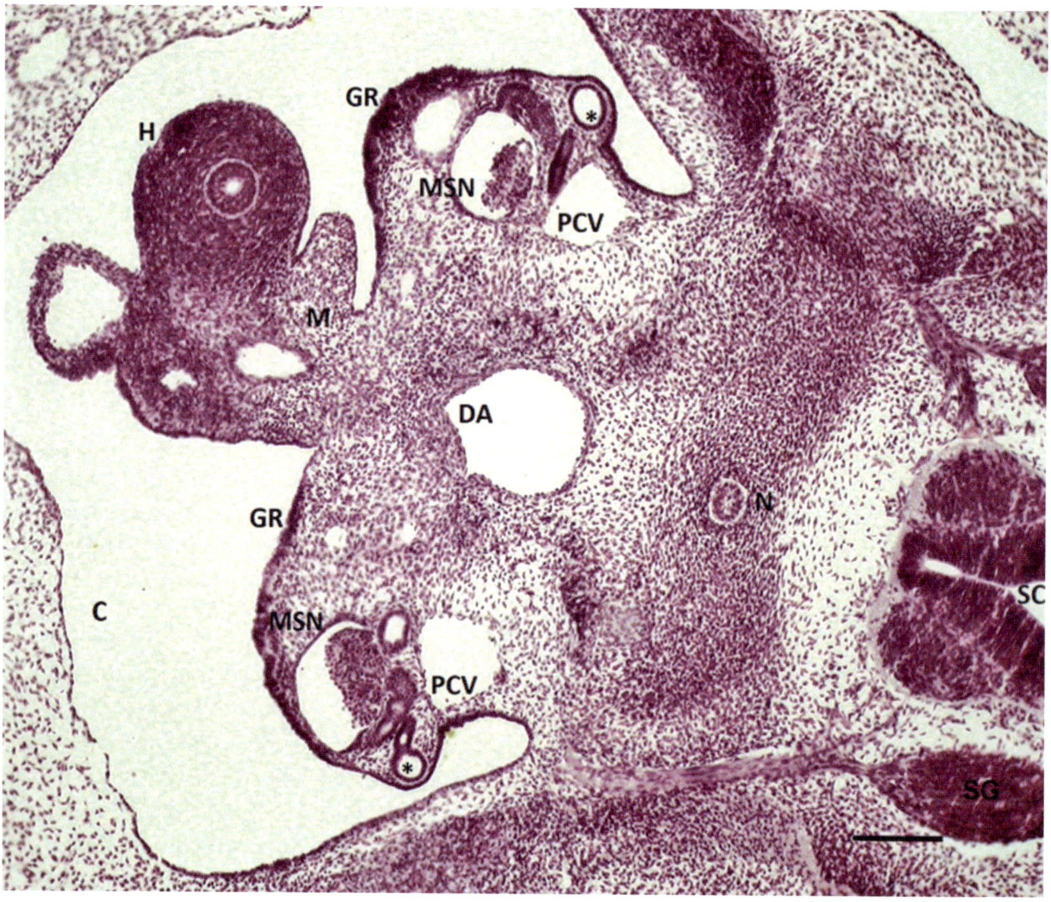

Fig. 4.1 Transverse section from a 5-week-old human embryo. The primordial germ cells migrate to the region of the future gonads. Cells from the coelomic (*C*) epithelium and the mesonephros proliferate, forming gonadal ridges (*GR*). Medially, the tubules of the mesonephros (*MSN*) enter the Wolffian (mesonephric) ducts (*asterisks*). *DA* dorsal aorta, *H* hindgut, *M* mesenterium, *N* notochord, *PCV* posterior cardinal vein, *SC* spinal cord, *SG* spinal ganglion (hematoxylin and eosin [H&E] stain; *scale bar*, 100 μm)

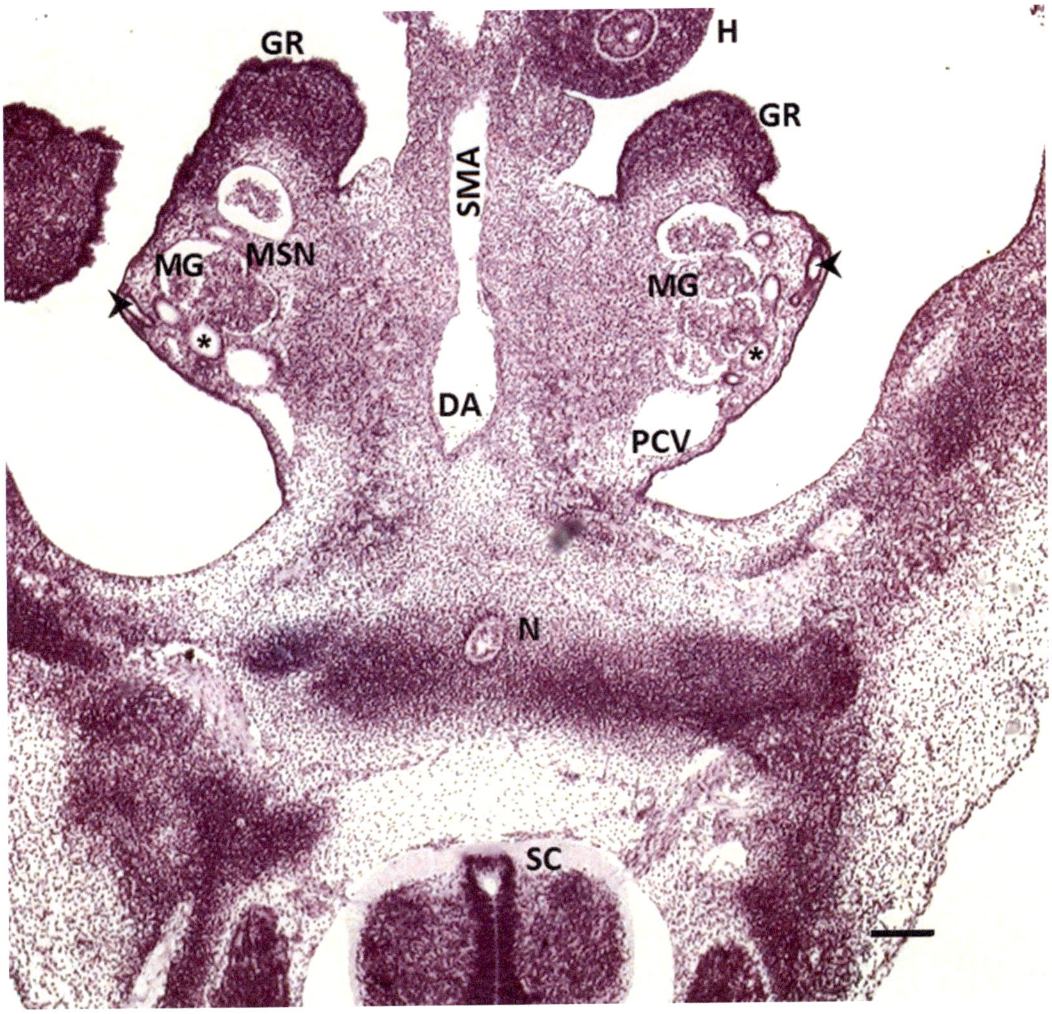

Fig. 4.2 Transverse section from a 5-week-old human embryo. The urogenital ridge is a transitory structure formed by the gonadal ridge (*GR*), mesonephros (*MSN*), and meso- and paramesonephric ducts. Branches of the aorta (*DA*) form the mesonephric glomeruli (*MG*), which drain into the posterior cardinal vein (*PCV*). The *arrowheads* point to the Müllerian (paramesonephric) ducts; the *asterisk* indicates the Wolffian (mesonephric) duct. *H* hindgut, *N* notochord, *SC* spinal cord, *SMA* superior mesenteric artery (H&E stain; *scale bar*, 100 μm)

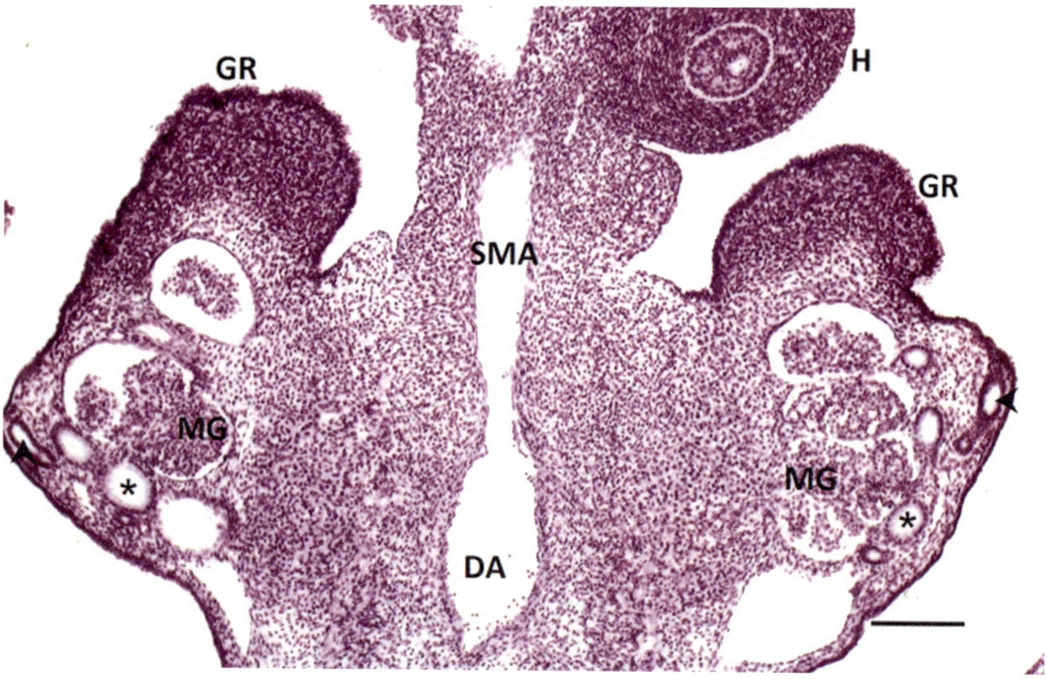

Fig. 4.3 Transverse section from the urogenital ridge structures in a 5-week-old human embryo. The superior mesenteric artery (*SMA*) originates from the dorsal aorta (*DA*). The *arrowheads* point to the Müllerian (paramesonephric) duct and the *asterisks* to the Wolffian (mesonephric) duct. *GR* gonadal ridge, *H* hindgut, *MG* mesonephric glomeruli (H&E stain; *scale bar*, 100 μm)

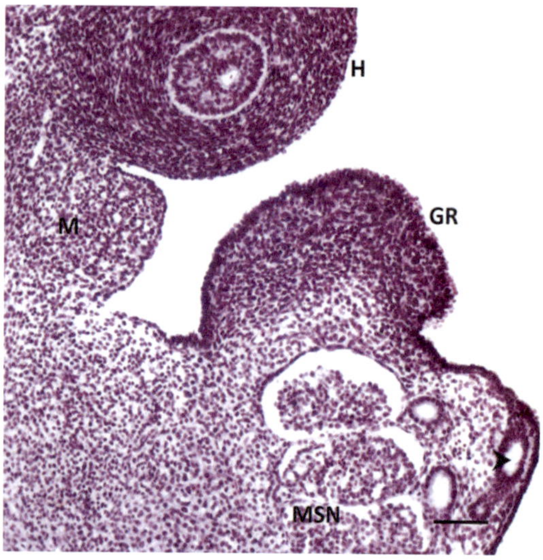

Fig. 4.4 Transverse section from a 5-week-old human embryo. Shown are the following structures of the urogenital ridge: the gonadal ridge (*GR*), mesonephros (*MSN*), and Müllerian (paramesonephric) duct (*arrowhead*). Medially is the hindgut (*H*) held by the mesenterium (*M*) (H&E stain; *scale bar*, 50 μm)

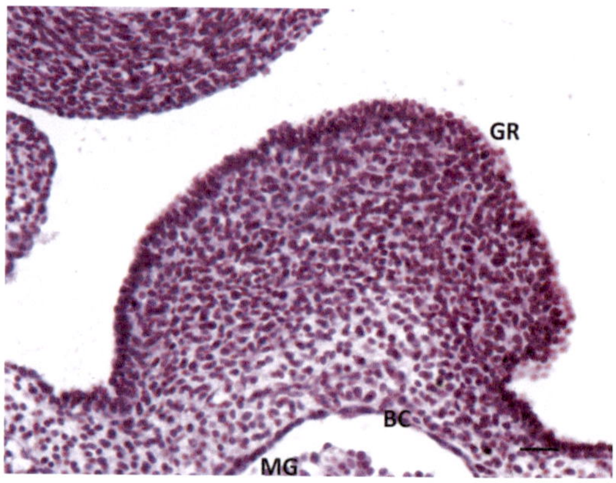

Fig. 4.5 Transverse section from a 5-week-old human embryo. The gonadal ridge (*GR*) appears in front of the mesonephric glomerulus (*MG*) within Bowman's capsule (*BC*) (H&E stain; *scale bar*, 100 μm)

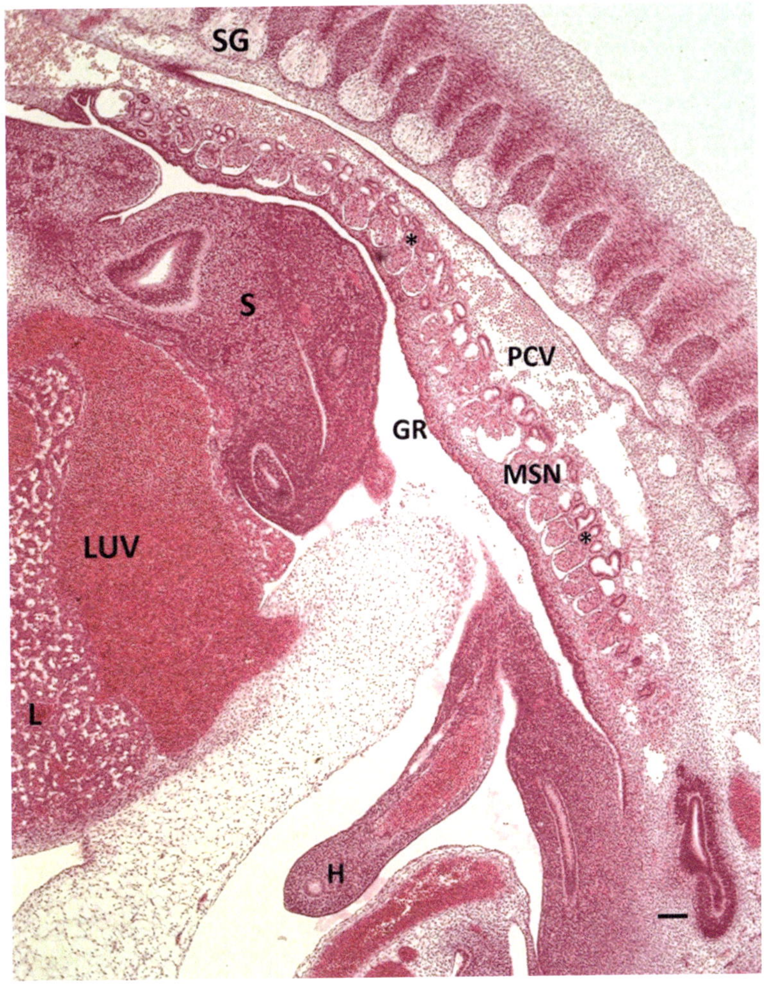

Fig. 4.6 Longitudinal section from a 6-week-old human embryo. The image shows the urogenital ridge, an elongated structure that lies on both sides of the midline between the lateral mesoderm and the root of the dorsal embryonic mesenterium. It consists of two main components: the nephrogenic cord, from which the urinary apparatus arises (marked here as the mesonephros [*MSN*]), and the gonadal ridge (*GR*) as the primordium of the gonad. The gonadal ridge extends from the upper thoracic region to the level of the cloaca, but the true gonad anlage develops only out of the middle area. The *asterisks* mark the Wolffian (mesonephric) duct. *H* hindgut, *L* liver, *LUV* left umbilical vein, *PCV* posterior cardinal vein, *S* stomach wall, *SG* spinal ganglion (H&E stain; *scale bar*, 100 μm)

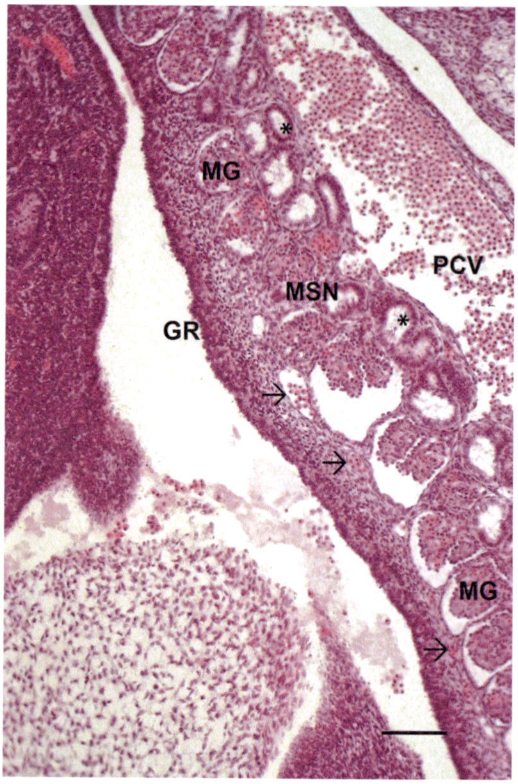

Fig. 4.7 Longitudinal section from a 6-week-old human embryo. The mesonephric glomeruli (*MG*) are a tuft of capillaries that originate in the lateral branches of the dorsal aorta and drain into the posterior cardinal vein (*PCV*). Note the newly formed blood islands (*arrows*) below the gonadal ridge (*GR*). Only the middle section will form the gonad. The tubules of the mesonephros (*MSN*) enter the Wolffian (mesonephric) duct (*asterisks*) (H&E stain; *scale bar*, 100 μm)

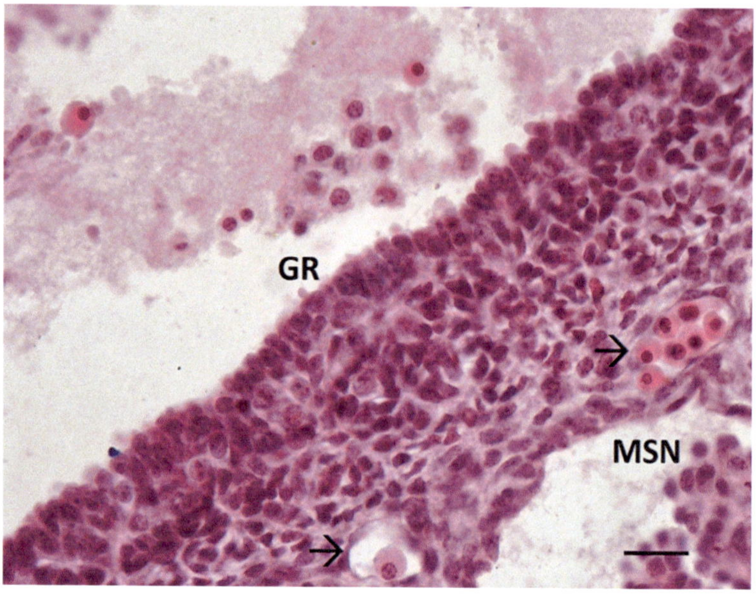

Fig. 4.8 Longitudinal section from a 6-week-old human embryo. Blood islands (*arrows*) between the gonadal ridge (*GR*) and the mesonephros (*MSN*) arise from mesoderm cells induced to form hemangioblasts, common precursors of vessel and blood cells (H&E stain; *scale bar*, 20 μm)

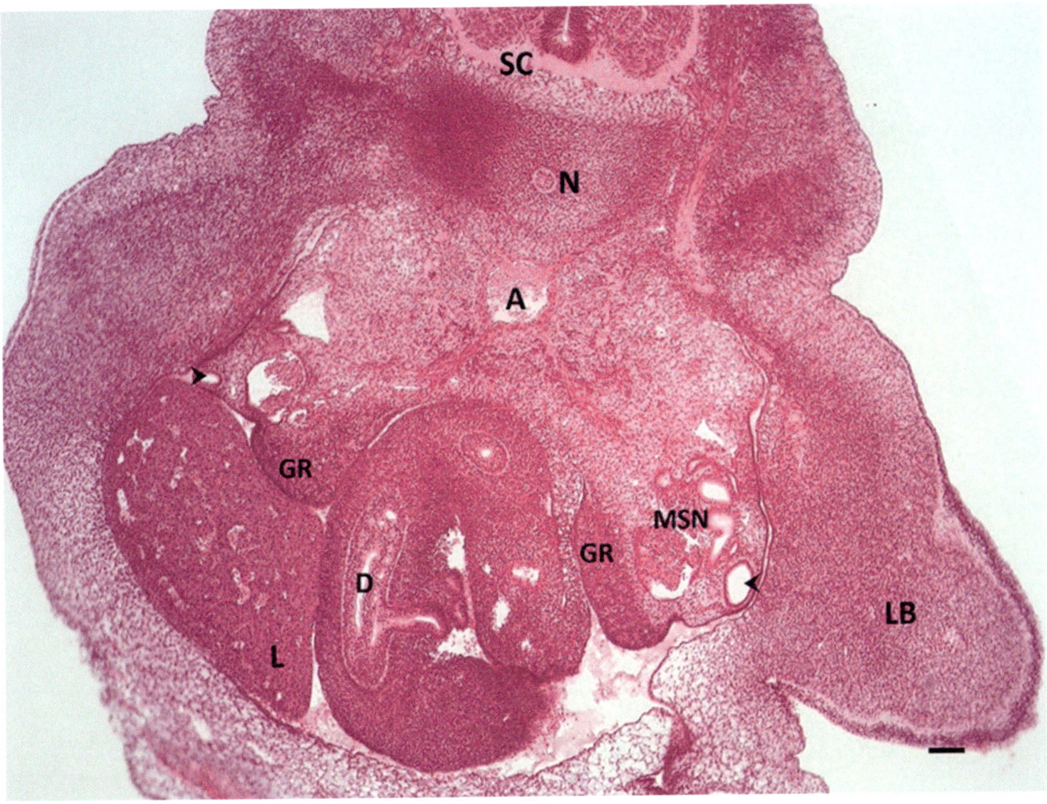

Fig. 4.9 Oblique section from a 6-week-old human embryo. The section is from the level of the lower limb bud (*LB*) primordium, showing the main structures of the lower abdomen: the liver (*L*), the duodenum (*D*), and a well-developed gonadal ridge (*GR*) rotating medially to the mesonephros (*MSN*). The *arrowheads* point to the Müllerian (paramesonephric) duct. *A* aorta, *N* notochord, *SC* spinal cord (H&E stain; *scale bar*, 100 μm)

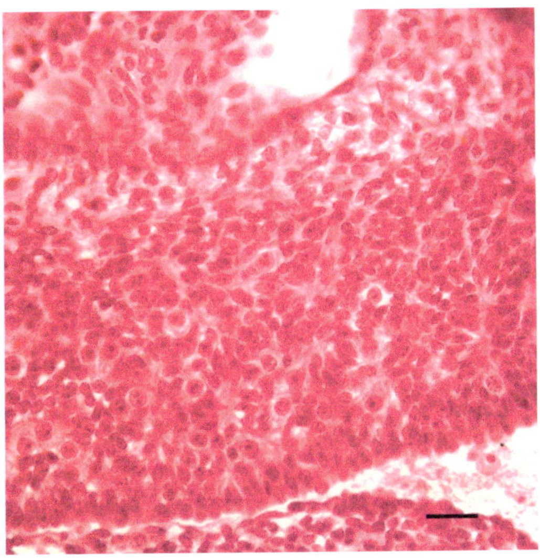

Fig. 4.10 Transverse section from the gonadal ridge of a 6-week-old human embryo (H&E stain; *scale bar*, 100 μm)

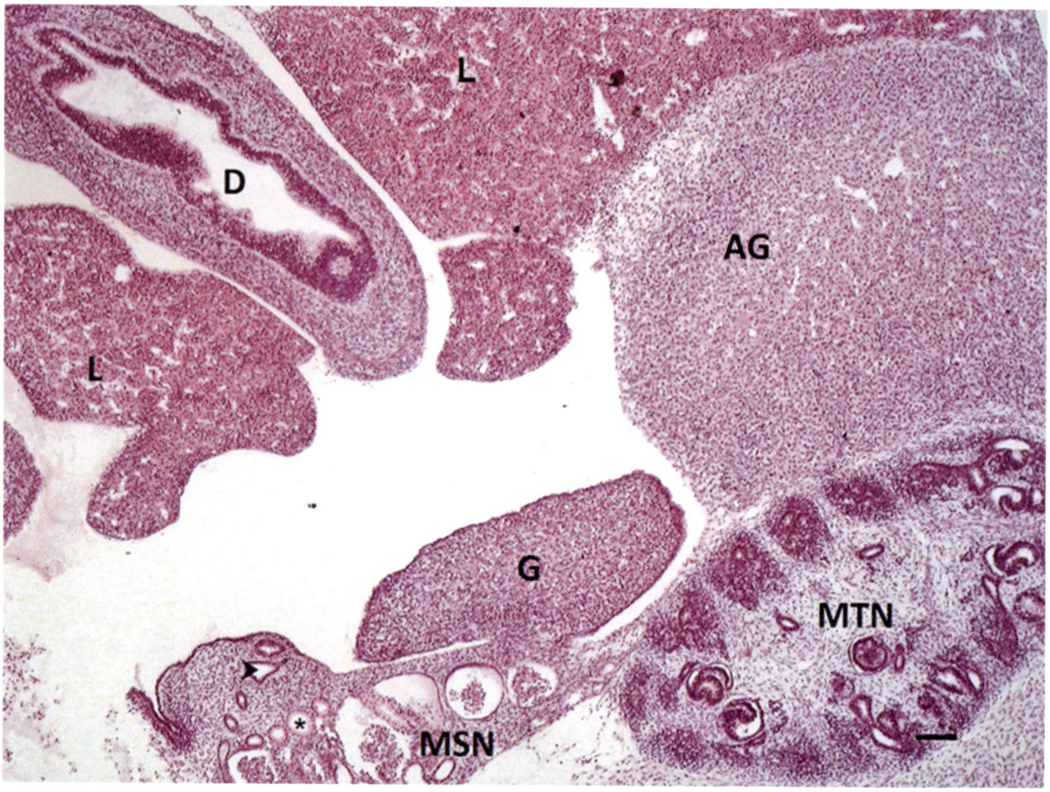

Fig. 4.11 Longitudinal section from the lower abdominal organs of a 7-week-old human embryo. Shown are the liver (*L*), adrenal gland (*AG*), duodenum (*D*), gonad (*G*), metanephros (*MTN*), mesonephros (*MSN*), Müllerian (paramesonephric) duct (*arrowhead*), and Wolffian (mesonephric) duct (*asterisk*) (H&E stain; *scale bar*, 100 μm)

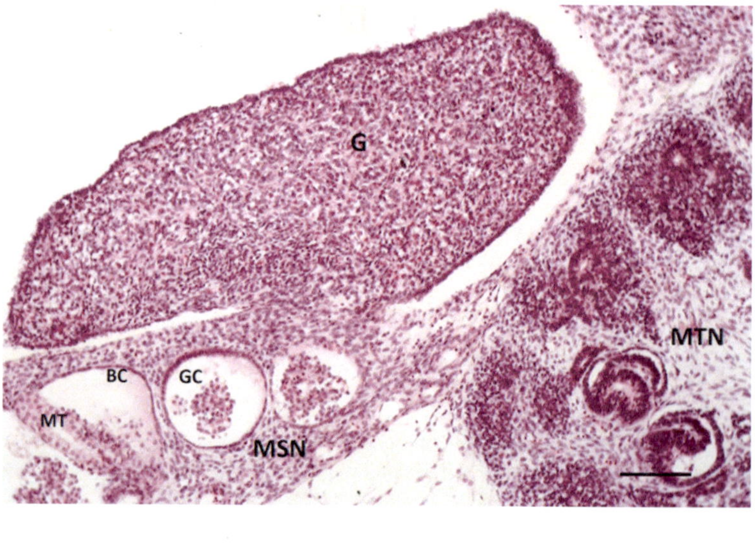

Fig. 4.12 Longitudinal section from a 7-week-old human embryo. The image shows the structures of the mesonephros (*MSN*): the tubules (*MT*), Bowman's capsule (*BC*), and the glomeruli in their cavity (*GC*), which are atrophying while the metanephros (*MTN*) is developing. The gonad (*G*) lies in front (H&E stain; *scale bar*, 100 μm)

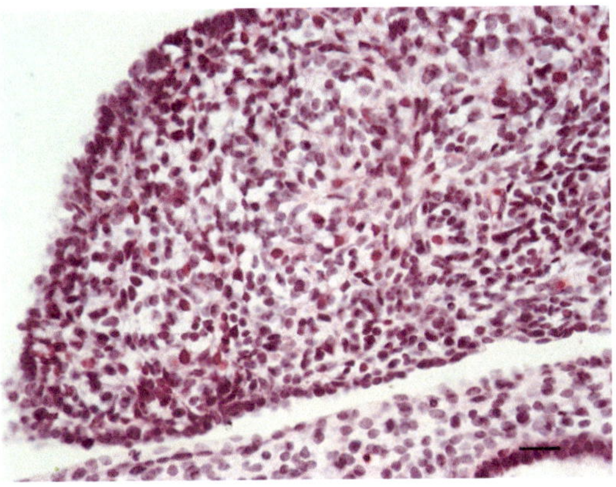

Fig. 4.13 Gonad of a 7-week-old human embryo in a longitudinal section (H&E stain; *scale bar*, 20 μm)

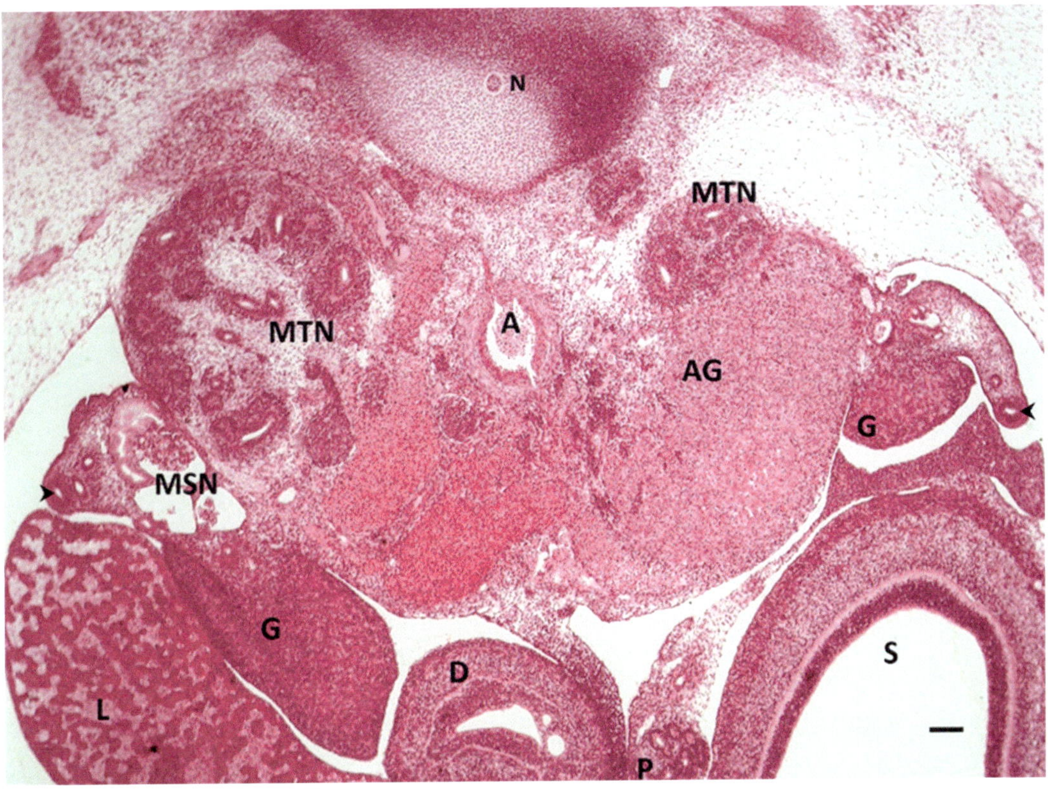

Fig. 4.14 Transverse section from the lumbar region of a 7-week-old human embryo. The metanephros (*MTN*) and adrenal glands (*AG*) are placed retroperitoneally. The gonads (*G*) rotate medially; the Müllerian (paramesonephric) ducts (*arrowheads*) will fuse later in development. *A* aorta, *D* duodenum, *L* liver, *MSN* mesonephros, *N* notochord, *P* pancreas, *S* stomach (H&E stain, *scale bar*, 100 μm)

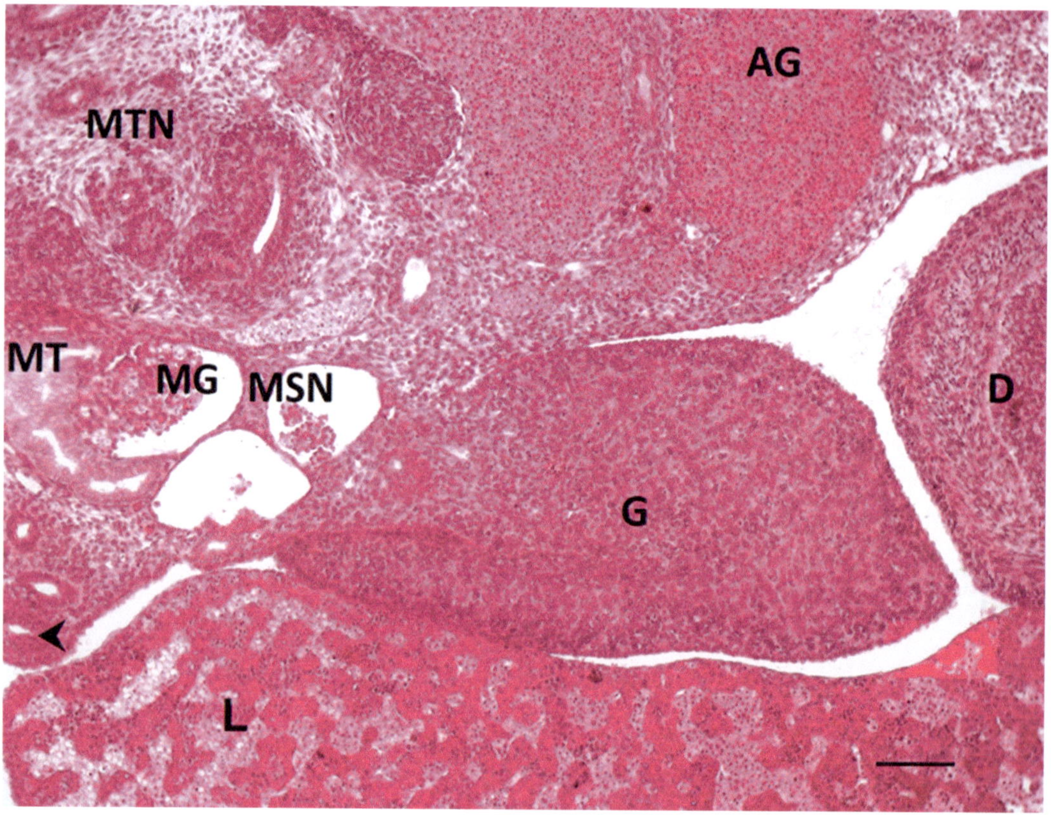

Fig. 4.15 Transverse section from a 7-week-old human embryo. There is communication among the gonad (*G*), metanephros (*MTN*), adrenal gland (*AG*), duodenum (*D*), liver (*L*), and structures of the mesonephros (*MSN*): the mesonephric glomerulus (*MG*) and mesonephric tubule (*MT*), with the Müllerian (paramesonephric) duct (*arrowhead*) positioned laterally (H&E stain; *scale bar*, 100 μm)

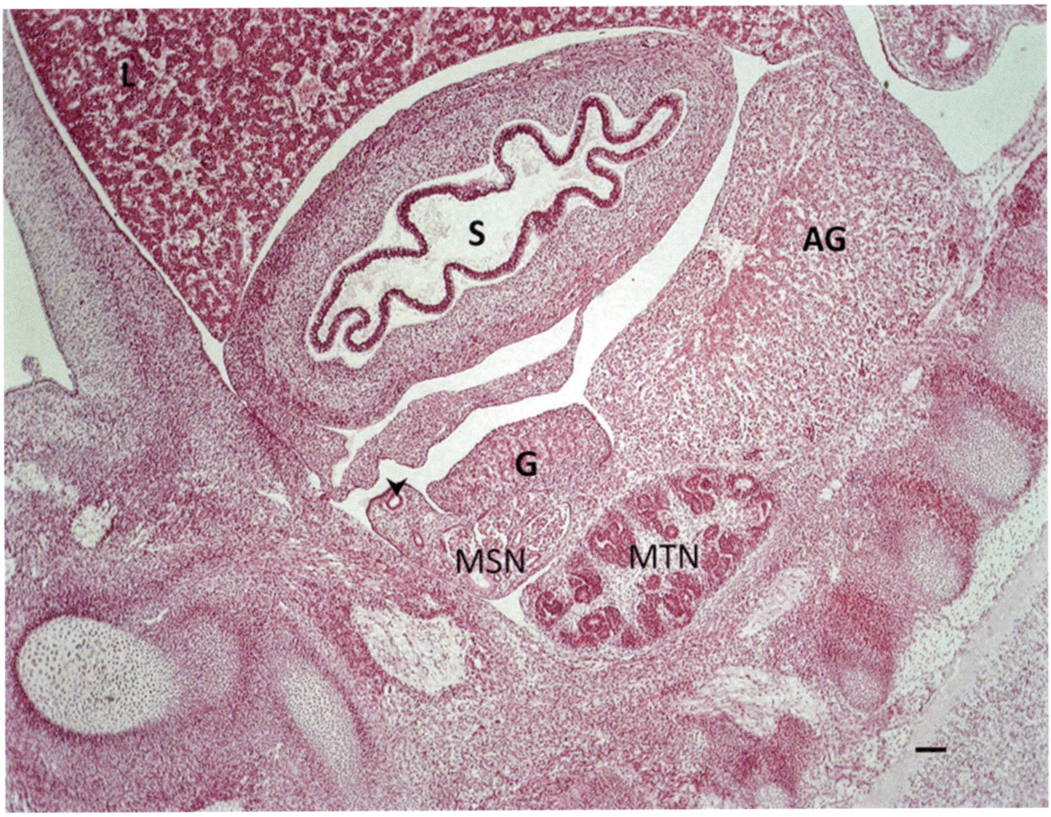

Fig. 4.16 Longitudinal section from an 8-week-old human embryo. The image shows the organs of the lower abdomen: the gonad (*G*), adrenal gland (*AG*), mesonephros (*MSN*), metanephros (*MTN*), liver (*L*), and stomach (*S*). The *arrowhead* points to the Müllerian (paramesonephric) duct (H&E stain; *scale bar*, 100 μm)

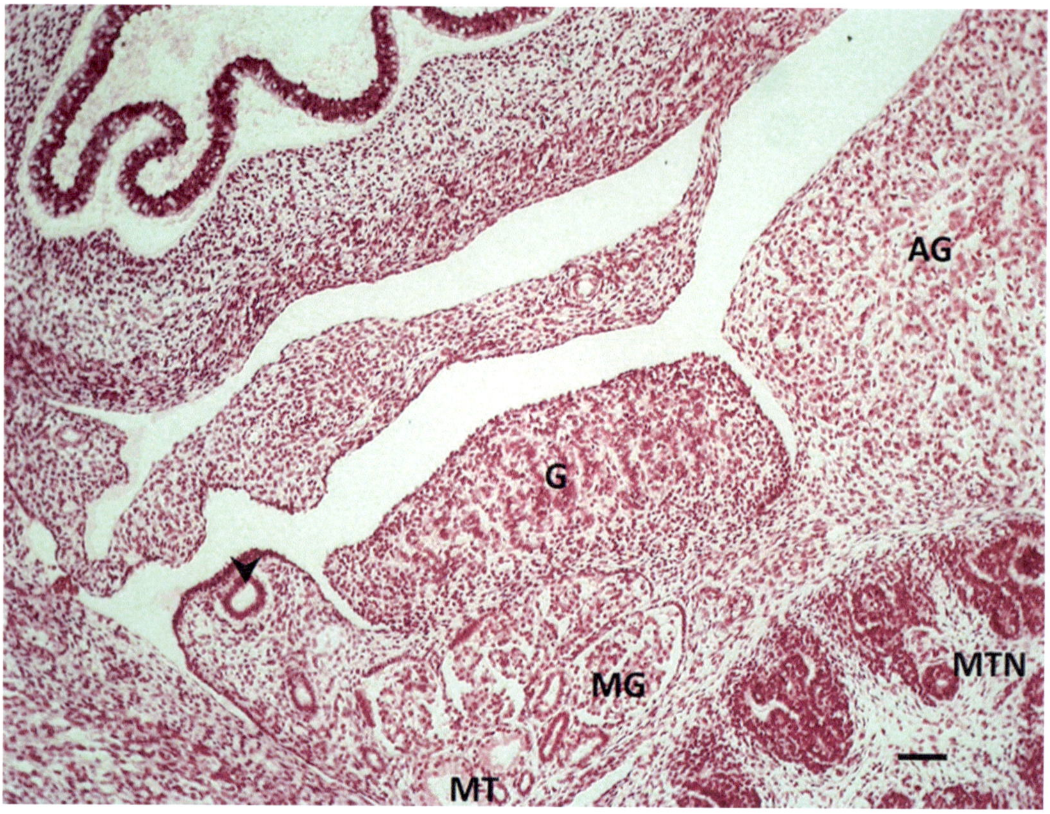

Fig. 4.17 The gonad (*G*) of an 8-week-old human embryo in a longitudinal plane. At this stage, the adrenal gland (*AG*) is much bigger than the metanephros (*MTN*), which will develop into the definite kidney. The mesonephric tubules (*MT*) and glomeruli (*MG*) are atrophying. The *arrowhead* points to the Müllerian (paramesonephric) duct (H&E stain; *scale bar*, 100 μm)

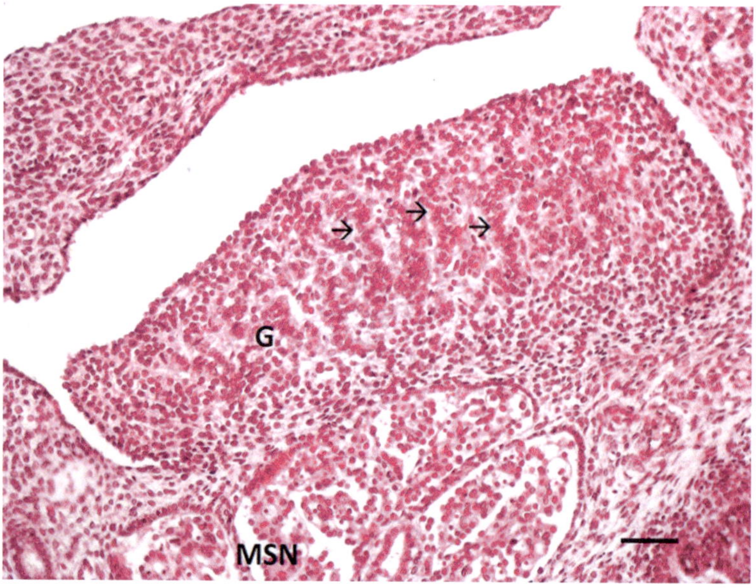

Fig. 4.18 Longitudinal section from the gonad (*G*) lying in front of the mesonephros (*MSN*) in an 8-week-old human embryo. The *arrows* point to the gonadal cords (H&E stain; *scale bar*, 50 μm)

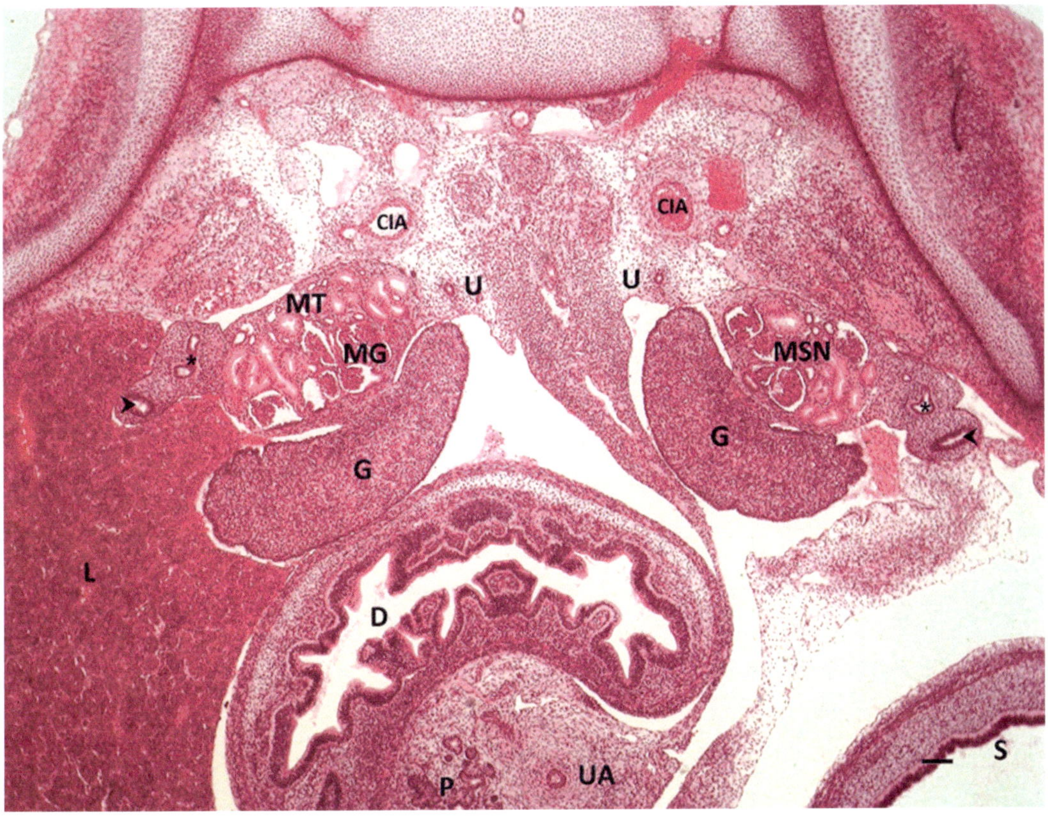

Fig. 4.19 Eight-week-old human embryo in a frontal section below the aortoiliac bifurcation. The urogenital ridge now clearly divides into a gonad (*G*) and the atrophying mesonephros (*MSN*). The ureter (*U*) descends from the more cranially placed metanephros. The *asterisk* marks the Wolffian (mesonephric) duct; the *arrowhead* points to the Müllerian (paramesonephric) duct. *CIA* common iliac artery, *D* duodenum, *L* liver, *MG* mesonephric glomeruli, *MT* mesonephric tubules, *P* pancreas, *S* stomach, *UA* umbilical artery (H&E stain; *scale bar*, 100 μm)

Fig. 4.20 Frontal section from an 8-week-old human embryo. The mesonephric (Wolffian) duct (*asterisk*) forms a transitory precursor of the adult excretory system with the S-shaped mesonephric tubules (*MT*). The medial end of the mesonephric tubule is closed and forms the Bowman's capsule that surrounds the glomerulus (*MG*). Lateral is the Müllerian or paramesonephric duct (*arrowhead*). The ureter (*U*) belongs to the metanephros (placed more cranially in this developmental stage). The gonad (*G*) is surrounded by the liver (*L*) and duodenum (*D*) (H&E stain; *scale bar*, 100 μm)

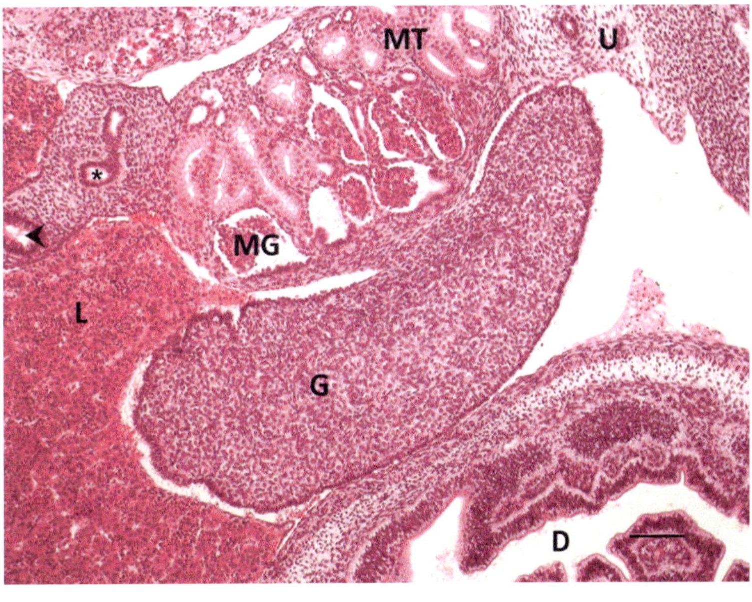

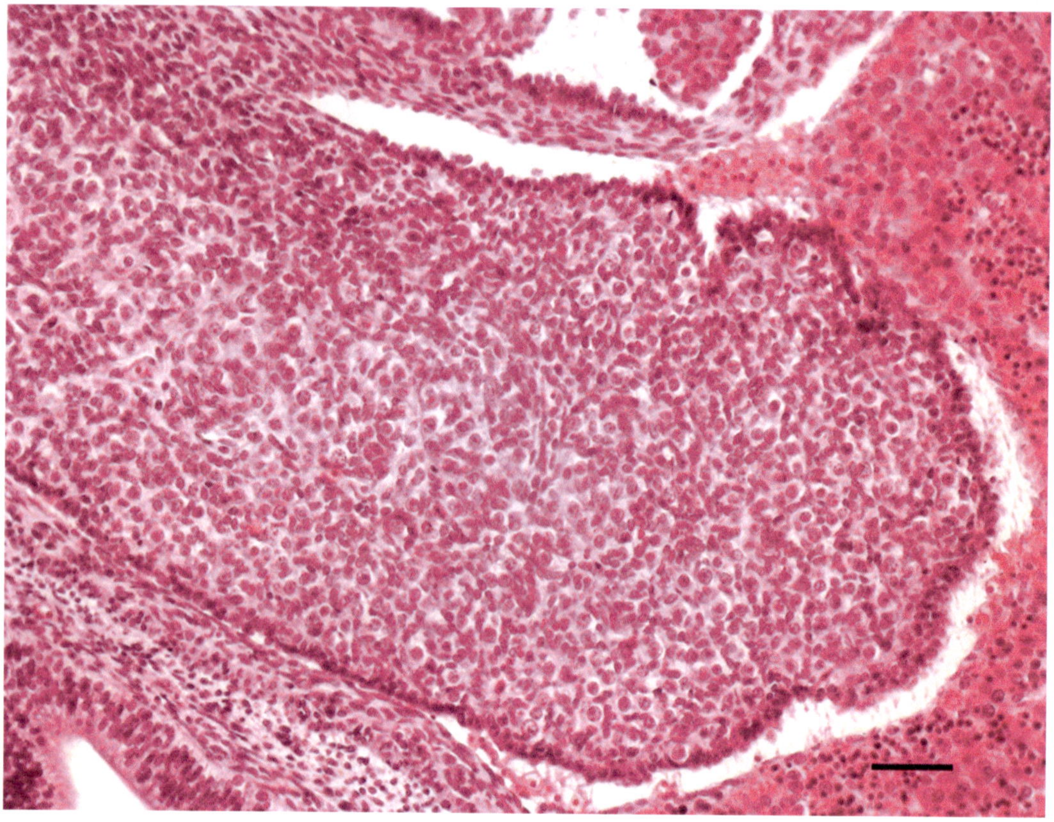

Fig. 4.21 Gonad of an 8-week-old human embryo in a frontal section showing the gonadal cords forming (H&E stain; *scale bar*, 50 μm)

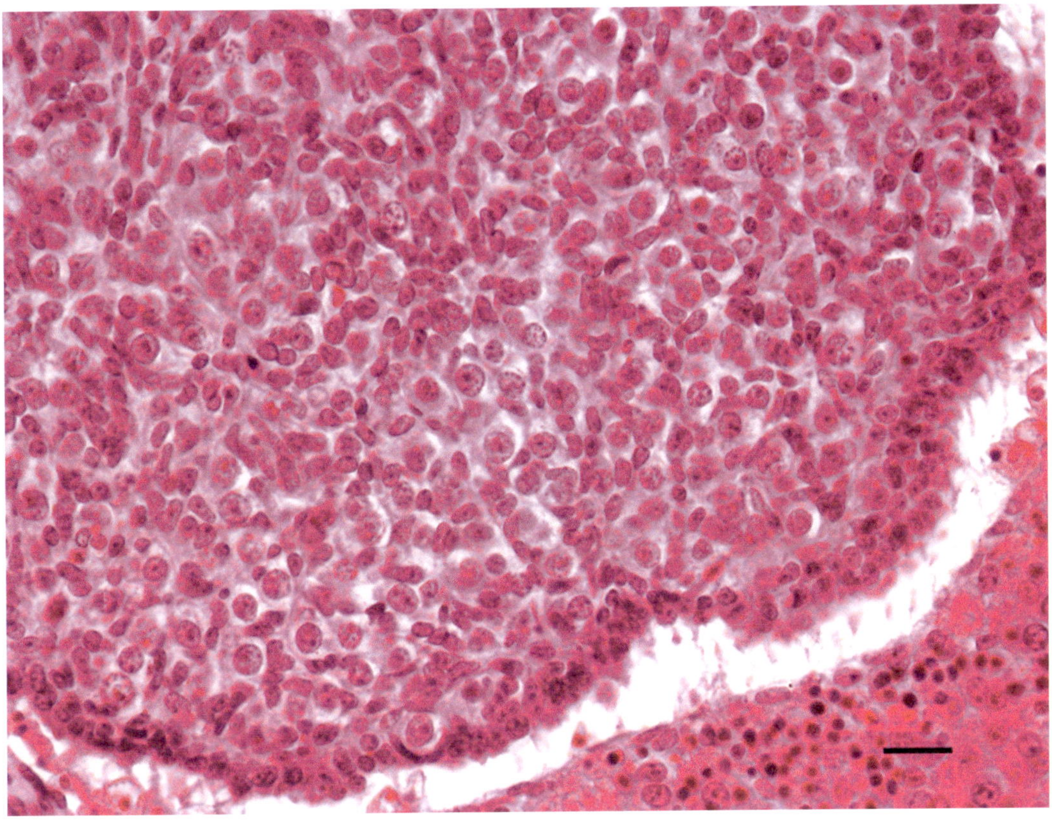

Fig. 4.22 Gonad of an 8-week-old human embryo (H&E stain; *scale bar*, 20 μm)

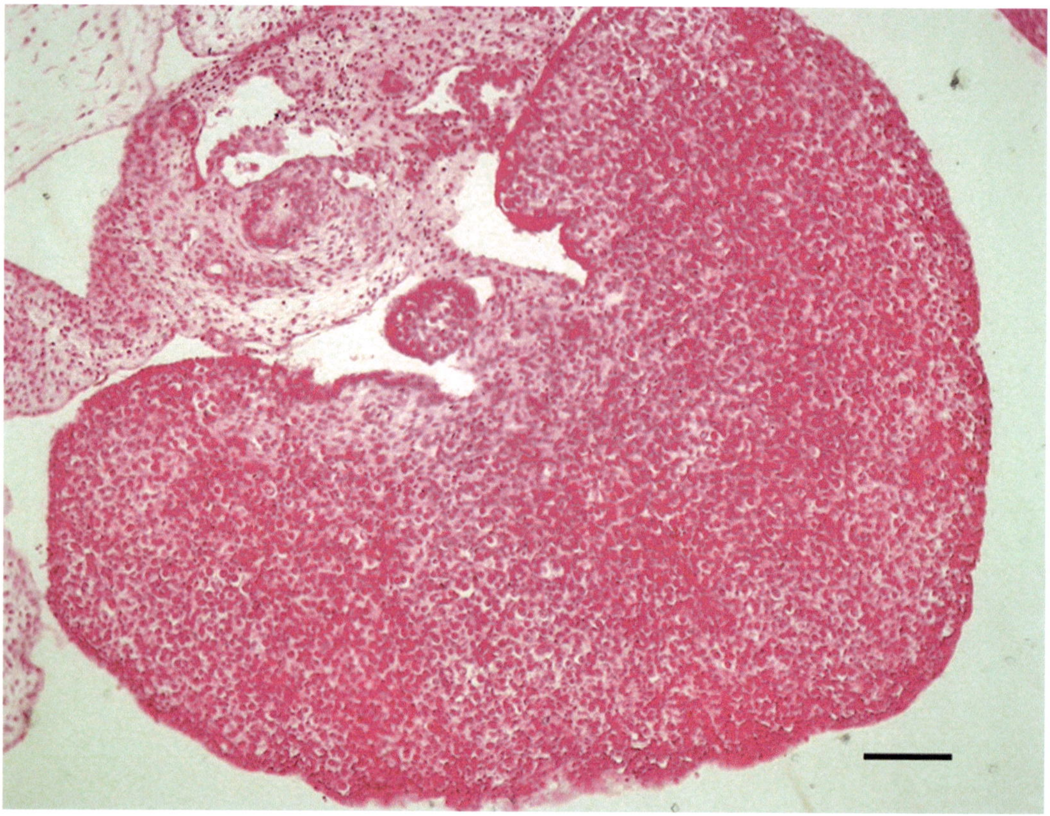

Fig. 4.23 Transverse section from a gonad and gonadal mesentery of a 10-week-old human fetus (H&E stain; *scale bar*, 100 μm)

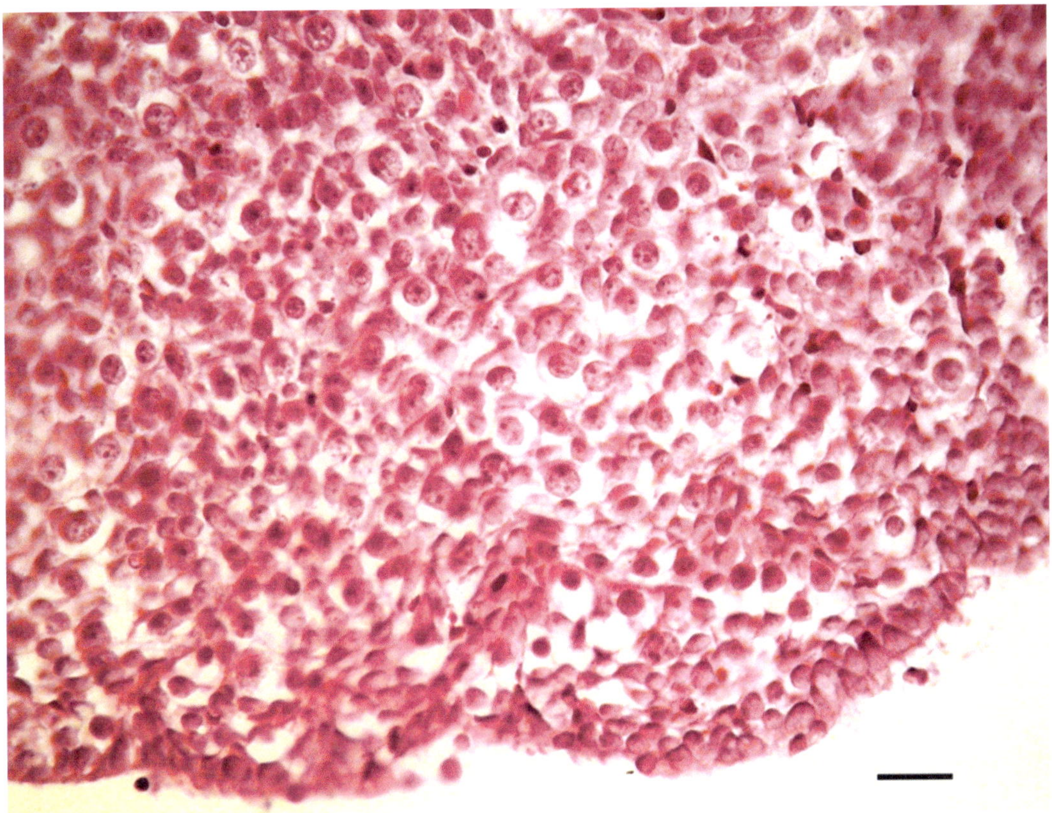

Fig. 4.24 Ten-week-old human fetal gonad (H&E stain; *scale bar*, 20 μm)

References

1. Moore KL, Persaud TVN. Chapter 12: The urogenital system. In: The developing human. 8th ed. Philadelphia: Saunders/Elsevier; 2008. p. 244–84.
2. Sizonenko PC. Human sexual differentiation. In: Bertrand J, Rappaport R, Sizonenko PC, editors. Pediatric endocrinology. Baltimore: Williams and Wilkins; 1993. p. 88–99.
3. Sadler TW. Chapter 15: Urogenital system. In: Sadler TW, editor. Langman's medical embryology. 11th ed. Philadelphia: Lippincott William and Wilkins; 2010. p. 235–63.
4. Duančić V. Development of the urogenital system. In: Švajger A, editor. Basic human embryology [in Croatian]. 9th ed. Zagreb: Medicinska knjiga Beograd-Zagreb; 1985. p. 129–47.

The Fetal Human Testis

5

Marina Kos and Tanja Leniček

At our pathology department, all spontaneously aborted fetuses are routinely autopsied and analyzed. During fetal autopsy, the position of the testes is assessed and noted, and both testes are taken for routine histopathologic analysis. They are fixed in 10 % buffered formalin, embedded in paraffin and routinely processed, cut in 5-μm sections, and stained with hematoxylin and eosin (H&E). If the microscopic features are within the normal limits for gestational age, the microscopic appearance is not described in detail in the autopsy report. If there is any deviation from the appearance expected for gestational age, it is described in detail. Although immunohistochemical (IHC) analysis is not done routinely, for the purpose of this chapter and for a better understanding of fetal development, certain markers were used (see the figures).

10–20 Weeks' Gestation

During the 11th gestational week, proliferation and maturation of Leydig cells begin, inducing the transformation of gonocytes to prespermatogonia and resulting in changes in microscopic appearance.

By 14 weeks' gestation, Leydig cells constitute half the volume of the testis, separating the tubules by a distance of one to four tubular diameters. The capsule consists of the tunica vaginalis (the flat to low cuboidal epithelium) and tunica albuginea (several horizontal layers of loose connective tissue). The tips of the sex cords, future seminiferous tubules, do not blend into the tunica albuginea, and the ends of their loops are not coiled.

At 14 weeks' gestation, the most numerous cells in the cords/tubules are Sertoli cells, which are perpendicular to the basement membrane and localized around the germ cells. Their main role is to maintain the germ cells and regulate their maturation. By 14–15 weeks' gestation, the transformation of gonocytes into fetal prespermatogonia takes place. The gonocytes migrate toward the basement membrane through the stage of intermediate cells, transforming into fetal prespermatogonia when they reach the basement membrane. Gonocytes that are the stem cells of the fetal male germ cells proliferate, producing increasing numbers of fetal prespermatogonia, and maintain their own number. Peritubular myoid cells appear before 15 weeks' gestation and later become more prominent.

20–40 Weeks' Gestation

During this period, Leydig cells of the fetal type undergo regression and dedifferentiation. They become smaller, vacuolated, and, in the end,

M. Kos, M.D., Ph.D. (✉) • T. Leniček, M.D.
Clinical Department of Pathology "Ljudevit Jurak",
Clinical Hospital Center "Sestre milosrdnice",
Zagreb, Croatia
e-mail: dr_marina_kos@yahoo.com

D. Ježek (ed.), *Atlas on the Human Testis*,
DOI 10.1007/978-1-4471-2763-5_5, © Springer-Verlag London 2013

spindle shaped, blending into the interstitium. Some of these cells also may undergo apoptosis. The tunica albuginea thickens steadily, and the tunica vasculosa forms, as do the interlobular septa that are invariably present after 25–28 weeks' gestation. The future seminiferous tubules increase in length and coiling, whereas the tubular diameter does not change much. The number of germ cells per tubular diameter remains constant throughout gestation. The septal, mediastinal, and extratesticular parts of the rete become more apparent.

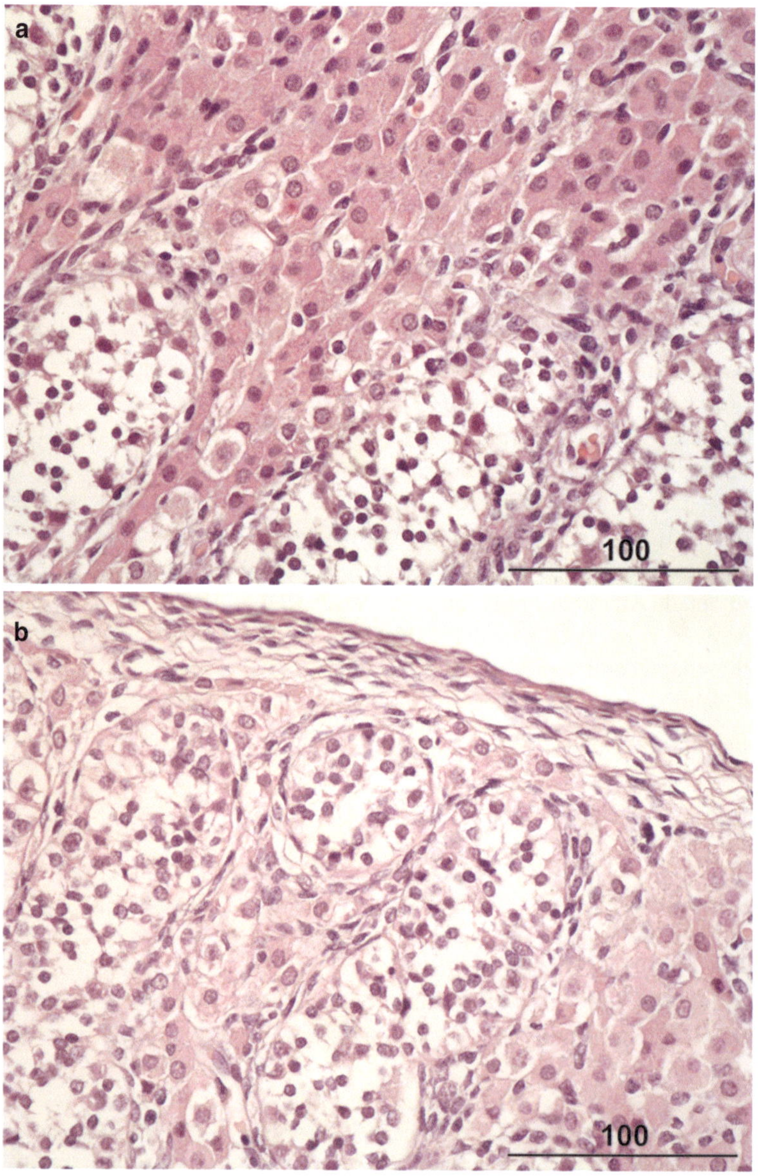

Fig. 5.1 Fetal testis at 17 weeks' gestation. (**a**) The Leydig cells still constitute a large portion of the testicular volume. (**b**) The columns of Leydig cells between the tubules become narrow, because the Leydig cells regress in size and number (H&E stain; *scale bar*, 100 μm)

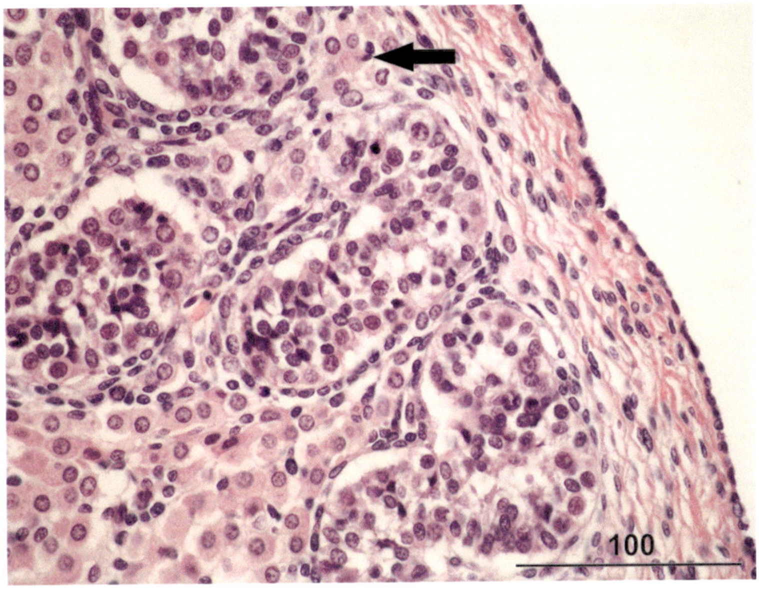

Fig. 5.2 Fetal testis at 17 weeks' gestation. The capsule is composed of visceral epithelium of the tunica vaginalis (on the surface) and the tunica albuginea, whose thickness ranges from 25 to 111 μm in the period from 17 to 20 weeks' gestation. Loose connective tissue deep in the tunica albuginea is beginning to form and later will become the tunica vasculosa. Between the tips of the tubules and the tunica albuginea of the capsule, there are very few Leydig cells (*arrow*) (H&E stain; *scale bar*, 100 μm)

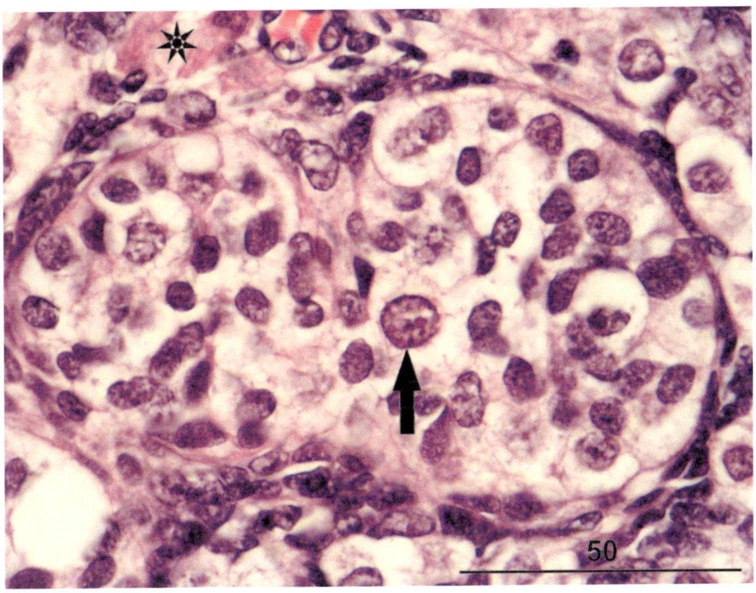

Fig. 5.3 Fetal testis at 17 weeks' gestation: a sex cord, a future seminiferous tubule. On light microscopy, differentiation among germ cell subtypes is not possible, and differentiating germ cells from Sertoli cells is rather difficult. Germ cells contain a large round, centrally placed nucleus; one or two large round nucleoli; and scanty cytoplasm (*arrow*). Sertoli cells have smaller, oval, and more basophilic nuclei. The median sex cord/tubular diameter is 59 μm (range, 43–87 μm). The median number of germ cells per tubular diameter is seven (range, 1–13), and the median number of Sertoli cells is 32 (range, 16–52). The tubular diameter and median number of germ and Sertoli cells do not change appreciably until the end of gestation. The tubules are surrounded by one or two layers of myoid and fibroblast cells, whereas the interstitium contains Leydig cells (*asterisk*) (H&E stain; *scale bar*, 50 μm)

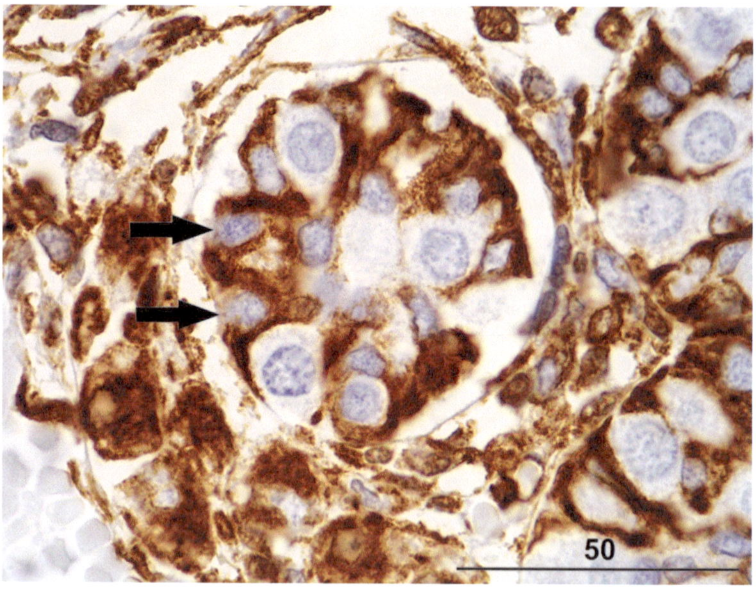

Fig. 5.4 Fetal testis at 17 weeks' gestation: a sex cord. The immature Sertoli cell has an oval to round nucleus and an inconspicuous nucleolus. Sertoli cells show cytoplasmic expression of vimentin (*arrows*), whereas the germ cells are negative (IHC stain for vimentin; *scale bar*, 50 μm)

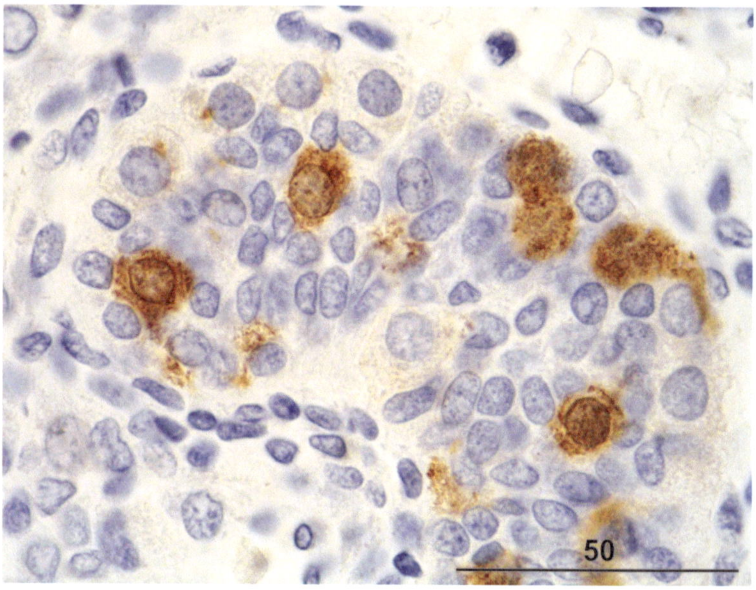

Fig. 5.5 Fetal testis at 17 weeks' gestation: a sex cord. Fetal germ cells, or gonocytes, express the cell membrane proto-oncogene receptor CD-117 (c-kit), placenta-like alkaline phosphatase (PLAP), and nuclear transcriptional regulator Oct3/4. These markers appear early in gestation, disappearing gradually at term or soon after birth. The Leydig, Sertoli, and interstitial cells do not express these markers, and they are not expressed in normal adult germ cells (IHC stain for CD-117; *scale bar*, 50 μm)

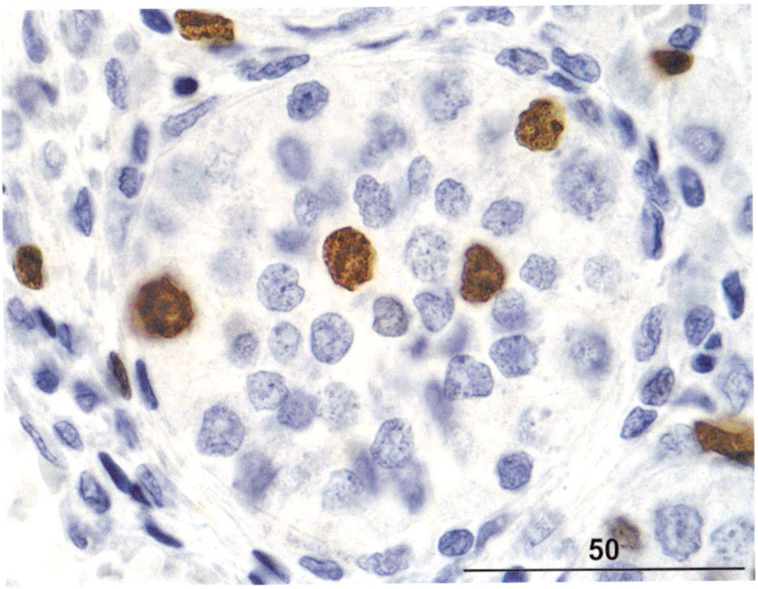

Fig. 5.6 Fetal testis at 17 weeks' gestation: a sex cord. Fetal germ cells reveal strong nuclear staining with Ki-67 throughout gestation. After birth, only a few cells show proliferative activity (IHC stain for Ki-67; *scale bar*, 50 μm)

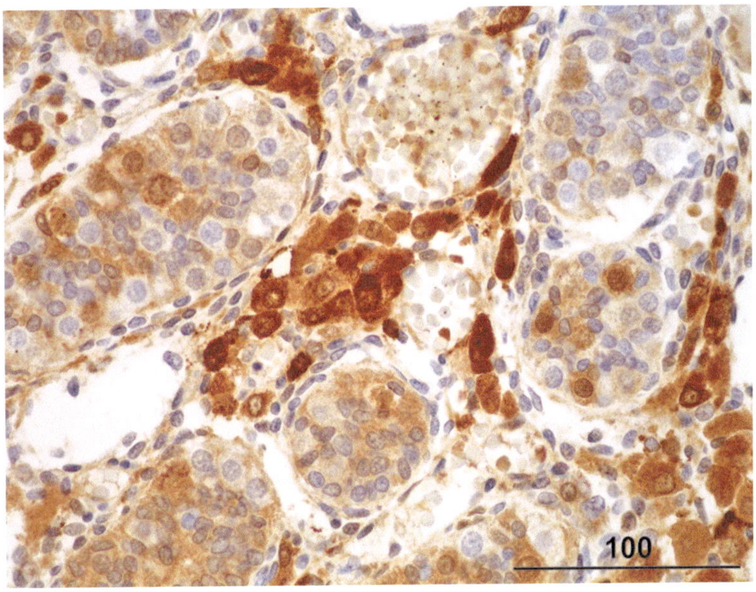

Fig. 5.7 Fetal testis at 17 weeks' gestation. Leydig cells show strong cytoplasmic calretinin immunoreactivity (IHC stain for calretinin; *scale bar*, 100 μm)

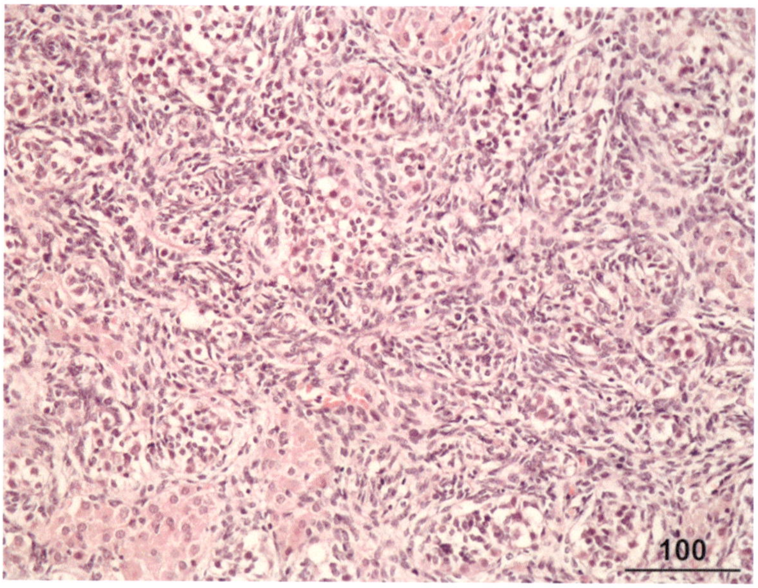

Fig. 5.8 Fetal testis at 17 weeks' gestation: rete testis. At this gestational age, it is difficult to distinguish between the rete compartments because the lumina are not formed yet (H&E stain; *scale bar*, 100 μm)

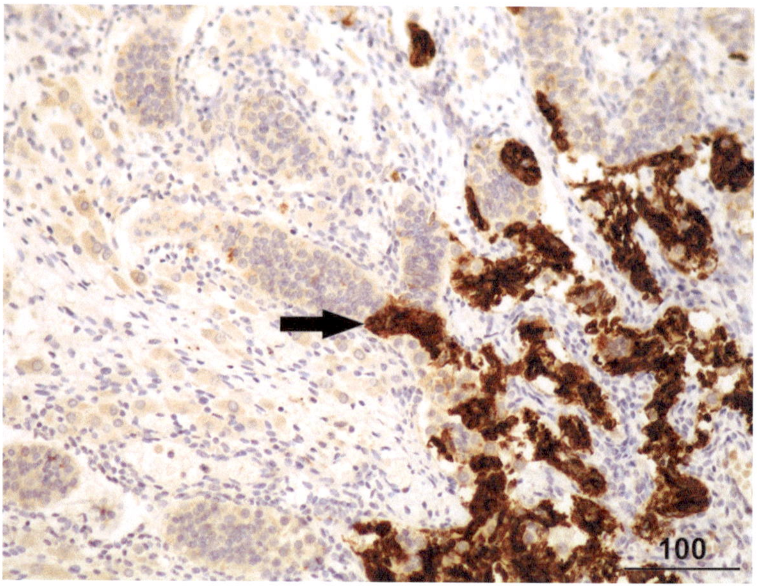

Fig. 5.9 Fetal testis at 17 weeks' gestation: rete testis. The hilar ends of the sex cords consist only of CK8-positive Sertoli cells (*arrow*) and end at the part of the rete that will become the mediastinal compartment after formation of the lumina. The rete epithelium also shows CK8 immunopositivity (IHC stain for CK8; *scale bar*, 100 μm)

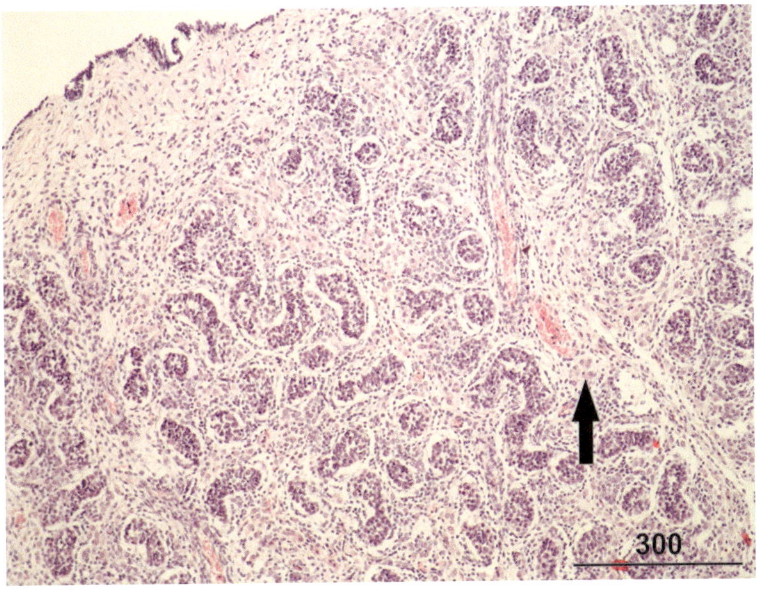

Fig. 5.10 Fetal testis at 25 weeks' gestation. The tunica albuginea thickens further and shows an outer, more condensed layer and an inner, loose, more vascular layer. The interlobular septa are formed and contain blood vessels and some Leydig cells (*arrow*). Leydig cells are still seen in the interstitium, but they rarely separate the tubules for more than one tubular diameter (H&E stain; *scale bar*, 300 μm)

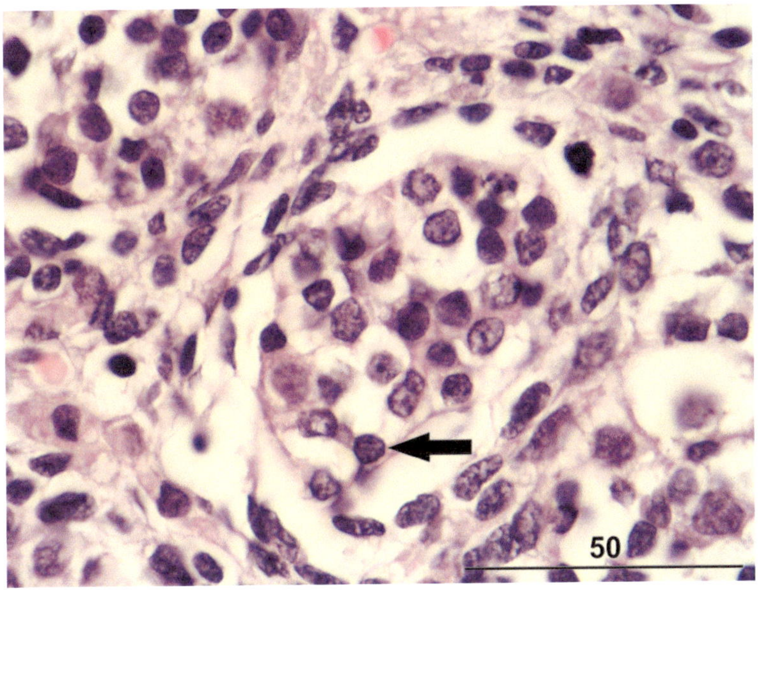

Fig. 5.11 Fetal testis at 25 weeks' gestation: a future seminiferous tubule. The lumen is ill-defined and contains Sertoli and germ cells. The fetal Sertoli cell has an oval to round nucleus and an inconspicuous nucleolus (*arrow*). Fetal Sertoli cells outnumber germ cells by a ratio of 7:1 and undergo active mitotic division during this period (H&E stain; *scale bar*, 50 μm)

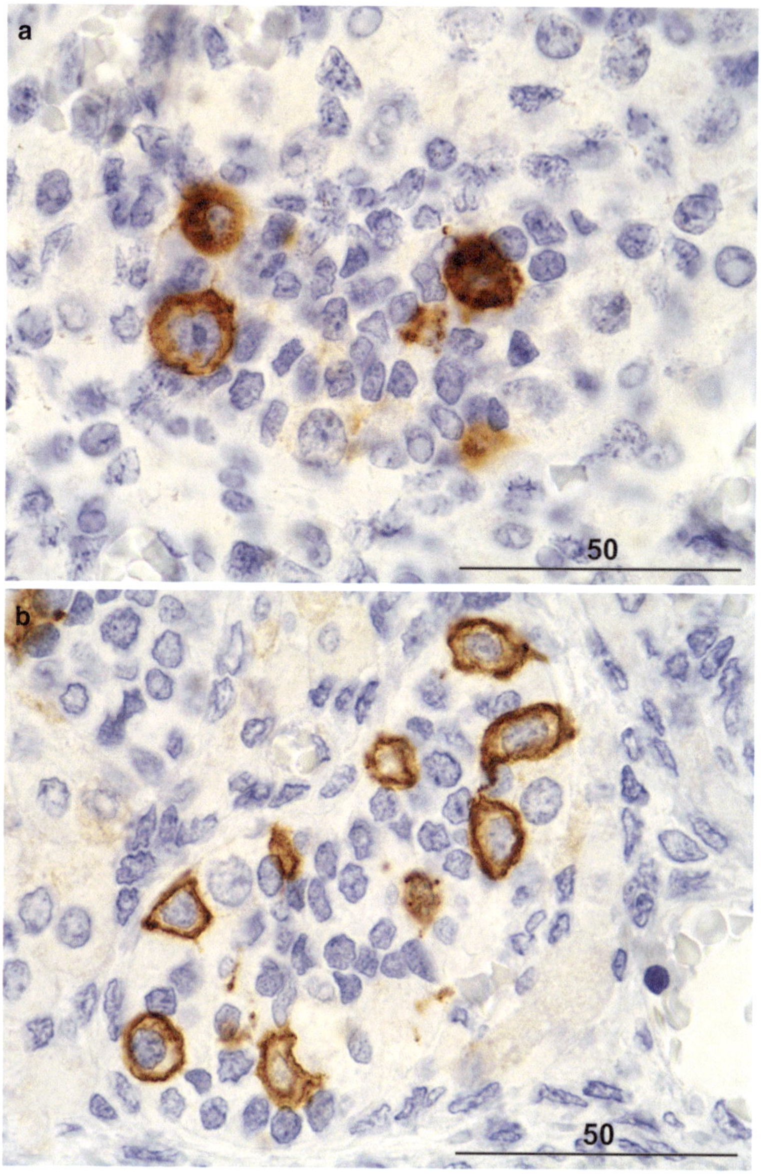

Fig. 5.12 Fetal testis at 25 weeks' gestation: a sex cord (future seminiferous tubule). With the advancing maturation of the fetal germ cells, at about 20 weeks of gestation, the expression of PLAP is the first to decrease, followed by Oct3/4 and c-kit. Because the rates of expression disappearance are different, it is possible to observe positive staining of germ cells for all three of these markers simultaneously, or for all combinations (IHC stain for PLAP (**a**) and CD-117 (**b**); *scale bar*, 50 μm)

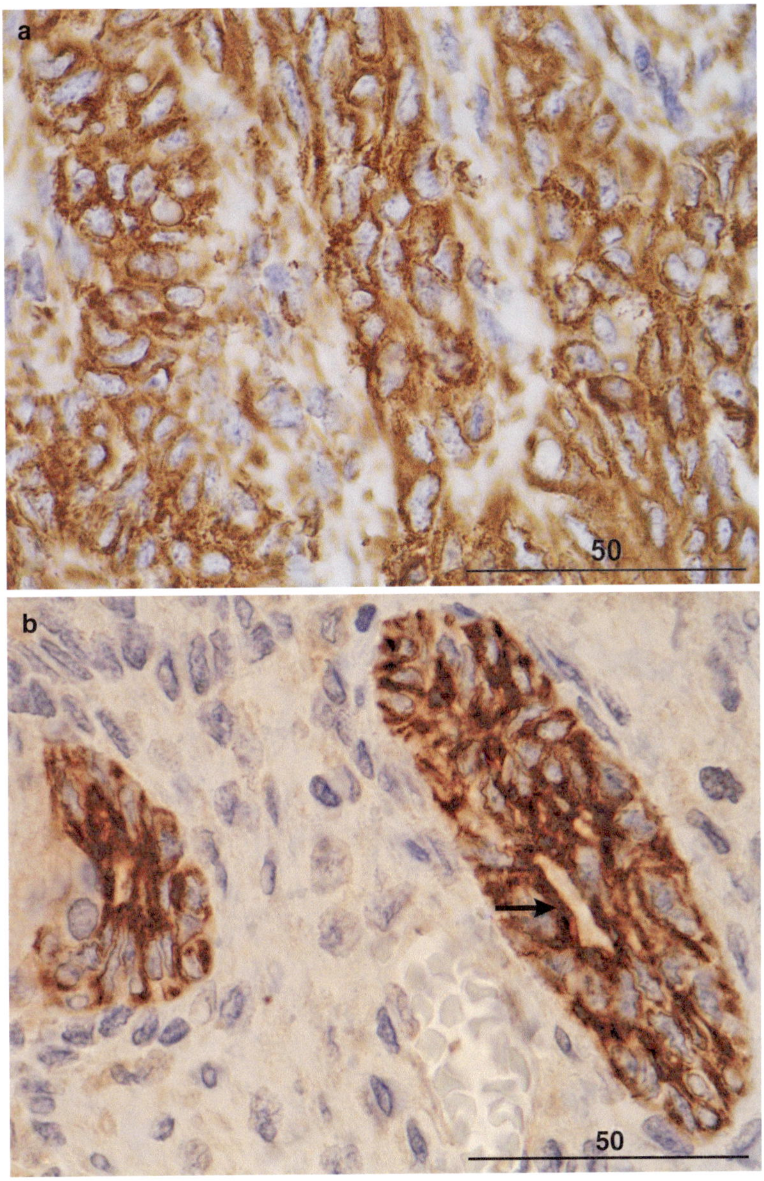

Fig. 5.13 Fetal testis at 25 weeks' gestation: the rete testis. The cytoplasm of the rete epithelium contains keratin and vimentin intermediate filaments. The keratins, primarily of low molecular weight, may be first identified at the tenth week of fetal life and precede the appearance of vimentin by 2–3 weeks. The lumina in the testicular rete begin to form at approximately 24 weeks' gestation, and CK8 immunoreactivity is stronger at the luminal border of the epithelium (*arrow*) (IHC stain for vimentin (**a**) and CK8 (**b**); *scale bar*, 50 μm)

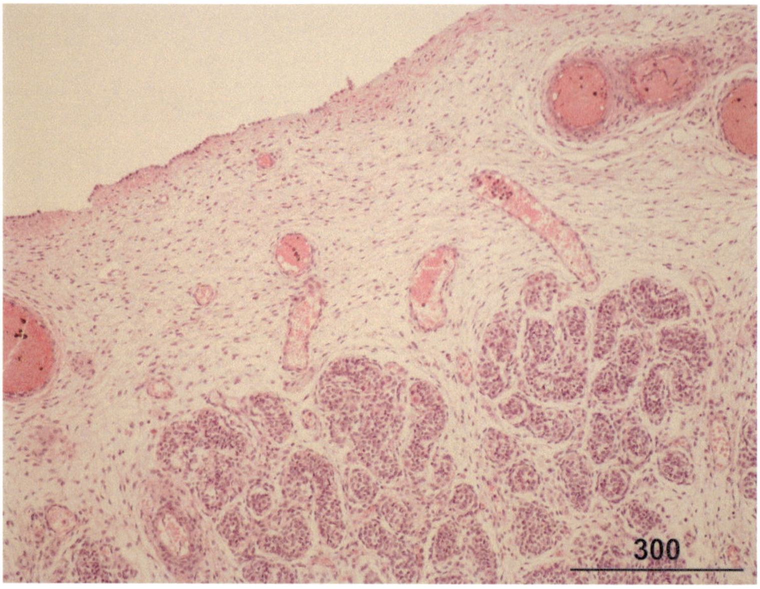

Fig. 5.14 Fetal testis at 30 weeks' gestation. The capsule has thickened even more because of the thickening of the tunica albuginea and the complete development of the tunica vasculosa. The tunica vasculosa forms the complete interlobular septa, connecting the subcapsular area to the hilum. The future seminiferous tubules appear very crowded because their length and coiling increase, reaching the maximum approximately at this gestational age (H&E stain; *scale bar*, 300 μm)

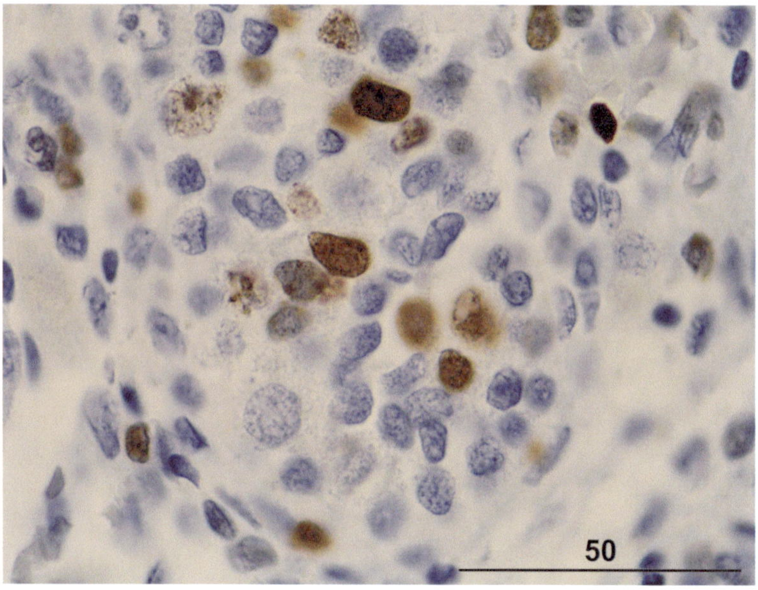

Fig. 5.15 Fetal testis at 30 weeks' gestation. At this stage, both germ and Sertoli cells show proliferative activity on nuclear staining with Ki-67, but after birth, they become mitotically inactive (IHC stain for Ki-67; *scale bar*, 50 μm)

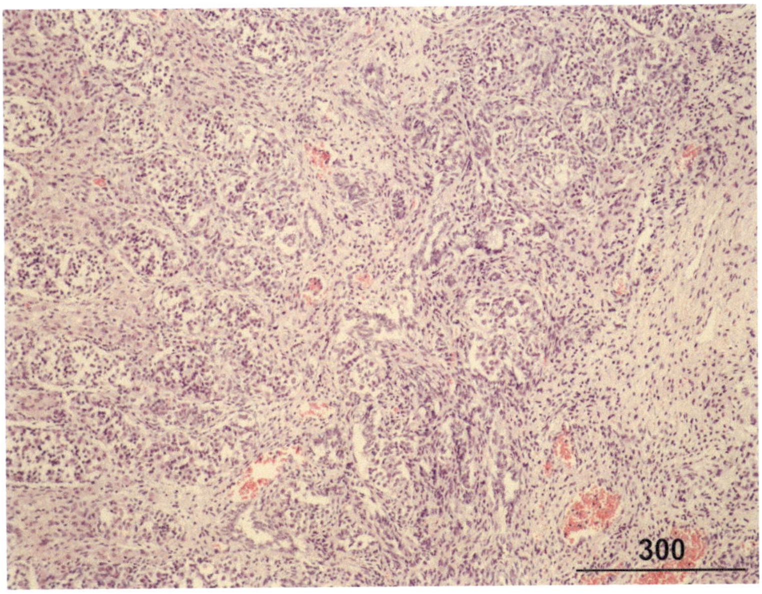

Fig. 5.16 Fetal testis at 30 weeks' gestation: rete testis. The rete testis consists of septal, mediastinal, and extratesticular compartments, the largest being the mediastinal compartment (H&E stain; *scale bar*, 300 μm)

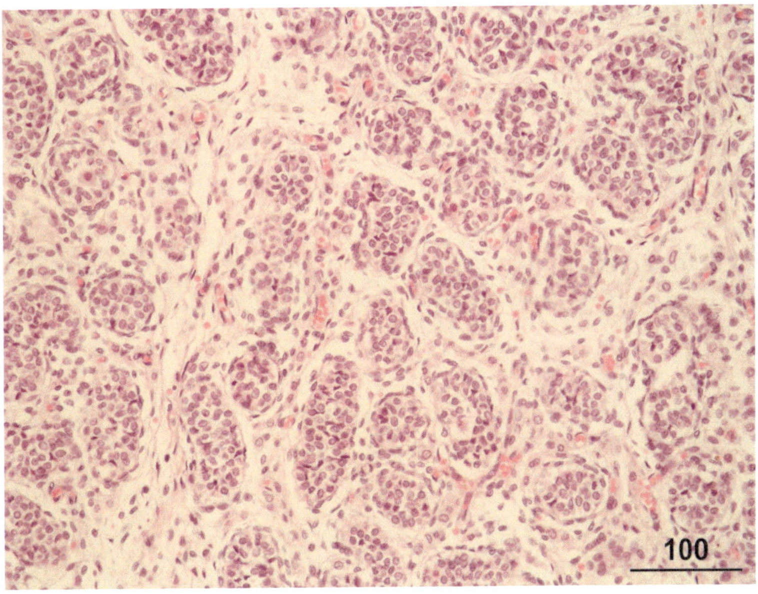

Fig. 5.17 Fetal testis at 38 weeks' gestation. Because Leydig cells are difficult to identify in the routine histologic section from an autopsy, they seem to be absent from some testes of near-term fetuses. However, a few spindle-shaped eosinophilic Leydig cells are found in the interstitium (*see* Fig. 5.18) (H&E stain; *scale bar*, 100 μm)

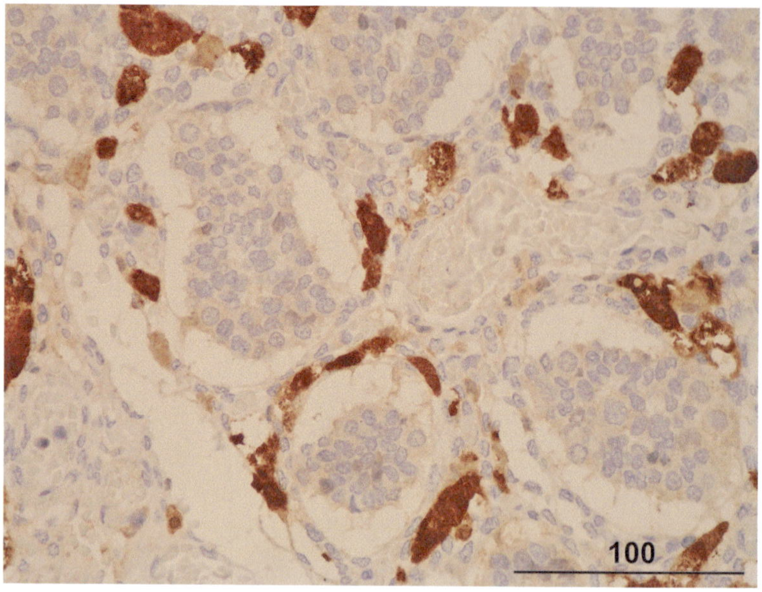

Fig. 5.18 Fetal testis at 38 weeks' gestation. Leydig cells show positive immunostaining for calretinin, making the distinction easier (IHC stain for calretinin; *scale bar*, 100 μm)

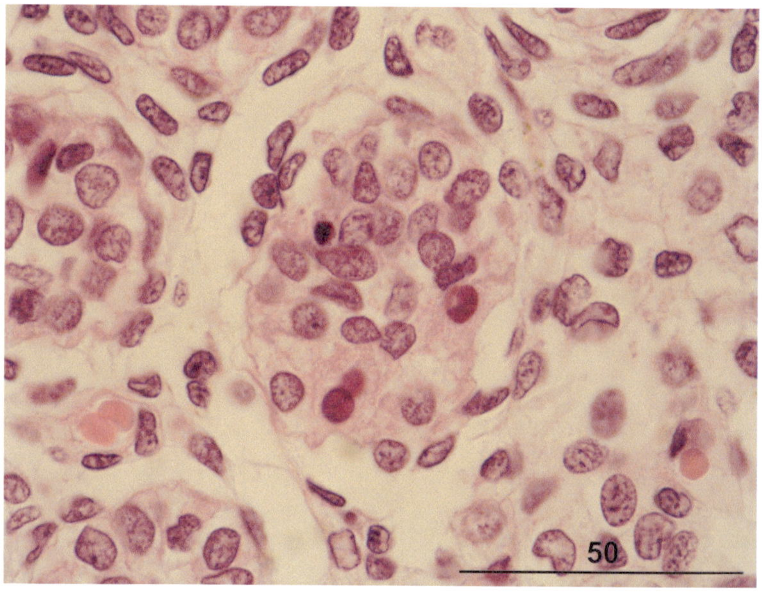

Fig. 5.19 Fetal testis at 38 weeks' gestation: a future seminiferous tubule. At birth, the germ cells average four per tubule cross section, two thirds of them being gonocytes. Apoptosis of germ cells is an important feature in their maturation (H&E stain; *scale bar*, 50 μm)

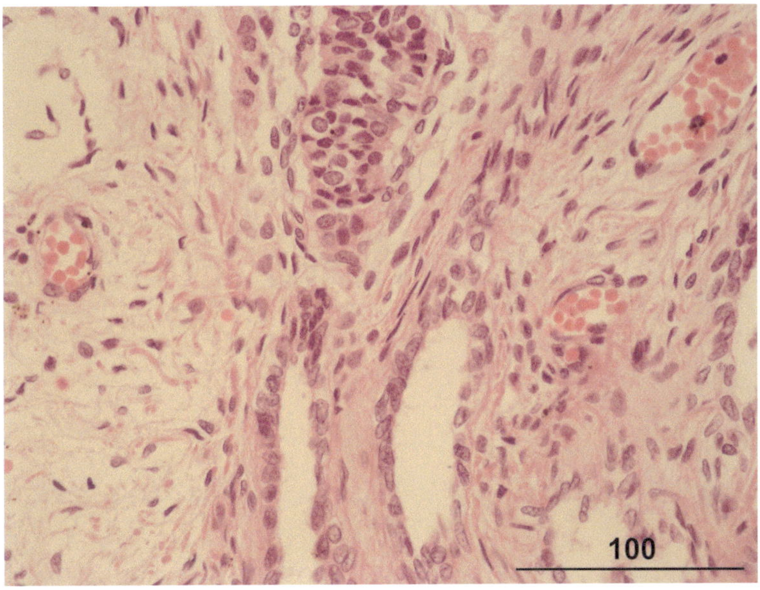

Fig. 5.20 Fetal testis at 38 weeks' gestation: the rete testis. The septal part of the rete testis is composed of the straight, short, narrow tubules (tubuli recti) connecting the hilar ends of the sex cords to the mediastinal compartment. The hilar ends of sex cords usually consist only of Sertoli cells and, in a plug-like form, protrude into the rete lumen. The formation of the lumina in the testicular rete is completed at term or after birth (H&E stain; *scale bar*, 100 μm)

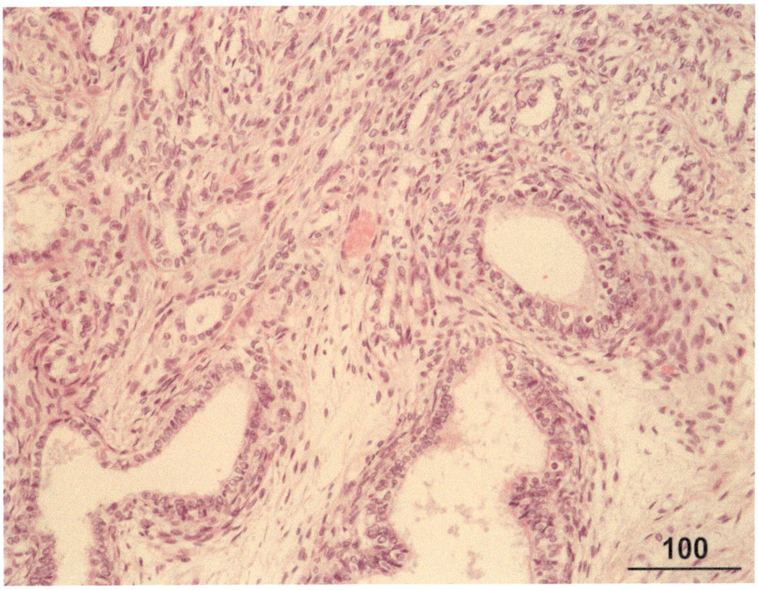

Fig. 5.21 Fetal testis at 38 weeks' gestation: the rete testis. Smaller tubular and cleft-like spaces lined with cuboidal or low columnar epithelium (*upper left*) are the extratesticular part of the rete, forming five to 12 tubules that join with the efferent ductules (*lower right*). The epithelium of efferent ductules appears pseudostratified because of ciliated and nonciliated columnar and basal cells (H&E stain; *scale bar*, 100 μm)

Recommended Literature

de Miguel M, Bethencourt F, Arenas M, Fraile B, Paniagua R. Intermediate filaments in the Sertoli cells of the aging human testis. Virchows Arch. 1997;431:131–8.

Dinges H, Zatloukal K, Schmid C, Mair S, Wirnsberger G. Co-expression of cytokeratin and vimentin filaments in rete testis and epididymis. An immunohistochemical study. Virchows Arch A Pathol Anat Histopathol. 1991;418:119–27.

Hittmair A, Zelger B, Obrist P, Dirnhofer S. Ovarian Sertoli-Leydig cell tumor: a SRY gene-independent pathway of pseudomale gonadal differentiation. Hum Pathol. 1997;28:1206–10.

Honecker F, Stoop H, de Krijger R, Chris Lau YF, Bokemeyer C, Looijenga LH. Pathobiological implications of the expression of markers of testicular carcinoma in situ by fetal germ cells. J Pathol. 2004; 203:849–57.

Huff DS. Chapter 9: Testis. In: Ernst LM, Ruchelli ED, Huff DS, editors. Color atlas of fetal and neonatal histology. New York: Springer; 2011. p. 121–41.

Muller J, Skakkebaek NE. Quantification of germ cells and seminiferous tubules by stereological examination of testicles from 50 boys who suffered sudden death. Int J Androl. 1983;6:143–56.

Sharpe R, McKinnell C, Kivlin C, Fisher JS. Proliferation and functional maturation of Sertoli cells, and their relevance to disorders of testis function in adulthood. Reproduction. 2003;125:769–84.

Suaraz-Quian CA, Martinez-Garcia F, Nistal M, Regadera J. Androgen receptor distribution in adult human testis. J Clin Endocrinol Metab. 1999; 84:350–8.

Svechnikov K, Landreh L, Weisser J, et al. Origin, development and regulation of human Leydig cells. Horm Res Paediatr. 2010;73:93–101.

Trainer TD. Chapter 37: Testis and excretory duct system. In: Mills SE, editor. Histology for pathologists. 3rd ed. Philadelphia: Lippincott Williams and Wilkins; 2007. p. 944–63.

Waters BL, Trainer TD. Development of the human fetal testis. Pediatr Pathol Lab Med. 1996;16:9–23.

6 The Testis in Childhood Between Birth and Puberty

Ewa Rajpert-De Meyts, Kristian Almstrup, John E. Nielsen, and Niels E. Skakkebæk

The testis of newborns and young infants is quite active endocrinologically [1, 2]. Serum levels of gonadotropins, testosterone, insulin-like factor 3 (INSL3), and inhibin B peak around age 2–3 months, when testosterone is close to a lower adult level [1, 3]. Despite this pronounced androgen activity in early infancy, the absence of expression of androgen receptors in Sertoli cells renders them unresponsive and prevents maturation of the testis [4]. In general, the histology of the infantile testis is similar to that of the third-trimester fetal testis. Leydig cells are noticeable during the postnatal hormone peak. The seminiferous tubules contain infantile spermatogonia and undifferentiated Sertoli cells, which produce large amounts of anti-Müllerian hormone (AMH) and inhibin B [1, 4, 5]. Both types of cells increase in number during the first 3 months of life [6, 7]. A few gonocyte-like cells may be seen during the first few months after birth. These cells express some of the embryonic germ cell markers (e.g., placental-like alkaline phosphatase [PLAP], M2A/D2-40/PDPN) and pluripotency markers, such as OCT4, NANOG, and AP2γ, although the expression is often weak compared with early fetal gonocytes [8, 9].

After the infantile mini-puberty, hormone levels decrease rapidly, and by the age of 6 months, both luteinizing hormone (LH) and testosterone are no longer measurable in serum by most of standard methods [1]. Leydig cells are fibrocyte-like, and very few adult-type Leydig cells can be found. Single gonocyte-like cells are rarely found after mini-puberty [8]. Their presence beyond the first year of life is a sign of testicular dysgenesis and is associated with a high risk of germ cell neoplasia (see Chap. 12). Single primary spermatocytes may be seen from the age of 4 years, but these cells apparently do not survive [10]. During childhood, the testis is not completely inactive; the tubular length and total volume of the gonad increase slightly from a volume of approximately 1 to 3 mL (measured by orchidometer) when puberty begins [11]. Sertoli cells remain immature and highly positive for AMH and podoplanin (M2A/D2-40/PDPN) until puberty, when both proteins are rapidly downregulated, whereas serum testosterone increases again [4, 5, 12, 13]. Leydig cells, which initiate testosterone production in response to sharply increased LH stimulation, increase in number and display the typical morphology of adult Leydig cells. This is associated with rapid growth of seminiferous tubules and the onset of spermatogenesis. The pubertal period during which the testis grows often lasts a couple of years and concludes when the gonad has reached an orchidometer-measured volume of 15–30 mL. The histologic pattern of the growing pubertal testis is often heterogeneous, with a mixture of fully differentiated and undifferentiated tubules.

E. Rajpert-De Meyts, M.D., D.M.Sc. (✉)
K. Almstrup, M.Sc., Ph.D. • J.E. Nielsen, M.Sc.
N.E. Skakkebæk, M.D., D.M.Sc.
Department of Growth and Reproduction, Copenhagen University, Rigshospitalet, Blegdamsvej 9, DK-2100 Copenhagen, Denmark
e-mail: erm@rh.dk

D. Ježek (ed.), *Atlas on the Human Testis*,
DOI 10.1007/978-1-4471-2763-5_6, © Springer-Verlag London 2013

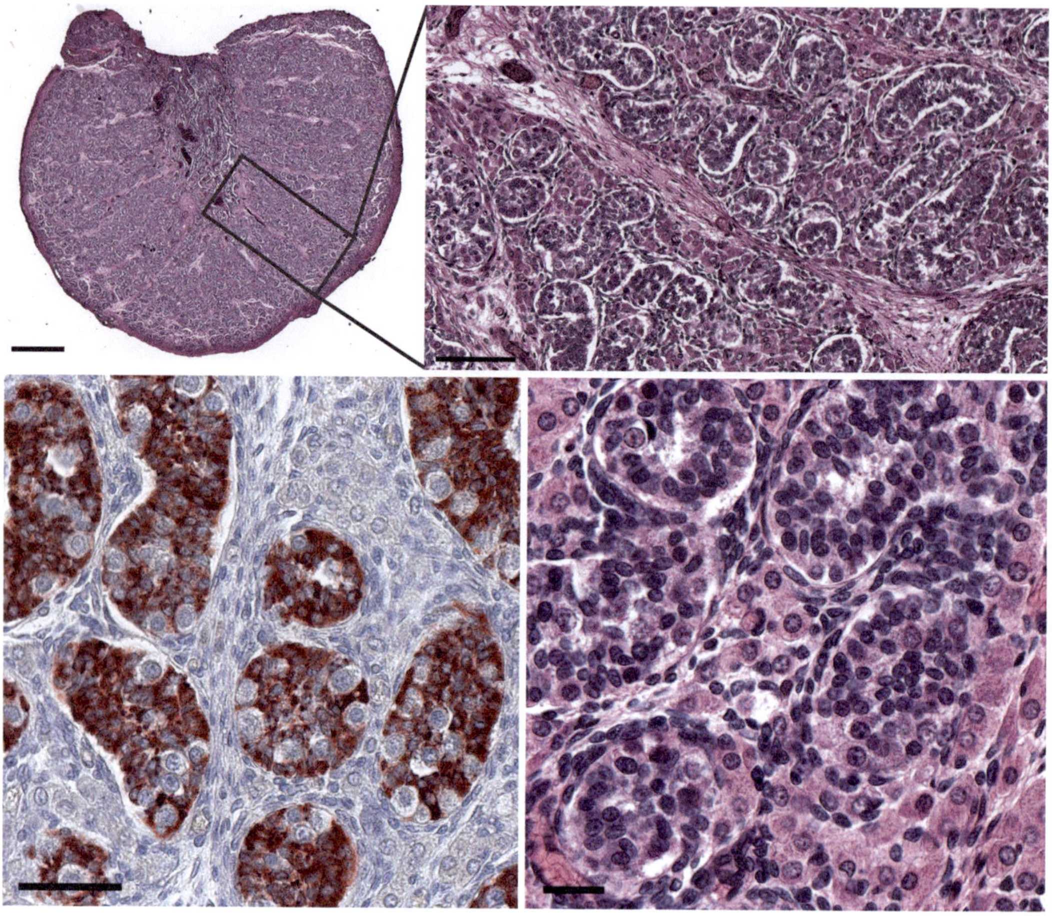

Fig. 6.1 Cross section of a testis of a newborn boy who died 1 day after birth. The parenchyma is divided by septa into lobules. Higher magnification of one lobule shows well-developed infantile seminiferous cords located in the interstitial tissue with abundant Leydig cells. As shown in detail in the two bottom images, the cords' membranes contain only Sertoli and germ cells (infantile spermatogonia) and are surrounded by a single layer of peritubular cells. The *left lower picture* shows intense expression of AMH, which in the testis is specific for immature Sertoli cells [5, 12] (Stieve's fixative, hematoxylin and eosin [H&E] staining, immunostaining for AMH. *Magnification bars*: *upper left*, 500 μm; *upper right*, 100 μm; *lower left*, 50 μm; *lower right*, 25 μm)

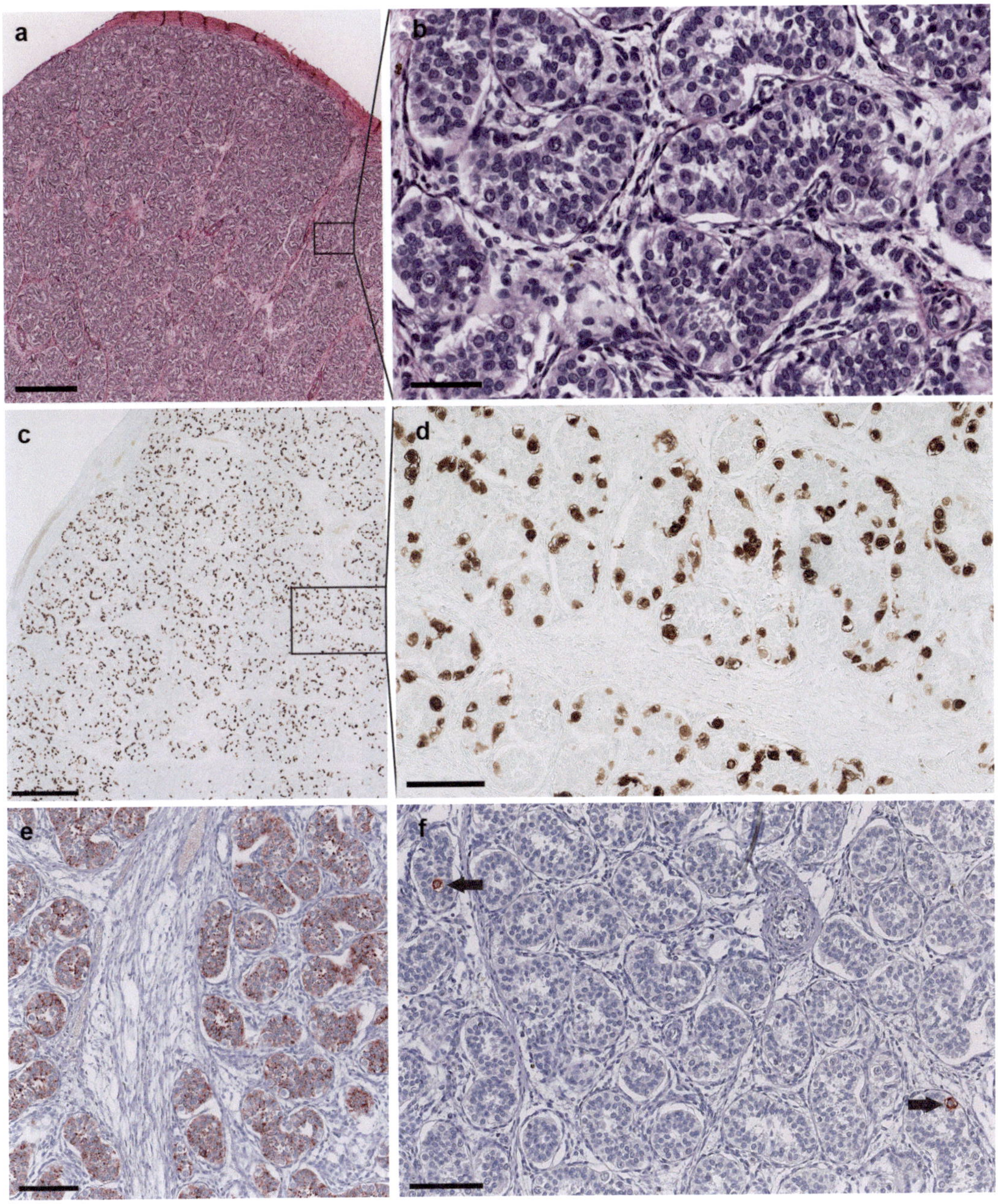

Fig. 6.2 Infantile testis at "mini-puberty" (specimen from a 3-month-old boy who died of severe pneumonia). (**a**) General view of the testis tissue. (**b**) The general structure of the testis parenchyma is similar to that of the neonatal testis, but relatively fewer Leydig cells are seen, whereas the number of germ cells is greater. (**c**, **d**) Germ cells marked *dark brown* by immunostaining for MAGEA4, which is a marker of spermatogonia [14, 15]. (**e**) Immature Sertoli cells are marked *red* by immunostaining for AMH. (**f**) The few gonocytes still present are visualized here by red immunostaining for PLAP (*arrows*) (*Magnification bars*: **a**, **c**: 500 μm; **b**: 50 μm; **d**–**f**: 100 μm)

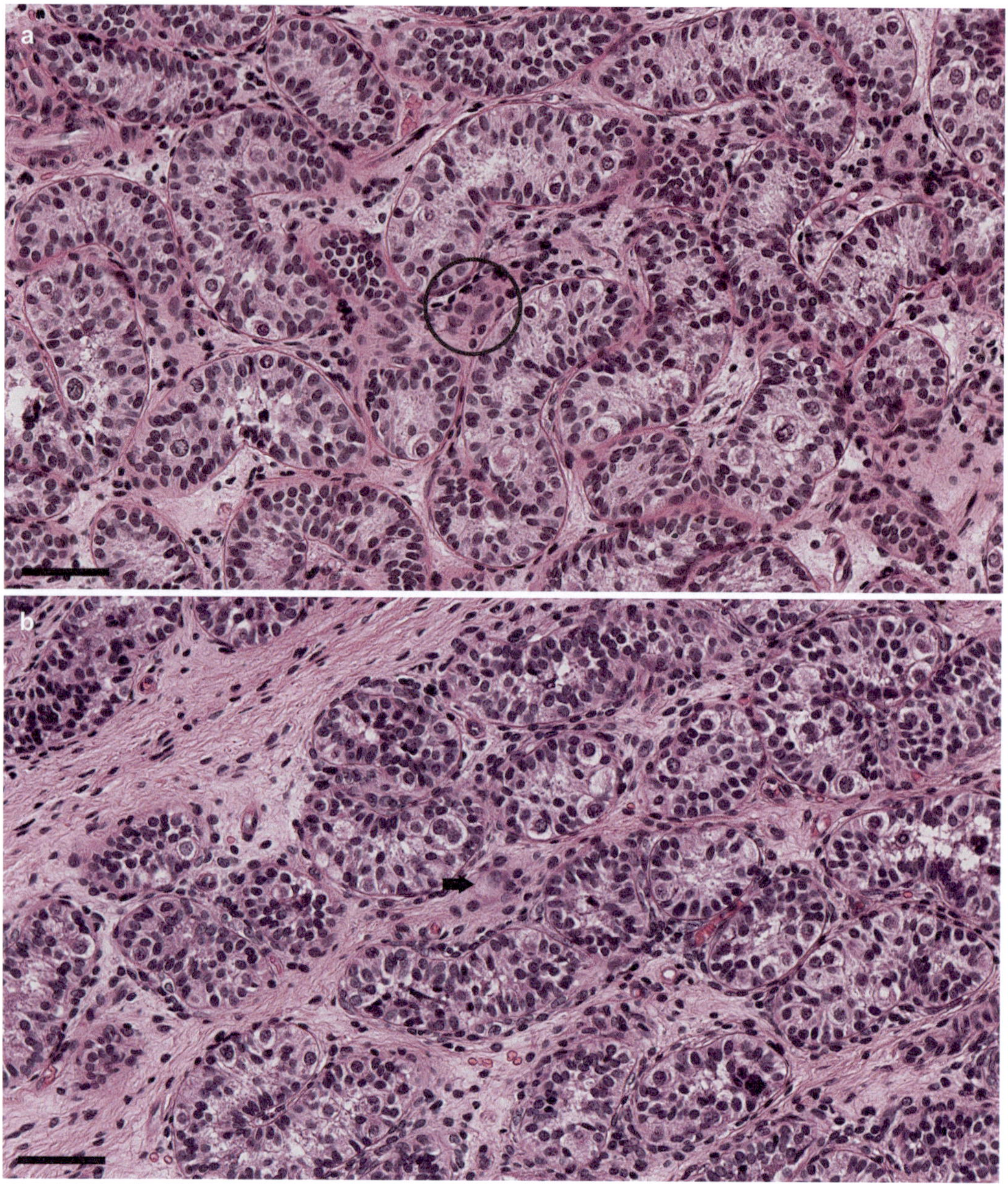

Fig. 6.3 Two examples of testis biopsies from boys aged 6.5 (**a**) and 7.5 years (**b**). Both boys had acute leukemia and underwent biopsy to control for leukemic infiltrates into the testes. Both testes show the same prepubertal pattern, with seminiferous tubules containing only immature Sertoli cells and spermatogonia. There is a relative paucity of Leydig cells in the interstitial compartment. Leydig cells are indicated by a *circle* (**a**) and an *arrow* (**b**) (Stieve's fixative, H&E staining. *Magnification bars*: 50 μm)

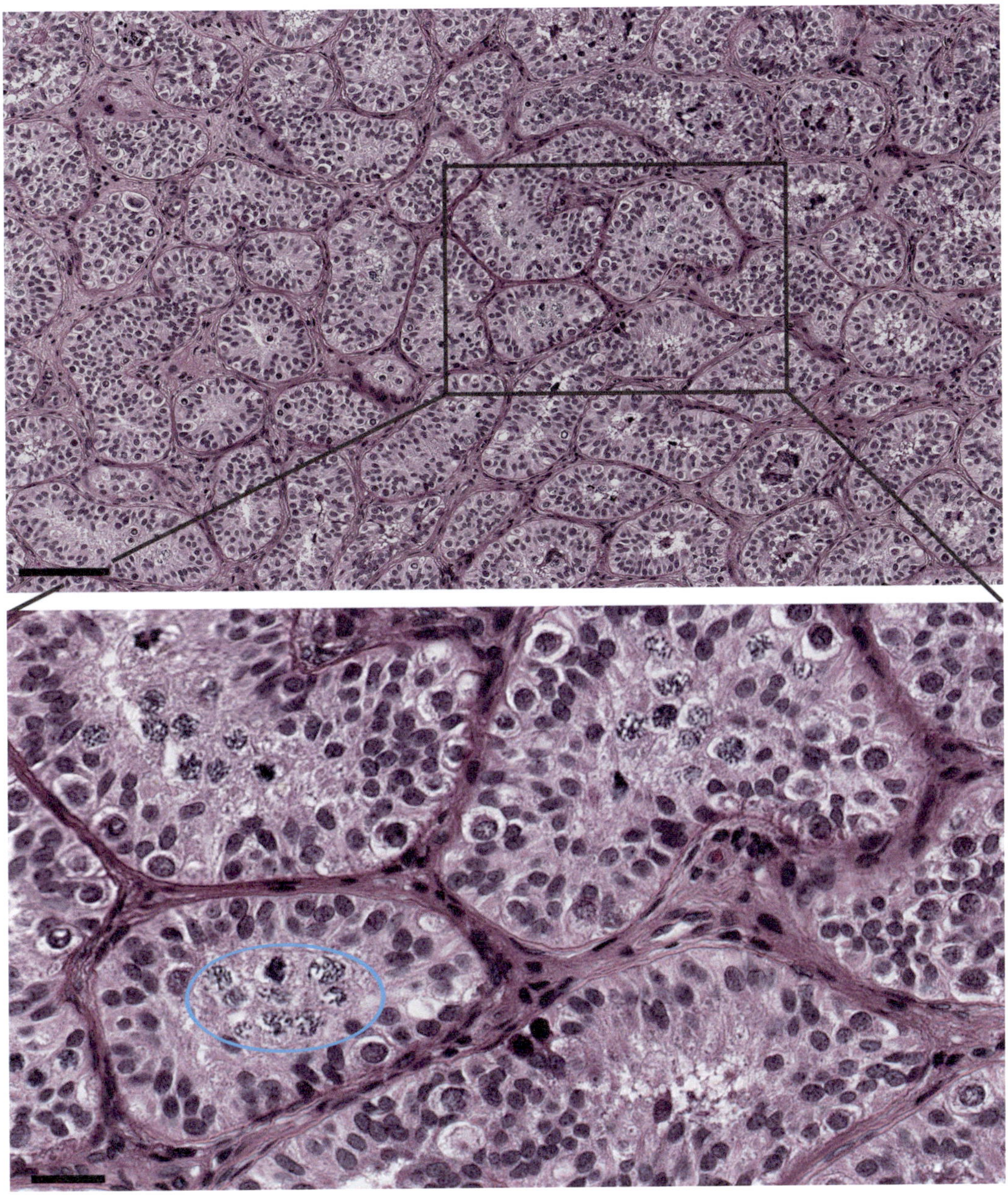

Fig. 6.4 Testis biopsy sample from an 11-year-old boy (obtained during follow-up for acute leukemia). The general testis histology resembles the pattern seen in the two younger boys (Fig. 6.3). Higher magnification reveals the presence of primary spermatocytes in a few seminiferous tubules. The spermatocytes do not mature beyond the pachytene stage and often deteriorate, as shown within the *blue ellipse* in one of the tubules in the bottom image. The Sertoli cells are not yet fully mature (Stieve's fixative, H&E staining. *Magnification bars*: *top*, 100 μm; *bottom*, 25 μm)

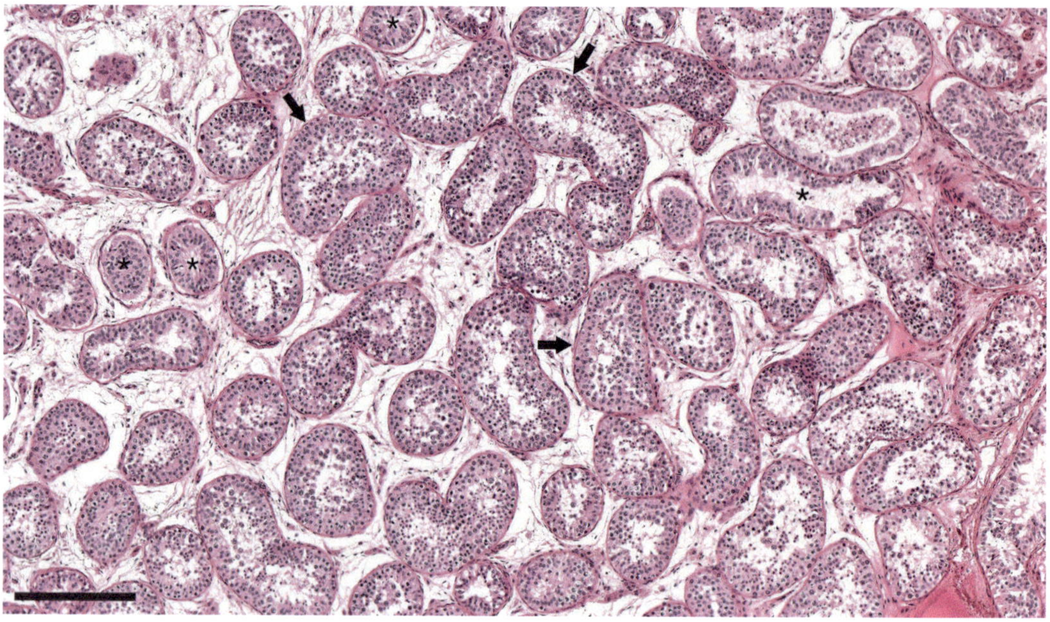

Fig. 6.5 Testis biopsy sample from a 14-year-old boy (who died accidentally) at the onset of puberty. A histologic overview reveals a patchy pattern; a few immature tubules without spermatids are marked with *asterisks*, whereas tubules with complete spermatogenesis and late spermatids (type Sd) are indicated by *arrows*. On ultrasound examination, such a testis may manifest an irregular echo pattern (H&E staining. *Magnification bar*: 250 μm)

References

1. Andersson AM, Toppari J, Haavisto AM, Petersen JH, Simell T, Simell O, Skakkebaek NE. Longitudinal reproductive hormone profiles in infants: peak of inhibin B levels in infant boys exceeds levels in adult men. J Clin Endocrinol Metab. 1998;83:675–81.
2. Chemes HE. Infancy is not a quiescent period of testicular development. Int J Androl. 2001;24:2–7.
3. Bay K, Andersson AM. Human testicular insulin-like factor 3: in relation to development, reproductive hormones and andrological disorders. Int J Androl. 2011; 34:97–109.
4. Rey RA, Musse M, Venara M, Chemes HE. Ontogeny of the androgen receptor expression in the fetal and postnatal testis: its relevance on Sertoli cell maturation and the onset of adult spermatogenesis. Microsc Res Tech. 2009;72:787–95.
5. Josso N, Picard JY, Rey R, di Clemente N. Testicular anti-Müllerian hormone: history, genetics, regulation and clinical applications. Pediatr Endocrinol Rev. 2006;3:347–58.
6. Cortes D, Müller J, Skakkebæk NE. Proliferation of Sertoli cells during development of the human testis assessed by stereological methods. Int J Androl. 1987; 10:589–96.
7. Müller J, Skakkebæk NE. Fluctuations in the number of germ cells during the late foetal and early postnatal periods in boys. Acta Endocrinol. 1984;105:271–4.
8. Jørgensen N, Giwercman A, Müller J, Skakkebæk NE. Immunohistochemical markers of carcinoma in situ of the testis also expressed in normal infantile germ cells. Histopathology. 1993;22:373–8.
9. Honecker F, Stoop H, de Krijger RR, Chris Lau YF, Bokemeyer C, Looijenga LH. Pathobiological implications of the expression of markers of testicular carcinoma in situ by fetal germ cells. J Pathol. 2004;203: 849–57.
10. Nistal M, Paniagua R. Occurrence of primary spermatocytes in the infant and child testis. Andrologia. 1984;16:532–6.
11. Zachmann M, Prader A, Kind HP, Häfliger H, Budliger H. Testicular volume during adolescence. Cross-sectional and longitudinal studies. Helv Paediatr Acta. 1974;29:61–72.
12. Rajpert-De Meyts E, Jørgensen N, Graem N, Müller J, Cate RL, Skakkebaek NE. Expression of anti-Müllerian hormone during normal and pathological gonadal development: association with differentiation of Sertoli and granulosa cells. J Clin Endocrinol Metab. 1999;84:3836–44.
13. Sonne SB, Herlihy AS, Hoei-Hansen CE, Nielsen JE, Almstrup K, Skakkebaek NE, et al. Identity of

M2A (D2-40) antigen and gp36 (Aggrus, T1A-2, podoplanin) in human developing testis, testicular carcinoma in situ and germ-cell tumours. Virchows Arch. 2006;449:200–6.
14. Aubry F, Satie A-P, Rioux-Leclercq N, Rajpert-De Meyts E, Spagnoli GC, Chomez P, et al. *MAGE-A4*, a germ-cell specific marker is differentially expressed in testicular tumours. Cancer. 2001;902:2778–85.
15. Takahashi K, Shichijo S, Noguchi M, Hirohata M, Itoh K. Identification of MAGE-1 and MAGE-4 proteins in spermatogonia and primary spermatocytes of testis. Cancer Res. 1995;55:3478–82.

7 Normal Morphology of the Human Testis and Epididymis

Davor Ježek, Viviana Kozina, and Andreja Vukasović

The aim of this chapter is to present the histologic architecture of the human testis and epididymis to clinicians who deal with spermatozoa (micro) surgical retrieval procedures, such as testicular sperm aspiration (TESA), testicular sperm extraction (TESE), percutaneous sperm aspiration (PESA), microsurgical epididymal aspiration (MESA), and similar techniques.

The testis is composed mainly of seminiferous tubules, coiled structures that are 150–250 μm in diameter and can reach more than 80 cm in length. The seminiferous tubules form the so-called tubular testicular compartment, whereas a loose connective tissue with androgen-producing Leydig cells creates the interstitial compartment. Both compartments are limited by a thick capsule of connective tissue, the tunica albuginea. On its surface, the tunica albuginea bears simple squamous epithelium (mesothelium) as a result of testicular descent, a process that in humans ends during the last 2 months of pregnancy. During testicular descent, the male gonad, initially attached to the posterior abdominal wall, travels through the inguinal canal and pulls down the evagination of the peritoneum, or processus vaginalis, as well as part of the abdominal muscles (cremasteric fascia and muscle). Thus, the testis reaches the scrotum by 33 weeks of gestation. The scrotal sac maintains a temperature of approximately 34–35 °C, which is needed for normal spermatogenesis. The tunica albuginea has connective tissue extensions, testicular septa, that incompletely divide the testicular parenchyma into testicular lobules.

The seminiferous tubules are lined with seminiferous epithelium and are encircled by lamina propria. The lamina propria is composed of five to seven layers of flattened peritubular (myoid) cells that occasionally contract. The seminiferous epithelium consists of supporting Sertoli cells and spermatogenic cells: spermatogonia, primary and secondary spermatocytes, and early and late spermatids. When early spermatids are released into the tubular lumen, they are considered mature spermatozoa. The seminiferous tubules end in a system of narrow straight ducts called *tubuli recti*. These straight ducts are devoid of spermatogenic cells and empty into flattened spaces called the *rete testis*. The rete testis is located in the thickening of the tunica albuginea, called the *mediastinum*. After passing through the straight ducts and rete testis, the spermatozoa reach the ductuli efferentes, 12–20 ducts with an irregular lumen. The efferent ducts fuse with the head (caput) of the epididymis.

D. Ježek, M.D., Ph.D. (✉) • V. Kozina, MEd
A. Vukasović, M.D.
Department of Histology and Embryology,
School of Medicine, University of Zagreb, Šalata 3,
HR-10000 Zagreb, Croatia
e-mail: davorjezek@yahoo.com, davor@mef.hr

D. Ježek (ed.), *Atlas on the Human Testis*,
DOI 10.1007/978-1-4471-2763-5_7, © Springer-Verlag London 2013

The epididymis consists of a highly coiled duct, the ductus epididymidis, which is 4–6 cm long and forms three distinctive regions of the epididymis: the head, the body, and the tail. In the tail of the epididymis, spermatozoa display forward motility. The epithelium of the epididymal duct is composed of principal columnar cells with stereocilia and basal cells. The epithelium contributes to the maturation of spermatozoa. The tail of the epididymis extends into the muscular tube, or vas deferens. The brief description of the testis and epididymis presented here is necessary for understanding both male fertility and infertility.

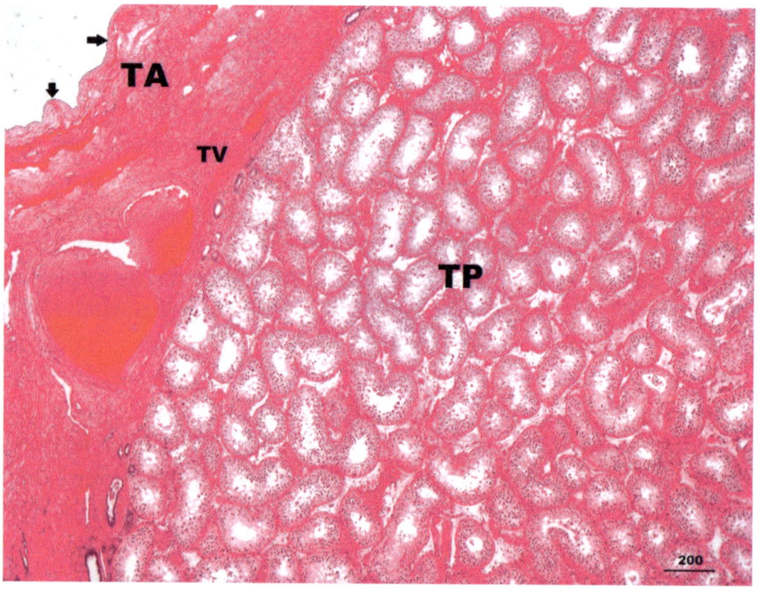

Fig. 7.1 Normal adult human testis. The testicular parenchyma (*TP*) is limited by a fibrous capsule, the tunica albuginea (*TA*). On its surface, the tunica bears a mesothelial lining (*arrows*) consisting of flattened simple squamous cells. The deeper part of the tunica albuginea is well vascularized and therefore is called the *tunica vasculosa* (*TV*) (Hematoxylin and eosin [H&E] stain, ×40; *scale bar*, 200 μm)

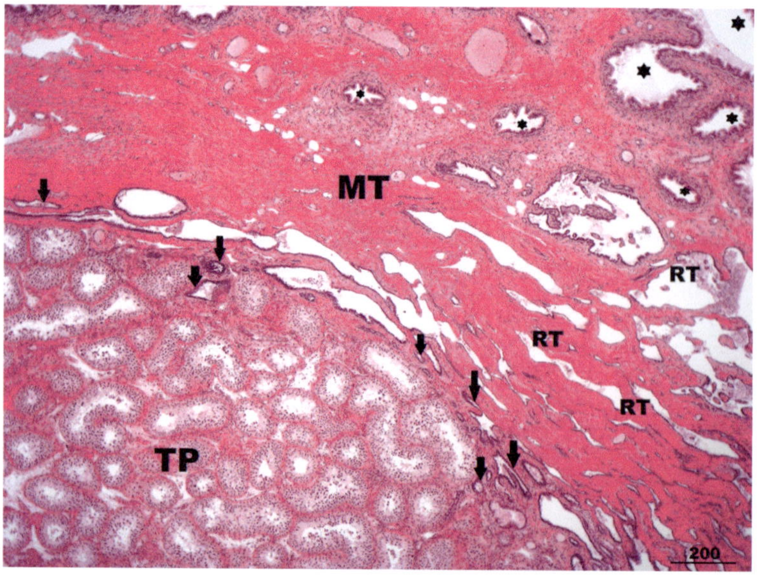

Fig. 7.2 Adult human testis with a normal structure. The testicular parenchyma (*TP*) consists of seminiferous tubules and gentle interstitial tissue. The seminiferous tubules empty into the system of narrow ducts, the tubuli recti (*arrows*). The tubuli recti open into many flattened channels within the mediastinum testis (*MT*). These channels are called the *rete testis* (*RT*). The rete testis fuses with 12–20 ducts of irregular lumen (*asterisks*) called *ductuli efferentes* (H&E stain, ×40; *scale bar*, 200 μm)

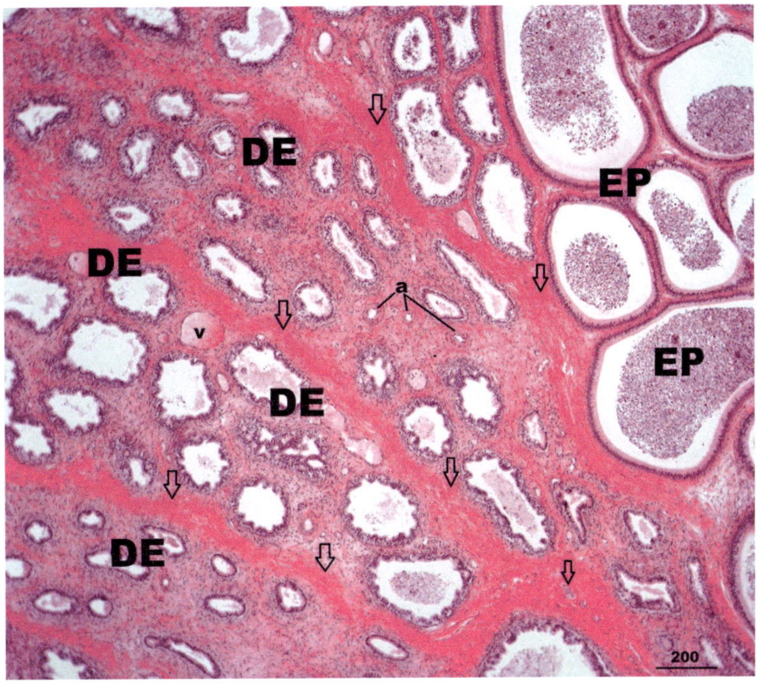

Fig. 7.3 Ductuli efferentes (*DE*) and part of the ductus epididymidis (*EP*). The efferent ducts are recognized by their irregular lumen and are surrounded by loose connective tissue bearing arteries (*a*) and veins (*v*). The ductus epididymidis has a more regular lumen filled with deposited spermatozoa. Dense connective tissue septa (*arrows*) run between the aforementioned structures (H&E stain, ×40; *scale bar*, 200 μm)

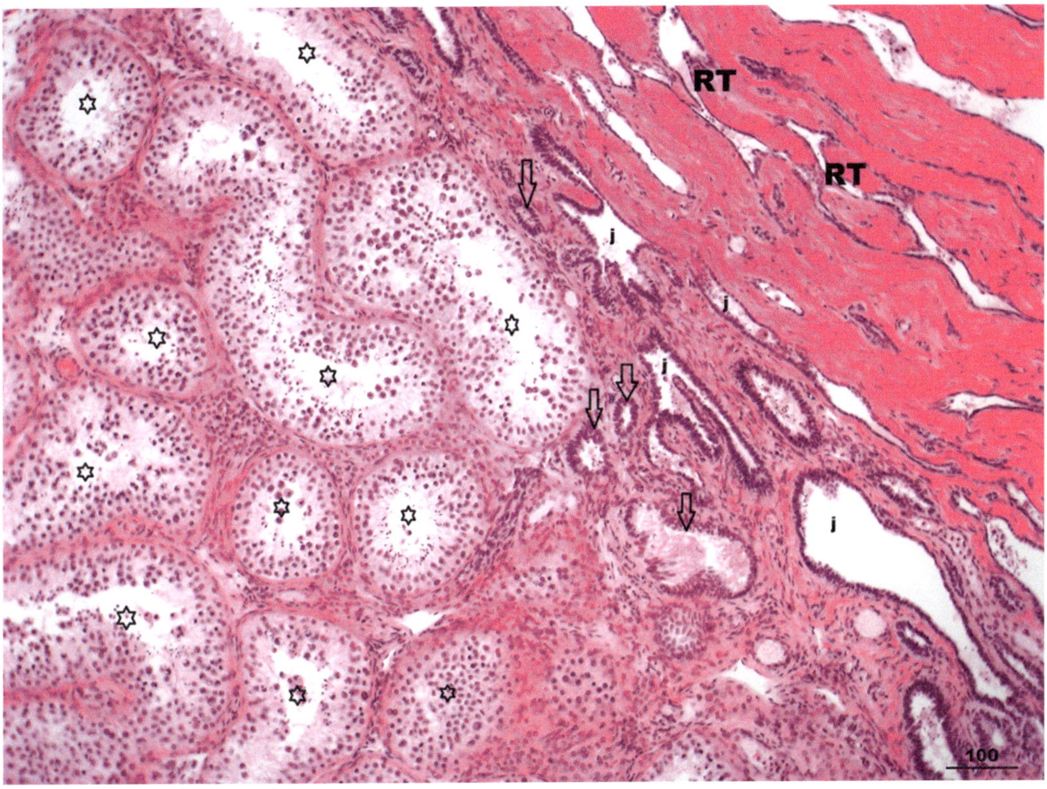

Fig. 7.4 Detail of a normal human testis. The seminiferous tubules (*asterisks*) empty into straight tubules (tubuli recti, *arrows*), a system of narrow ducts lined with columnar (proximal part) and cuboidal epithelium (distal part). At certain points, the straight tubules form a junction (*j*) with the rete testis (*RT*) (H&E stain, ×100; *scale bar*,100 μm)

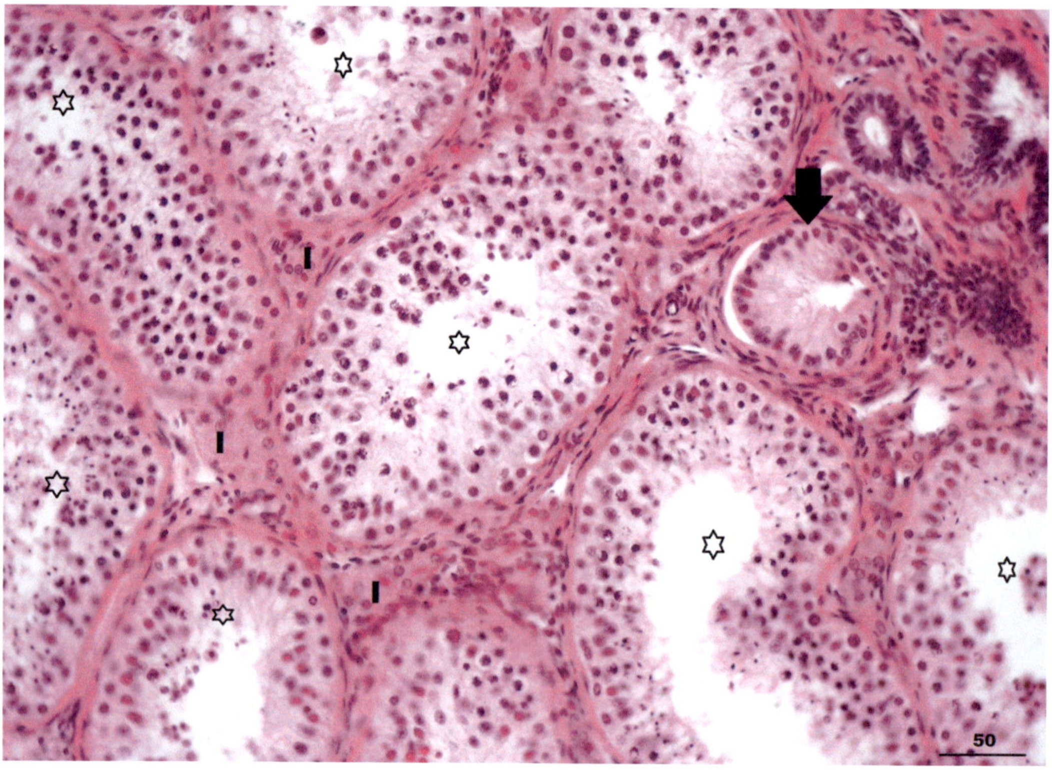

Fig. 7.5 Testicular parenchyma. The major part is occupied by seminiferous tubules (*asterisks*) lined with seminiferous epithelium (Sertoli and spermatogenic cells). Between the tubules, there is loose interstitial connective tissue (*I*). The most distal part of the tubule (*arrow*) is devoid of spermatogenic cells. Only modified Sertoli cells remain (H&E stain, ×200; *scale bar*, 50 μm)

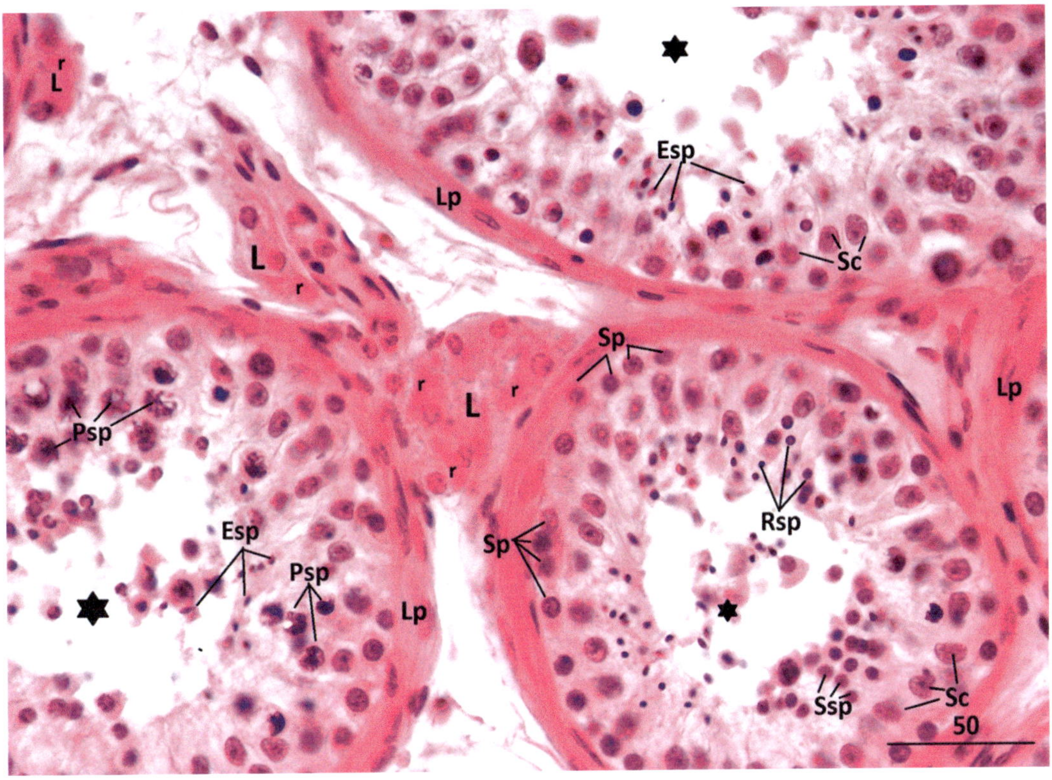

Fig. 7.6 Seminiferous tubules (*asterisks*) with accompanying interstitial tissue. Each tubule, limited by five to seven rows of peritubular (myoid) cells forming the lamina propria (*Lp*), is lined with seminiferous epithelium. Sertoli cells (*Sc*) divide the epithelium into basal and apical compartments. Spermatogonia (*Sp*) rest on the basement membrane in the basal compartment. Within the apical (meiotic) compartment, other types of spermatogenic cells are visible: primary spermatocytes (*Psp*), secondary spermatocytes (*Ssp*), round spermatids (*Rsp*), and elongated spermatids (*Esp*). When released into the tubular lumen, elongated spermatids are considered definitely formed spermatozoa. The interstitial loose connective tissue is rich in androgen-producing Leydig cells (*L*). Within some of these cells are Reinke's crystals (*r*) (H&E stain, ×400; *scale bar*, 50 μm)

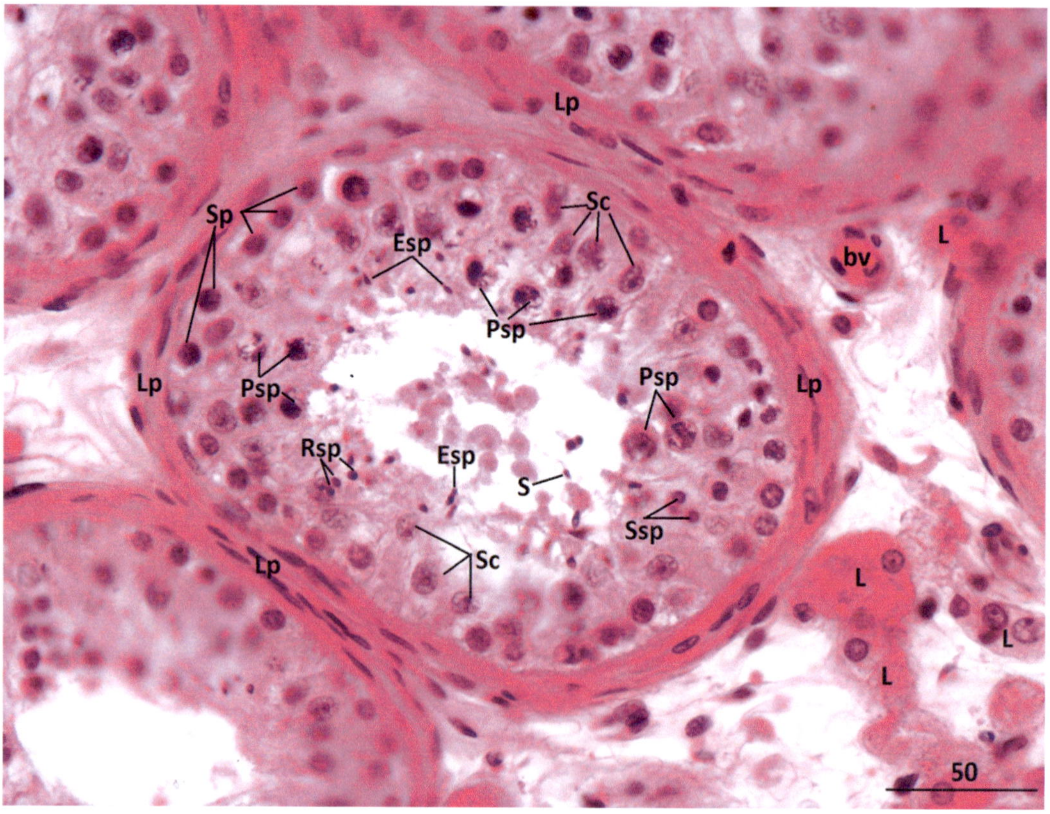

Fig. 7.7 Transverse section through several seminiferous tubules. The lamina propria (*Lp*) bears elongated peritubular cells. Sertoli cells (*Sc*) are recognized by their pear-shaped or ovoid nucleus with a prominent nucleolus. Spermatogonia (*Sp*) form heterogeneous groups of cells that either undergo permanent division (spermatogonia A, dark and pale) or enter meiotic prophase (spermatogonia B). Primary spermatocytes (*Psp*) arise from spermatogonia B and are subjected to the first meiotic division. They give rise to secondary spermatocytes (*Ssp*) that have a short life span and divide into haploid round (immature) spermatids (*Rsp*). Round spermatids dramatically change their cell shape, transforming into elongated (mature) spermatids (*Esp*). Occasionally, one may spot mature spermatozoa (*S*) released into the tubular lumen. Apart from Leydig cells (*L*), interstitial tissue bears many small blood vessels (*bv*) (H&E stain, ×400; *scale bar*, 50 μm)

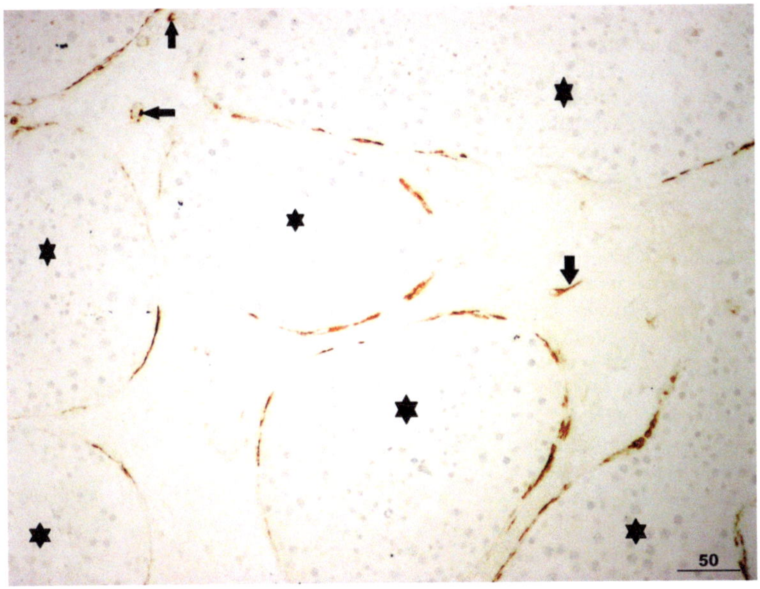

Fig. 7.8 Normal human testis: immunohistochemical (*IHC*) expression of α-smooth actin. Within the lamina propria of the seminiferous tubules (*asterisks*), peritubular (myoid) cells are positive. These cells have some of the properties of smooth muscle cells and, therefore, occasionally contract, enhancing the transport of spermatozoa toward the tubuli recti and rete testis. Smooth muscle cells of small blood vessels within the interstitium (*arrows*) also are positive (IHC stain [diaminobenzidine and hematoxylin counterstain], ×200; *scale bar*, 50 μm)

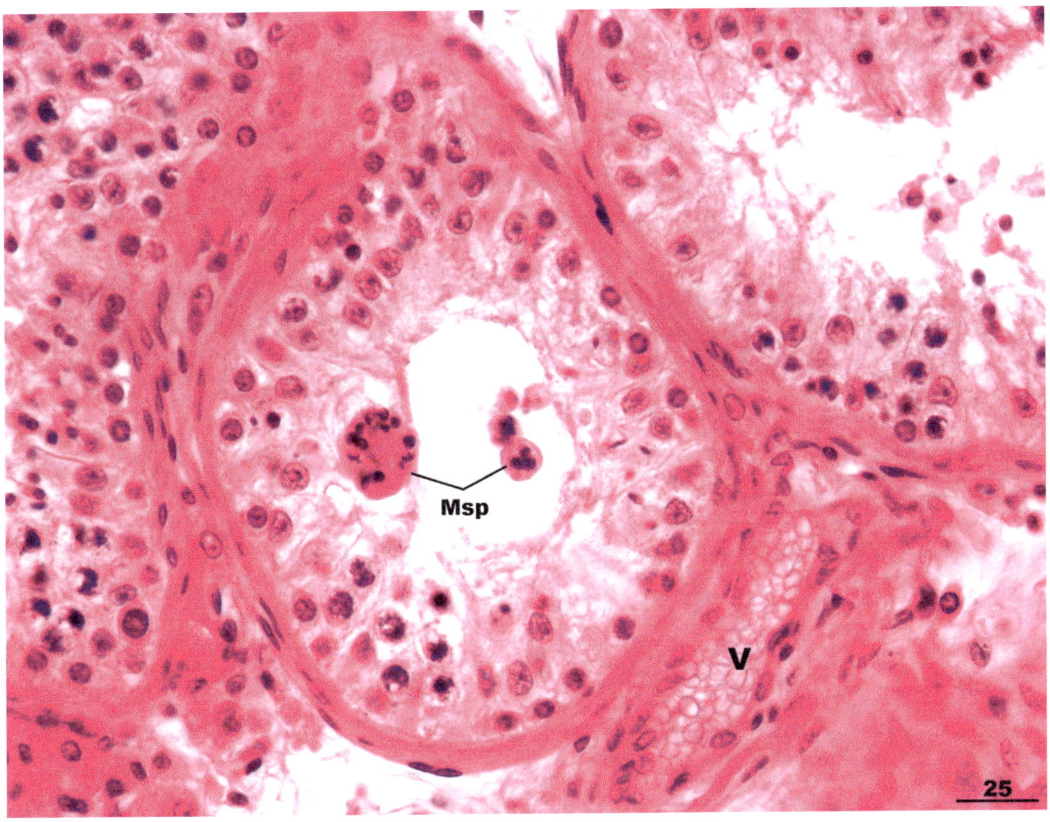

Fig. 7.9 Detail of part of the testicular parenchyma. Normal spermiogenesis results in the separation of spermatids from each other and their gradual transformation into spermatozoa. In this case, the spermatids fail to become individual cells and form megalospermatids (*Msp*), which occasionally may be encountered in normal seminiferous tubules. Interstitial tissue is rich in small vessels, in this case a vein (*v*) (H&E stain, ×400; *scale bar*, 25 μm)

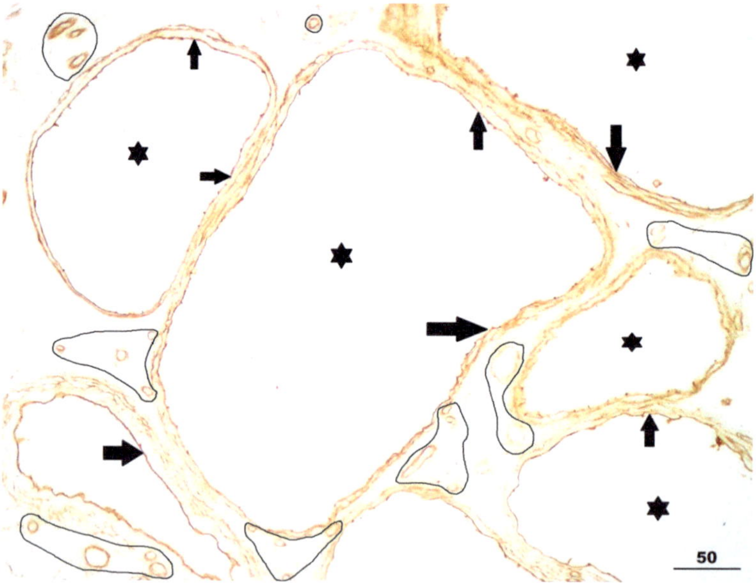

Fig. 7.10 Testicular parenchyma: expression of collagen type IV. The basement membrane of seminiferous epithelium (*arrows*) is a coproduct of Sertoli and peritubular cells. Each peritubular cell within the lamina propria of seminiferous tubules (*asterisks*) has its own basement membrane. The same goes for smooth muscle cells of small blood vessels (*encircled areas*) in the interstitial compartment (IHC stain [diaminobenzidine and hematoxylin counterstain], ×200; *scale bar*, 50 μm)

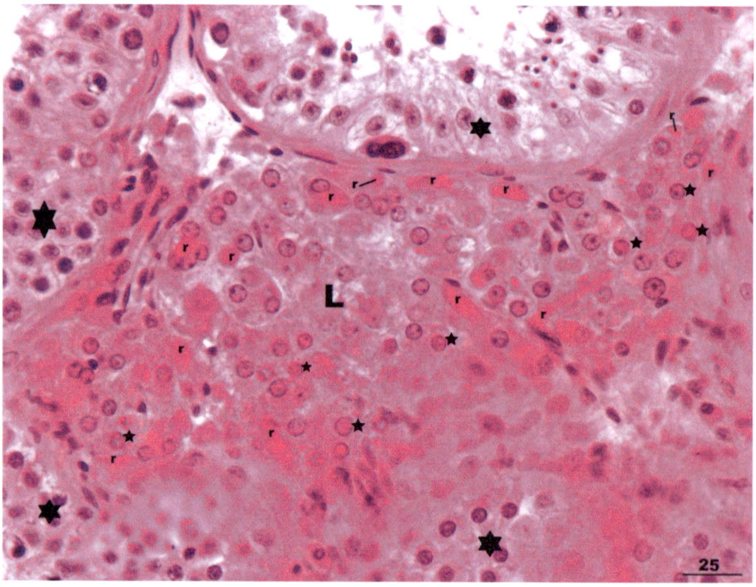

Fig. 7.11 Cluster of interstitial Leydig cells (*L*) situated between neighboring seminiferous tubules (*asterisks*). Leydig cells have a regular round or oval nucleus, sometimes with a prominent nucleolus. The nucleus is surrounded by abundant eosinophilic cytoplasm. Some interstitial cells bear in their cytoplasm Reinke's crystals (*r*) of different sizes. The nuclei of Leydig cells also might have the same crystal but much smaller (*arrows*) (H&E stain, ×400; *scale bar*, 25 μm)

Fig. 7.12 Testicular parenchyma: expression of testosterone. Between seminiferous tubules (*asterisks*), clusters of Leydig cells are positive (*encircled areas*) (IHC stain [diaminobenzidine and hematoxylin counterstain], ×400; *scale bar*, 50 μm)

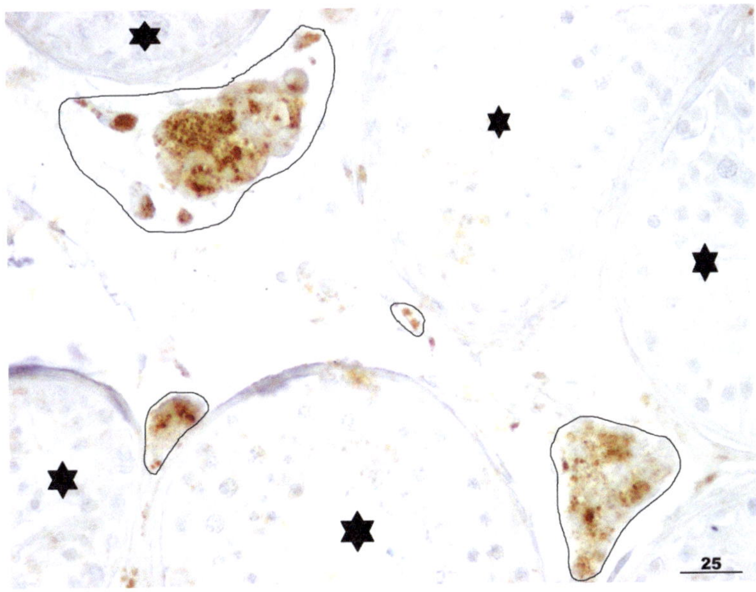

Fig. 7.13 Expression of macrophage marker KiM1P1. The loose connective tissue of the interstitial testicular compartment consists of all types of connective tissue cells, including macrophages. In this image, the positive cells are situated exclusively in the interstitium; none is found in the seminiferous tubules (*asterisks*), unlike in pathologic conditions (IHC stain [diaminobenzidine and hematoxylin counterstain], ×400; *scale bar*, 25 μm)

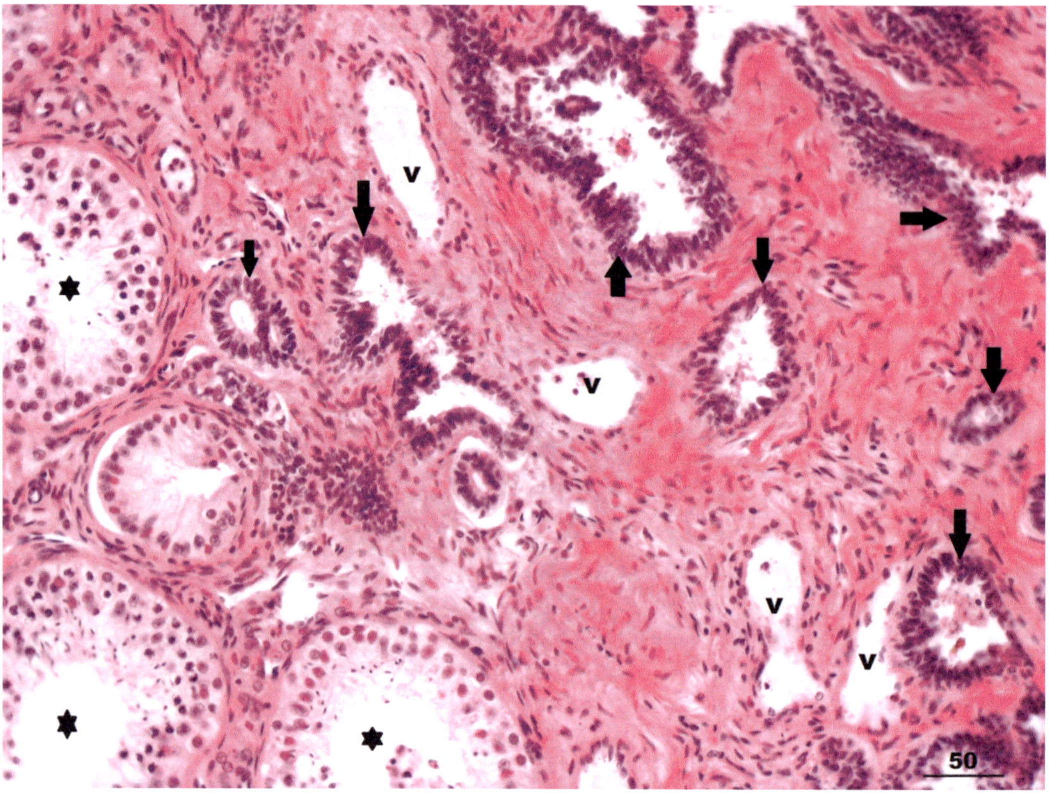

Fig. 7.14 Seminiferous tubules (*asterisks*) emptying into straight tubules (*arrows*). These tubules lack spermatogenic cells. Sertoli cells within this part of the tubules are mostly columnar. In the connective tissue, several veins (*v*) are visible (H&E stain, ×200; *scale bar*, 50 μm)

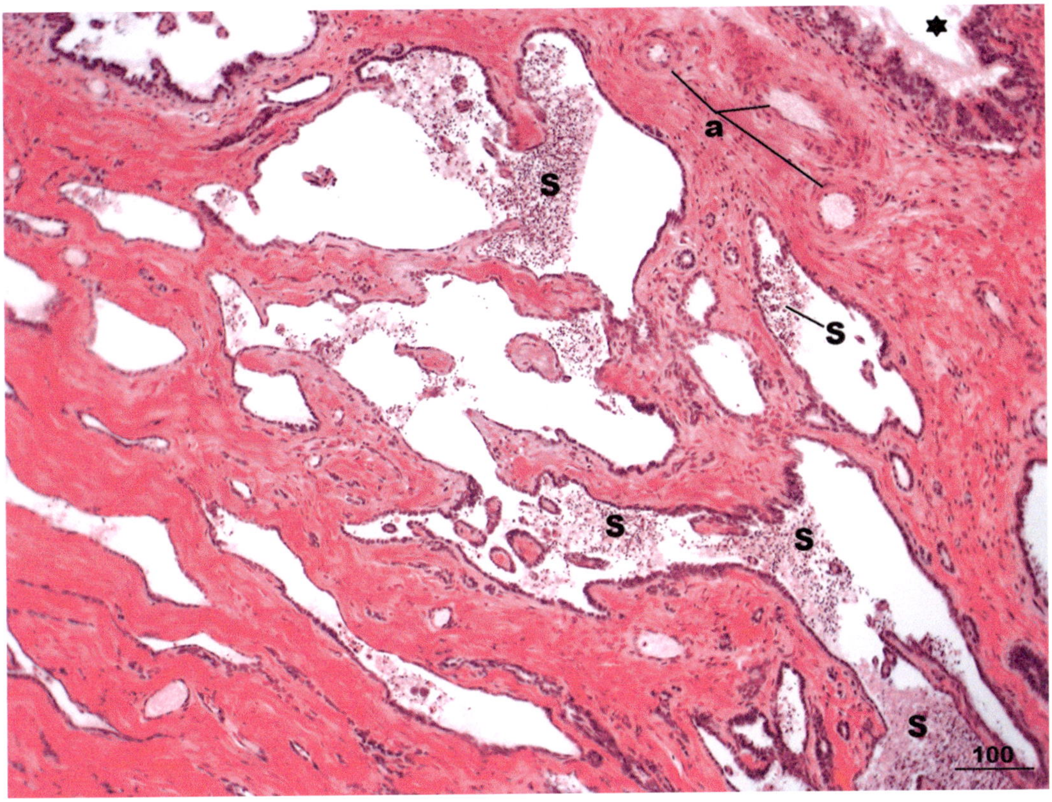

Fig. 7.15 Mediastinum testis. The mediastinum testis comprises a system of narrow and, sometimes, wide spaces lined with cuboidal epithelium. This system is called the *rete testis*. The wide parts contain abundant accumulations of spermatozoa (*S*) in their lumen. In the connective tissue of the mediastinum testis, there are many blood vessels (*a* [arteries]). Part of the efferent duct (*asterisk*) also is seen (H&E stain, ×100; *scale bar*, 100 μm)

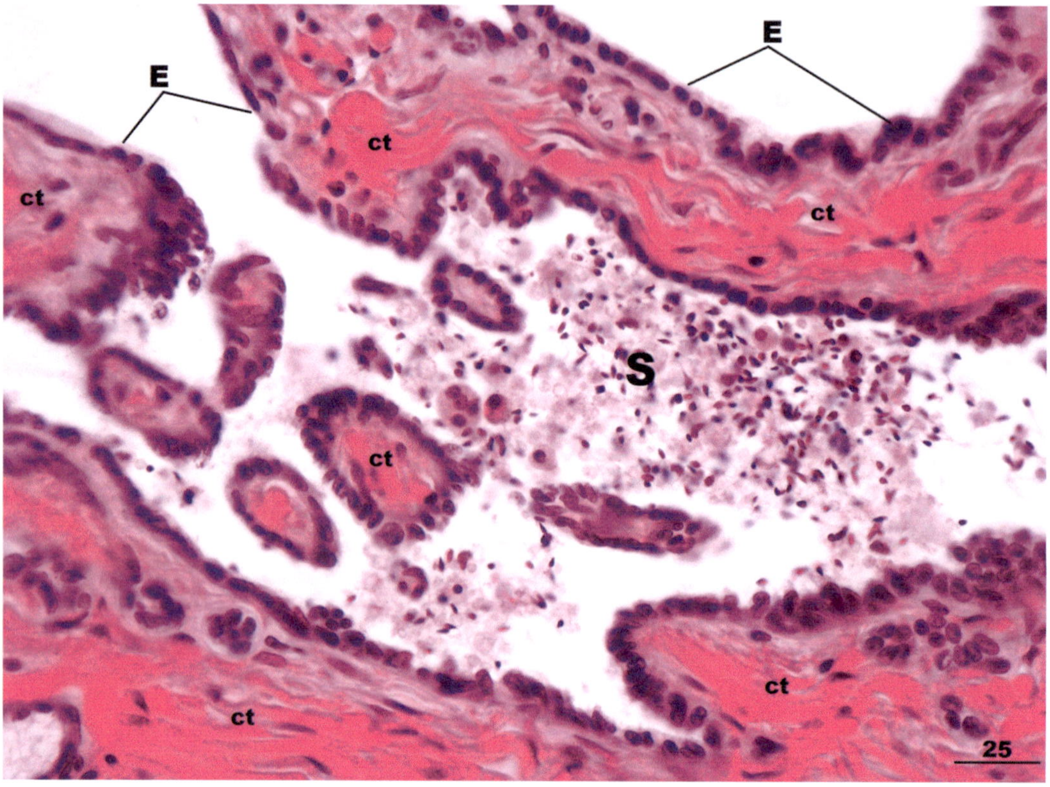

Fig. 7.16 Detail of the rete testis. The rete testis spaces are lined with cuboidal epithelium (*E*). Rich connective tissue (*ct*) provides support for the epithelium. In the spaces, one occasionally may see accumulations of spermatozoa (*S*) (H&E stain, ×400; *scale bar*, 25 μm)

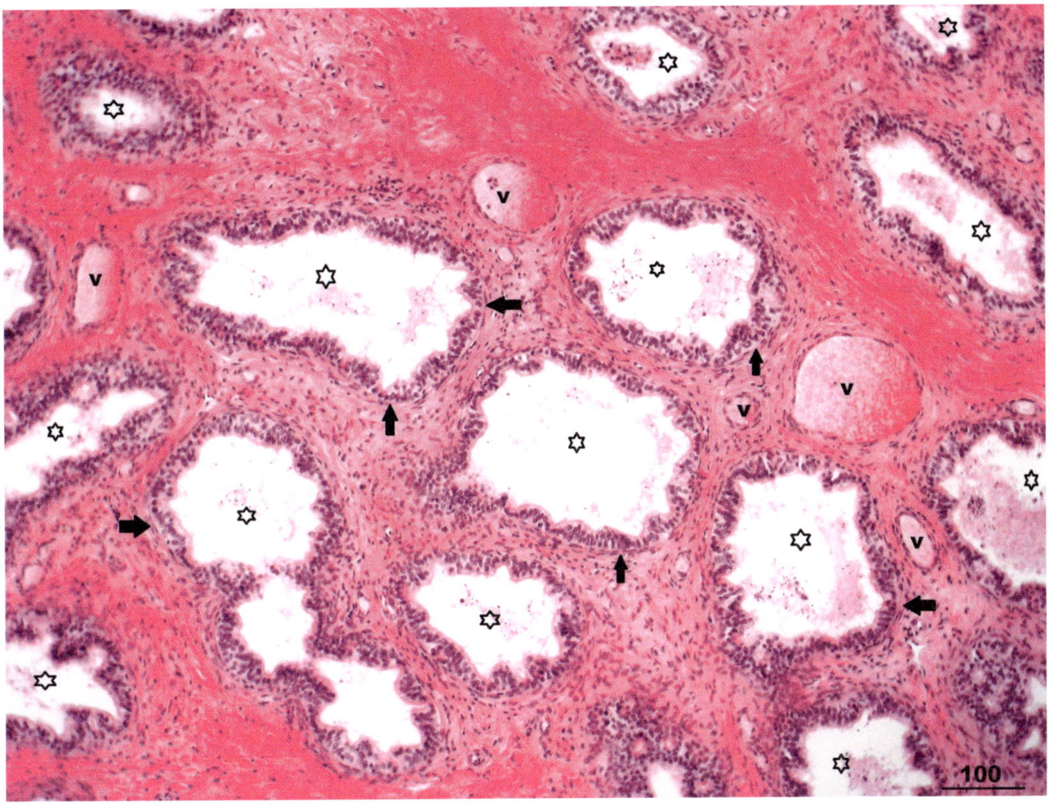

Fig. 7.17 Efferent ducts. Efferent ducts (ductuli efferentes, *asterisks*) are recognized by their rather irregular lumen. The epithelium consists of columnar and basal cells surrounded by loose connective and smooth muscle tissue (*arrows*). Several veins (*v*) are visible (H&E stain, ×100; *scale bar*, 100 μm)

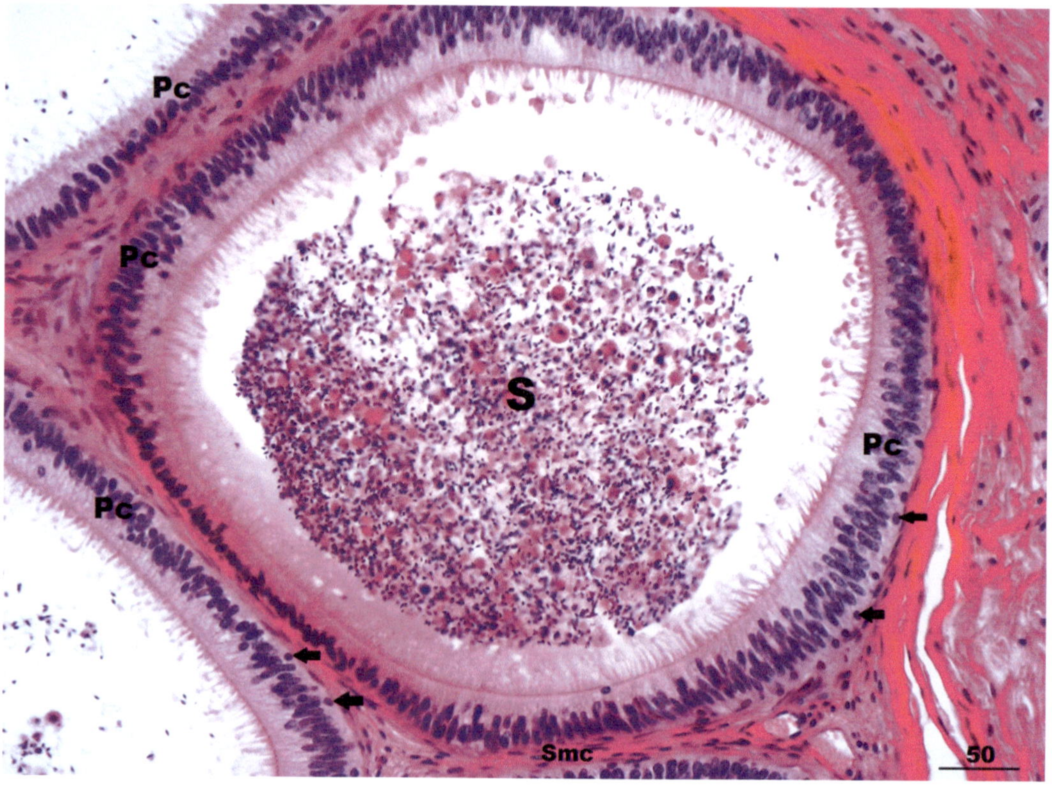

Fig. 7.18 Cross section through part of the ductus epididymidis. A regular lumen is filled with spermatozoa (*S*), some monocytes/macrophages and occasionally lymphocytes. These accumulations of spermatozoa may be used for PESA (percutaneous epididymal sperm aspiration) or MESA (microsurgical epididymal sperm aspiration). The duct itself is lined with columnar principal cells (*Pc*) and basal cells (*arrows*). A circular smooth muscle cell layer (*Smc*), seen here in the head region, is accompanied by the outer longitudinal layer in the body and tail regions (H&E stain, ×200; *scale bar*, 50 μm)

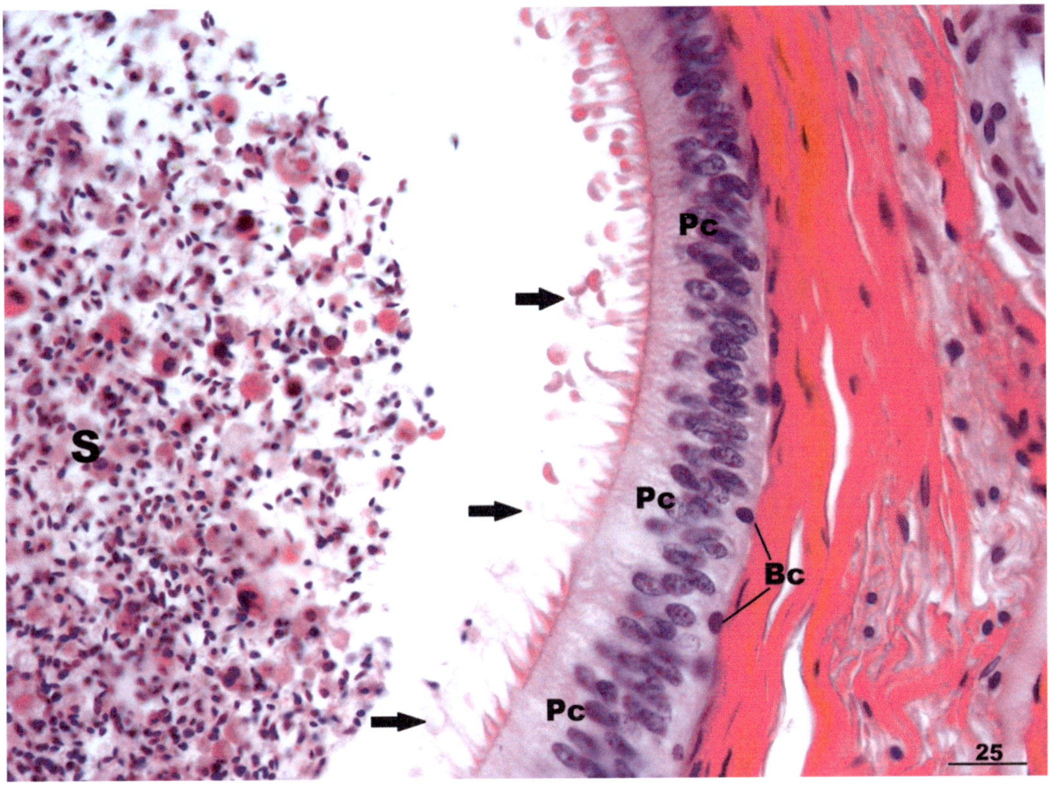

Fig. 7.19 Detail of the ductus epididymidis. On their apical surface, columnar principal cells (*Pc*) demonstrate extensions called *stereocilia* (*arrows*). As a result of fixation and dehydration of the specimen, only a few spermatozoa are in contact with the stereocilia, whereas the rest of the deposited spermatozoa (*S*) fill the lumen of the duct. *Bc* basal cells (H&E stain, ×400; *scale bar*, 25 μm)

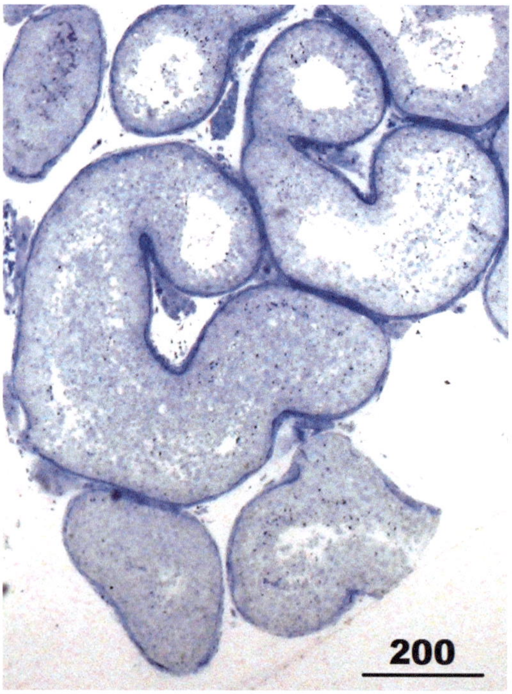

Fig. 7.20 Semithin section of a testicular biopsy specimen. As the result of a specific histologic procedure and plastic embedding, the seminiferous tubules and the interstitial tissue are preserved almost at their original size, with a minimum of shrinkage (Toluidine blue stain, ×40; *scale bar*, 200 μm)

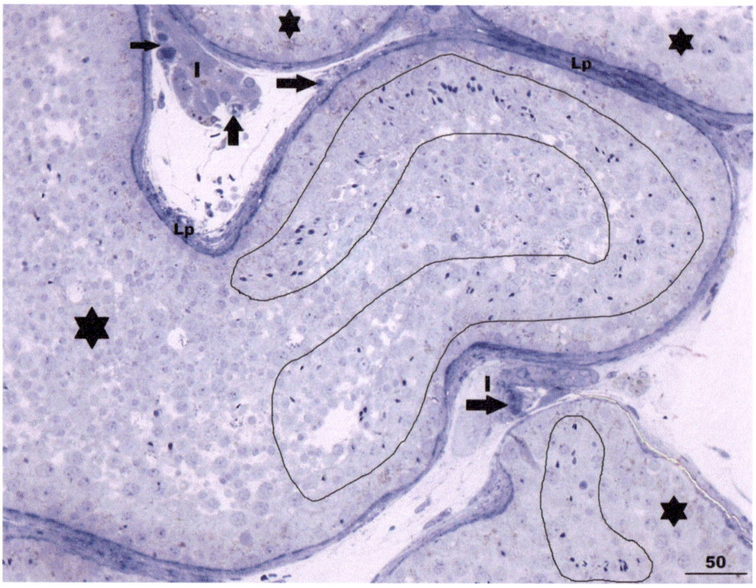

Fig. 7.21 Semithin section: detail of the previous figure. Seminiferous tubules (*asterisks*) with full spermatogenesis are limited by the lamina propria (*Lp*). The interstitial tissue (*I*) contains clusters of Leydig cells and some blood vessels (*arrows*). Because of the helical cycle of the seminiferous epithelium, only some areas in the tubule contain elongated spermatids and spermatozoa (*encircled areas*) (Toluidine blue stain, ×100; *scale bar*, 50 μm)

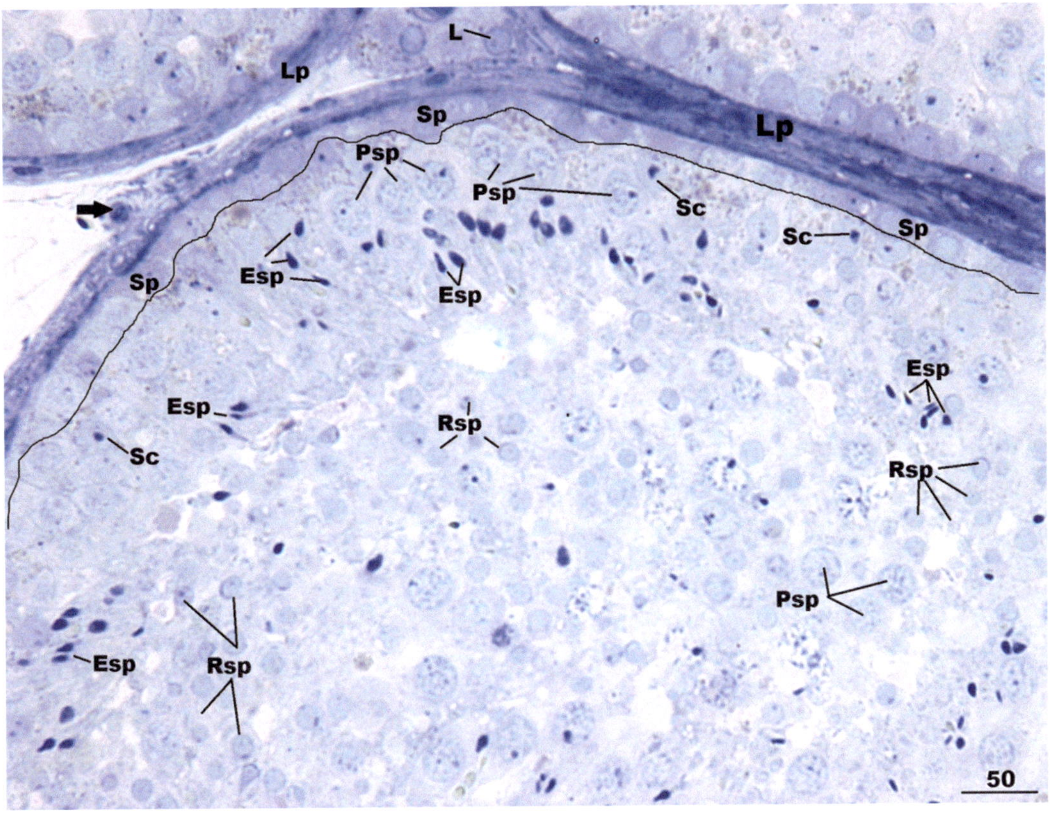

Fig. 7.22 Regular spermatogenesis in a semithin section. The *line* denotes a border between the basal and apical compartments of the seminiferous epithelium. Within the basal compartment, spermatogonia (*S*p) rest on the basement membrane. Adjacent to the basement membrane is the lamina propria (*Lp*) with some capillaries (*arrow*). Sertoli cells (*S*c) extend through both the basal and apical compartments. Most spermatogenic cells—primary spermatocytes (*Psp*), round spermatids (*Rsp*), and elongated spermatids (*Esp*)—are located within the apical compartment. Leydig cells (*L*) rest between neighboring seminiferous tubules (Toluidine blue stain, ×200; *scale bar*, 50 μm)

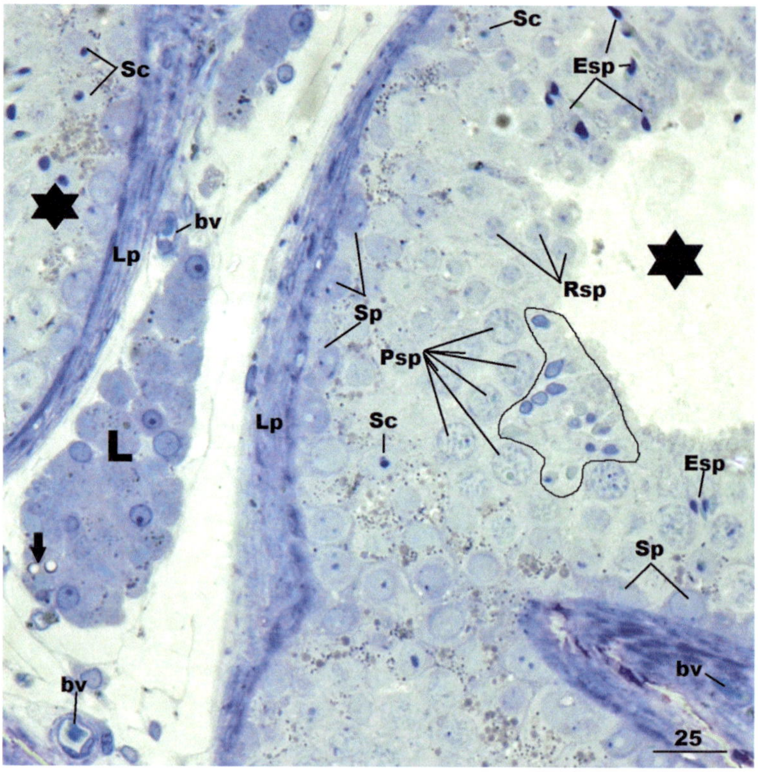

Fig. 7.23 Semithin section of normal testicular parenchyma. Parts of two seminiferous tubules (*asterisks*) and accompanying interstitium are visible. The seminiferous tubules, limited by lamina propria (*Lp*), are lined by Sertoli cells (*Sc*) and various types of spermatogenic cells: spermatogonia (*Sp*), primary spermatocytes (*Psp*), round or immature spermatids (*Rsp*), and elongated or mature spermatids (*Esp*). During the process of spermiogenesis, round spermatids are transformed into elongated spermatids (*encircled area*). Within the interstitium, Leydig cells (*L*) sometimes contain lipid droplets and vacuoles (*arrow*). Small blood vessels/capillaries (*bv*) also are visible (Toluidine blue stain, ×400; *scale bar*, 25 μm)

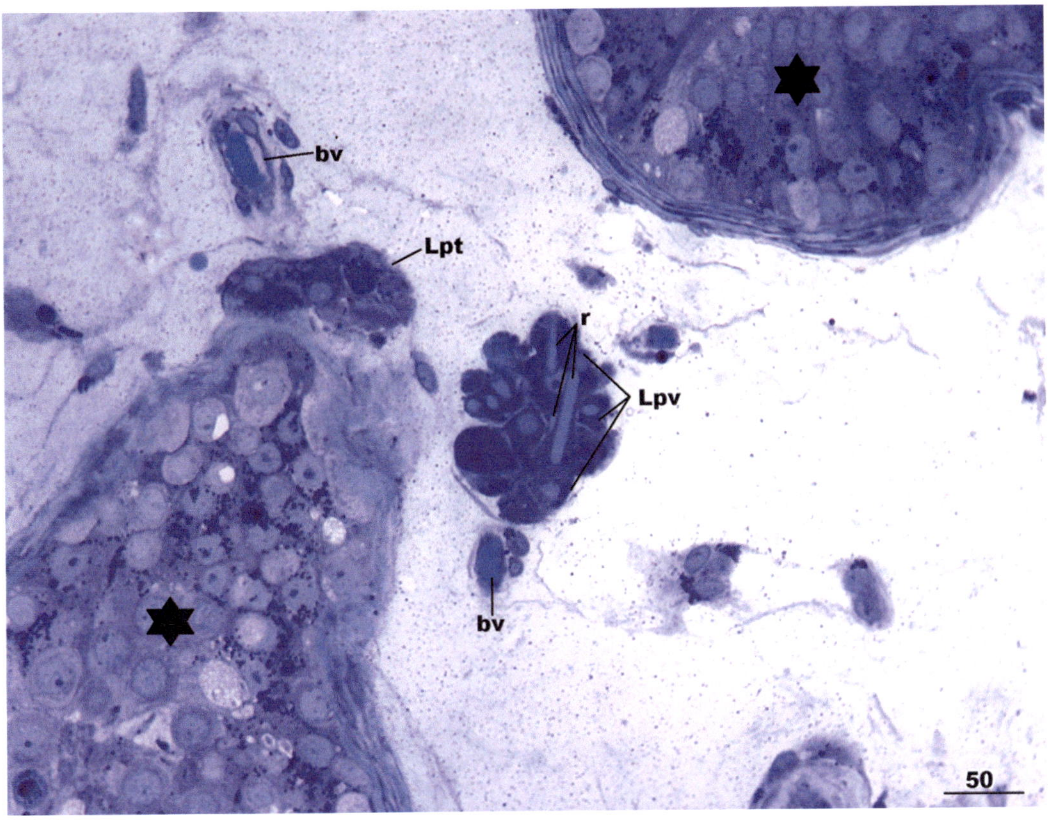

Fig. 7.24 Semithin section of testicular interstitial tissue. Leydig interstitial cells may be subdivided into two groups: peritubular cells (*Lpt*), which are adjacent to the seminiferous tubules (*asterisks*), and perivascular cells (*Lpv*), which are situated near blood vessels (*bv*). In this case, the perivascular group of Leydig cells contains several Reinke's crystals (*r*) (Toluidine blue stain, ×200; *scale bar*, 50 μm)

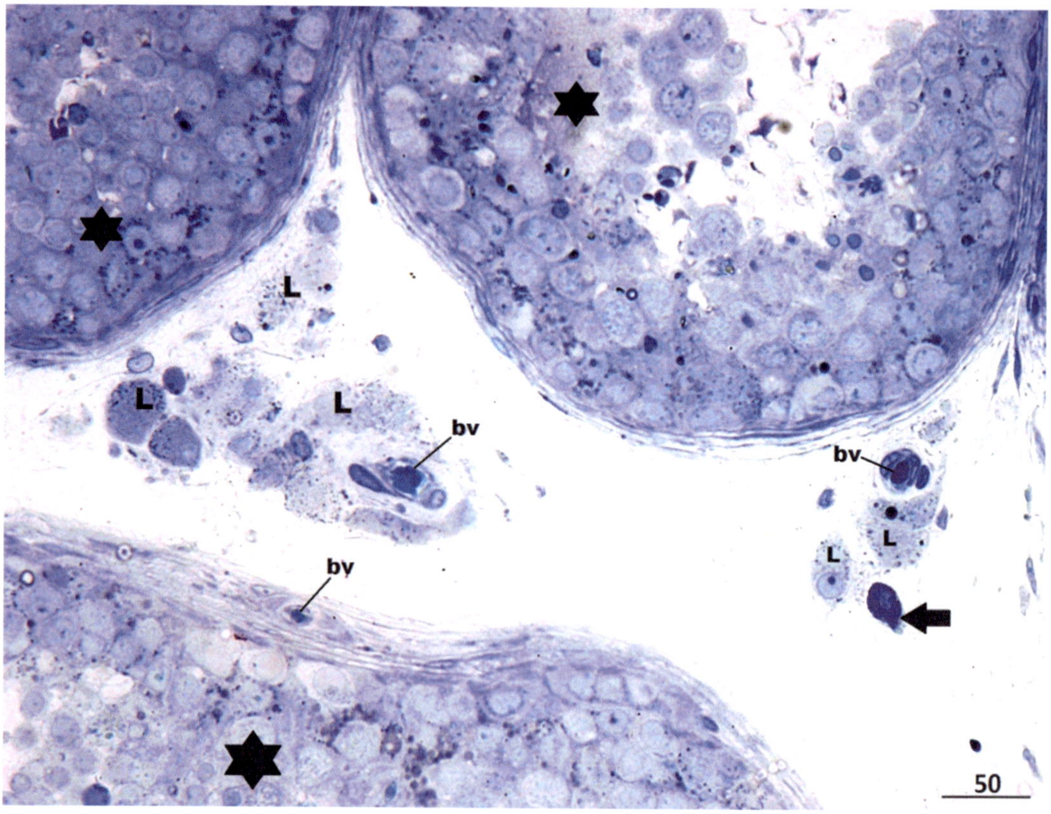

Fig. 7.25 Detail from a semithin section of testicular parenchyma. Apart from Leydig cells (*L*), loose connective tissue comprises all types of connective tissue cells, such as mast cells (*arrow*). The *asterisks* mark seminiferous tubules. *Bv* blood vessels (Toluidine blue stain, ×400; *scale bar*, 25 μm)

Recommended Literature

Bay K, Main KM, Toppari J, Skakkebæk NE. Testicular descent: INSL3, testosterone, genes and the intrauterine milieu. Nat Rev Urol. 2011;8:187–96.

Cerilli LA, Kuang W, Rogers D. A practical approach to testicular biopsy interpretation for male infertility. Arch Pathol Lab Med. 2010;134:1197–204.

Cheng CY, Mruk DD. A local autocrine axis in the testes that regulates spermatogenesis. Nat Rev Endocrinol. 2010;6:380–95.

Mital P, Hinton BT, Dufour JM. The blood-testis and blood-epididymis barriers are more than just their tight junctions. Biol Reprod. 2011;84:851–8.

Rajender S, Avery K, Agarwal A. Epigenetics, spermatogenesis and male infertility. Mutat Res. 2011;727: 62–71.

Spiller CM, Koopman P. Cell cycle control of germ cell differentiation. Results Probl Cell Differ. 2011;53: 269–308.

Svechnikov K, Landreh L, Weisser J, Izzo G, Colón E, Svechnikova I, Söder O. Origin, development and regulation of human Leydig cells. Horm Res Paediatr. 2010;73:93–101.

Damage of Spermatogenesis

8

Martin Bergmann and Davor Ježek

To perform a thorough histologic analysis or to interpret an evaluation report, an understanding of the process of normal spermatogenesis is necessary. Spermatogenesis is regarded as normal when most of the tubules in a given section represent one or more of the six stages of spermatogenesis (I–VI). These stages of spermatogenesis are a characteristic association of germ cells and represent several waves of spermatogenesis occurring simultaneously within the seminiferous epithelium. In the human, a so-called multistage arrangement is found. The structural preservation of testicular tissue fixed with Bouin's or Stieve's solution and embedded in paraffin is adequate to identify all cell types. Furthermore, it allows the use of modern histologic techniques, such as immunohistochemistry (IHC) and in situ hybridization, to find evidence of gene expression on the protein and mRNA levels, such as the correct histone–protamine exchange during spermiogenesis, or the state of differentiation of Sertoli and peritubular cells and the proliferation pattern of spermatogonia. If the seminiferous epithelium reveals a reduced number of elongated spermatids and defined stages of spermatogenesis cannot be identified, the term *hypospermatogenesis* should be applied. There are numerous defects, including multinuclear spermatids or spermatogonia and, most often, the occurrence of different spermatogenetic defects, such as maturation arrest at different levels (early round spermatids, primary spermatocytes, or spermatogonia), Sertoli cell–only syndrome, and the total loss of seminiferous epithelium, leaving only thickened lamina propria (tubular shadows) in the adjacent seminiferous tubules of a given section (mixed atrophy). Leydig cells may show a diffuse hyperplasia or focal nodular accumulation. Testicular intraepithelial neoplasia (TIN, syn. carcinoma in situ), which is known to be the precursor of seminomatous and nonseminomatous germ cell tumors—except for spermatocytic seminoma—may be recognized by its large size, clear cytoplasm, and large irregularly outlined nuclei with numerous nucleoli of the cells. The nature of these atypical spermatogonia must be determined by immunohistochemical staining of the membrane-bound placenta-like alkaline phosphatase (PLAP). The interstitial localization of these cells indicates an early invasive state of seminoma.

Besides paraffin sections, semithin sections are widely used in some andrologic centers. Semithin sections (0.70–1-μm thick) enable a fine analysis of the seminiferous epithelium and interstitial tissue, a distinction between residual bodies and late spermatid heads, and detection of carcinoma in situ cells. Therefore, this chapter also describes the various degrees of spermatogenesis damage observed in semithin sections.

M. Bergmann, Ph.D.
Institute for Veterinary Anatomy,
Histology and Embryology, Justus Liebig University,
Frankfurter Straße 98, Giessen 35392, Germany
e-mail: martin.bergmann@vetmed.uni-giessen.de

D. Ježek, M.D., Ph.D. (✉)
Department Histology and Embryology, School of Medicine, University of Zagreb,
Šalata 3, Zagreb 10000, Croatia
e-mail: davor@mef.hr; davorjezek@yahoo.com

D. Ježek (ed.), *Atlas on the Human Testis*,
DOI 10.1007/978-1-4471-2763-5_8, © Springer-Verlag London 2013

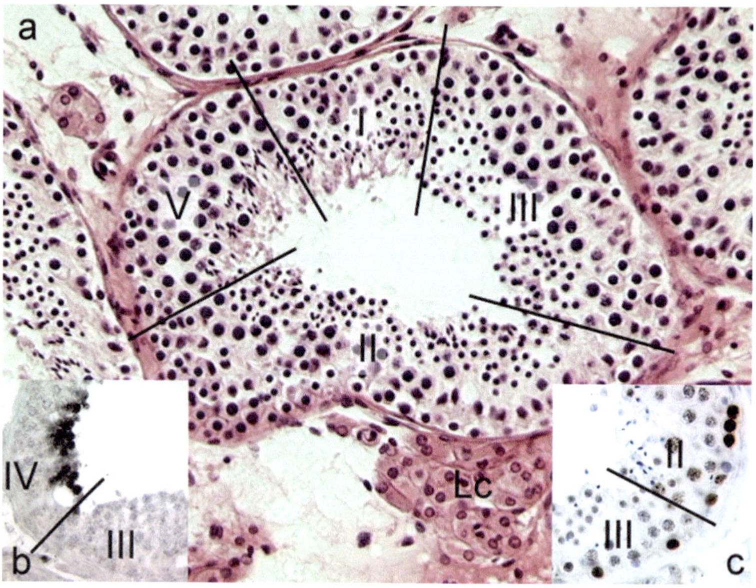

Fig. 8.1 Normal spermatogenesis. (**a**) Seminiferous tubule showing different stages (I–V) of spermatogenesis (multi-stage arrangement). (Hematoxylin and eosin [H&E] stain; primary magnification, ×20.) (**b**) In situ hybridization showing protamine 1 mRNA expression in step 4 spermatids of stage IV, indicating the initiation of histone–protamine exchange leading to the nuclear condensation and formation of sperm heads. Note the absence of mRNA in step 3 spermatids of stage III. (Primary magnification, ×40.) (**c**) Ki-67 (S-phase) IHC showing labeled spermatogonia in stages II and III (Hematoxylin counterstain; primary magnification, ×40), *Lc* Leydig cells

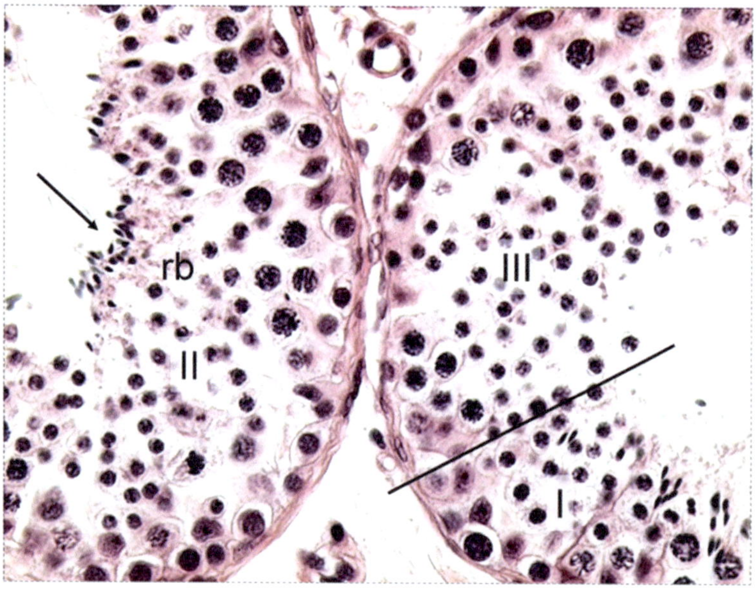

Fig. 8.2 Normal spermatogenesis. The left seminiferous tubule shows sperm release (*arrow*) and residual bodies (*rb*) at the end of stage II. The right tubule shows stages I and III (H&E stain; primary magnification, ×40)

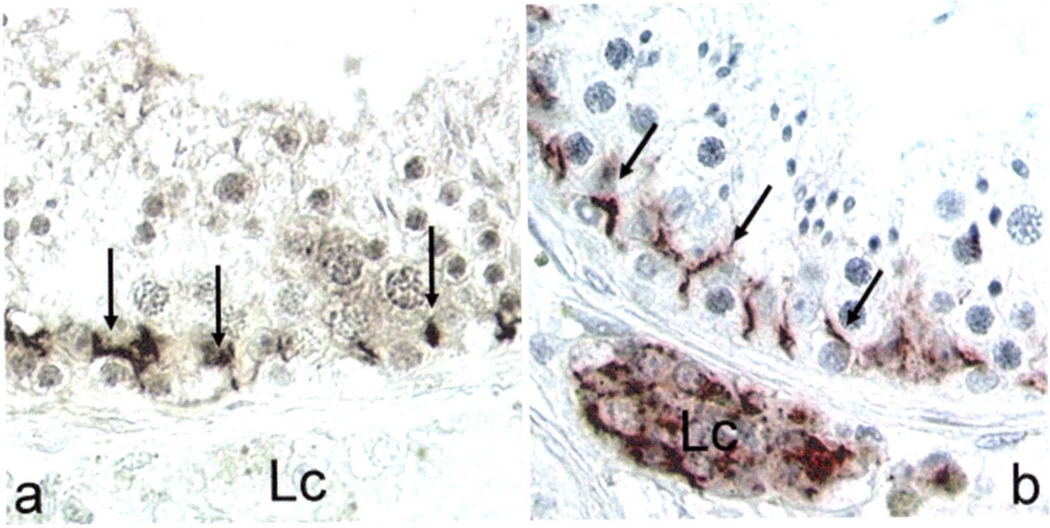

Fig. 8.3 Normal spermatogenesis. (**a**) A component of the inter–Sertoli cell junctional complex indicating an intact blood–testis barrier on ZO_2 IHC (*arrows*). (Primary magnification, ×40.) (**b**) A component of the inter–Sertoli cell junctional complex indicating an intact blood–testis barrier on gap junction protein Cx43 IHC (*arrows*). Note, in contrast to the ZO_2 image, interstitial Leydig cells (*Lc*) also are labeled (Hematoxylin counterstain; primary magnification, ×40)

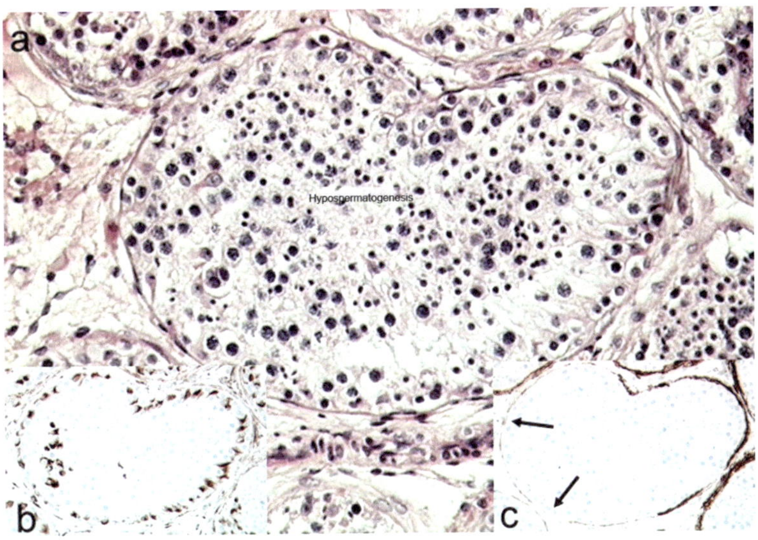

Fig. 8.4 Hypospermatogenesis (*hyp*). (**a**) Seminiferous tubule showing only qualitatively intact spermatogenesis up to elongated spermatids, but no clear stage arrangement. (H&E stain; primary magnification, ×20) (**b**) Vimentin IHC of the same tubule showing normal Sertoli cells. (Hematoxylin counterstain; primary magnification, ×20.) (**c**) Actin IHC of the same tubule. Note the absence of immunoreactivity in parts of the lamina propria (*arrows*), indicating dedifferentiation of myofibroblasts that cannot be seen by routine H&E staining (Hematoxylin counterstain; primary magnification, ×20)

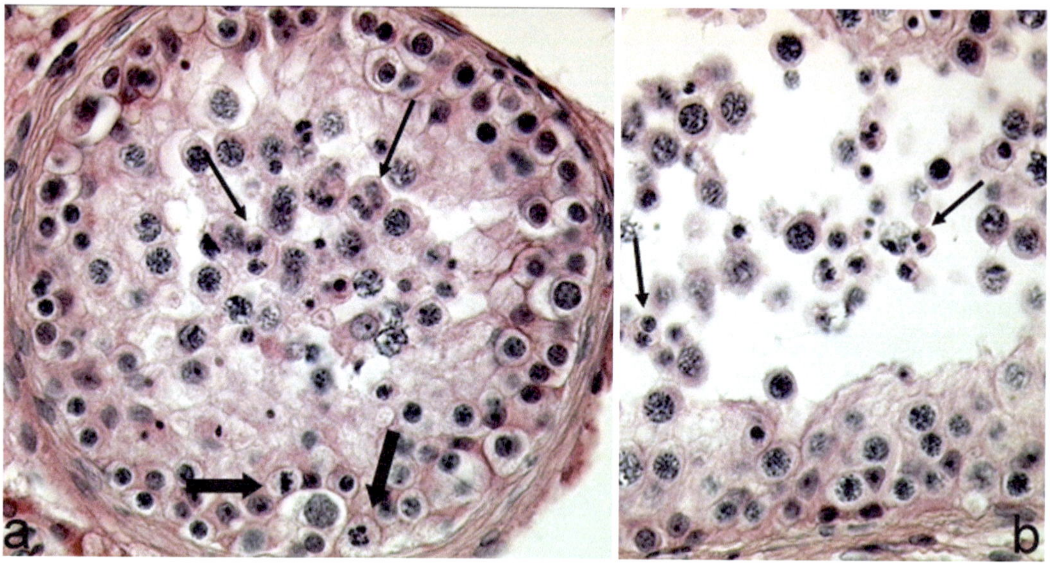

Fig. 8.5 Hypospermatogenesis and maturation arrest. (**a**) Seminiferous tubule showing hypospermatogenesis. Note the multinuclear spermatids (*arrows*), indicating defects in spermiogenesis. Mitotic figures (*bold arrows*) indicate ongoing spermatogonial renewal. (**b**) Seminiferous tubule showing maturation arrest at the level of early round spermatids. Note the numerous multinuclear spermatids and premature sloughing of spermatids into the tubular lumen (*arrows*) (H&E stain; primary magnification, ×40)

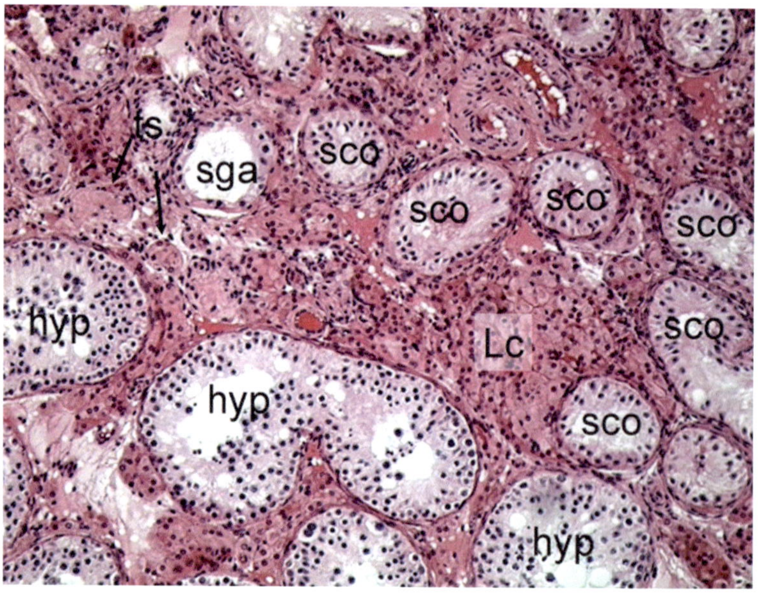

Fig. 8.6 Mixed atrophy. This histologic section shows various defects of spermatogenesis in the adjacent seminiferous tubules, such as hypospermatogenesis (*hyp*), arrest at the level of spermatogonia (*sga*), Sertoli cell–only syndrome (*sco*), and total loss of germ and Sertoli cells, leaving only thickened lamina propria (tubular shadows, *ts*, *arrows*). Note the additional diffuse Leydig cell hyperplasia (*Lc*) (H&E stain; primary magnification, ×10)

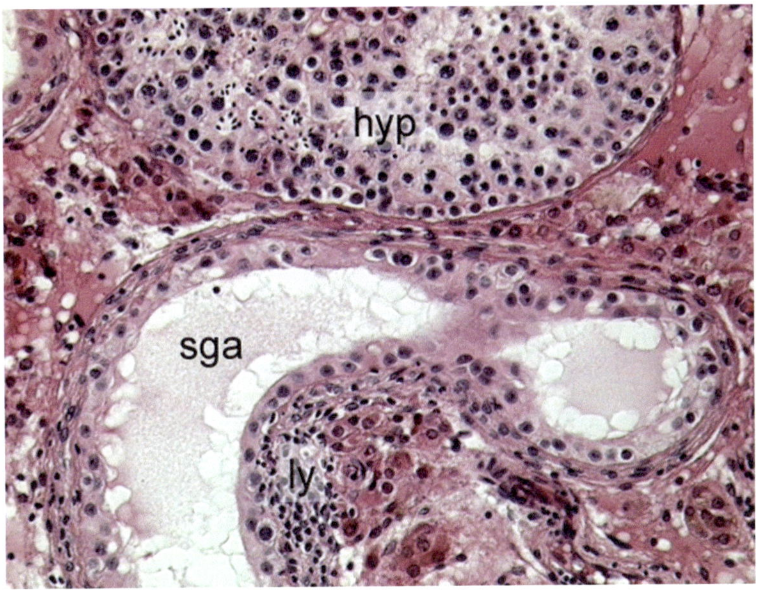

Fig. 8.7 Mixed atrophy. This histologic section shows a seminiferous tubule with hypospermatogenesis (*hyp*) adjacent to a tubule with an arrest of spermatogenesis at the level of spermatogonia (*sga*). Note the additional focal interstitial infiltrates of lymphocytes (*ly*) (H&E stain; primary magnification, ×40)

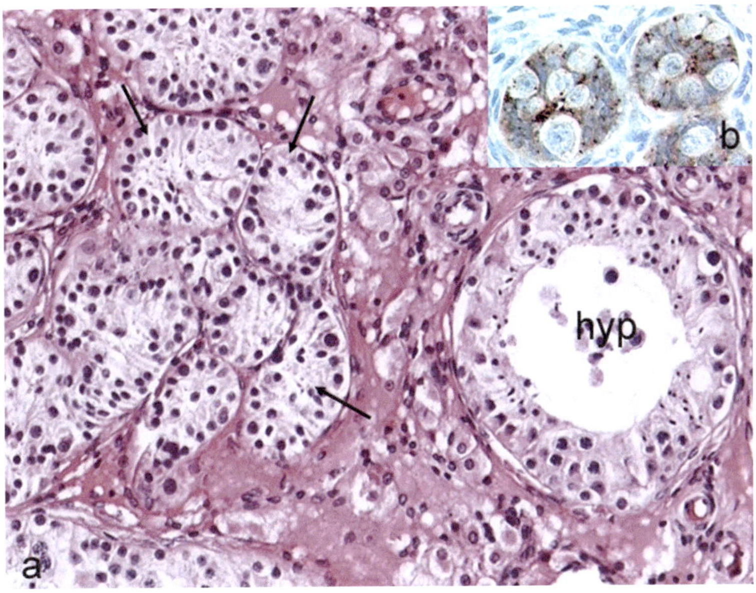

Fig. 8.8 Mixed atrophy. (**a**) Histologic section showing a seminiferous tubule with hypospermatogenesis (*hyp*) adjacent to immature seminiferous cords (*arrows*) consisting of immature Sertoli cells and spermatogonia. (H&E stain; primary magnification, ×40.) (**b**) Anti-Müllerian hormone immunostaining indicates a persistent fetal stage of Sertoli cells within the seminiferous cords (Hematoxylin counter-stain; primary magnification,×40)

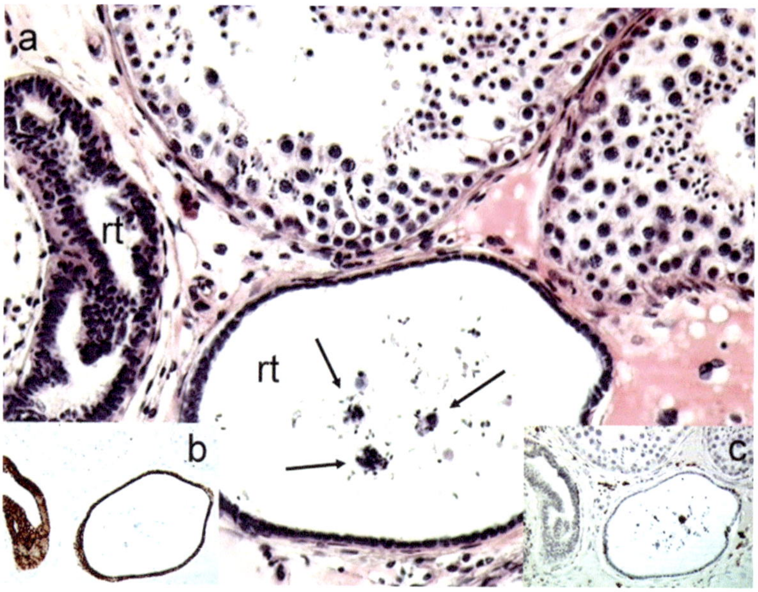

Fig. 8.9 Rete testis. (**a**) Histologic section showing seminiferous tubules with intact spermatogenesis adjacent to the rete testis (*rt*). Note the macrophages within the dilated area of the rete testis phagocytozing spermatozoa (*arrows*). (H&E stain; primary magnification, ×20.) (**b**) Cytokeratin 18 IHC shows labeling of the rete testis epithelium. (Hematoxylin counterstain; primary magnification, ×20.) (**c**) CD68+ IHC shows interstitial macrophages, as well as macrophages within the rete testis, phagocytosing spermatozoa (Hematoxylin counterstain; primary magnification, ×20)

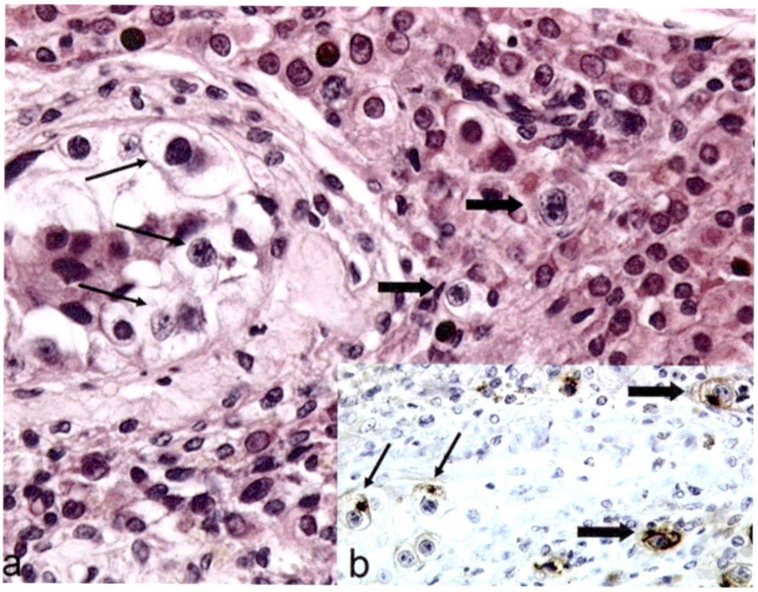

Fig. 8.10 Tumor development. (**a**) Histologic section showing preinvasive atypical germ cells within a seminiferous tubule (*arrows*) as well as within the interstitium (*bold arrows*). (H&E stain; primary magnification, ×40.) (**b**) Consecutive section of *panel* **a**. PLAP IHC shows PLAP-positive cells within the seminiferous tubule, indicating preinvasive testicular intraepithelial neoplasia as well as an invasive stage, indicated by PLAP-positive cells within the interstitium (Hematoxylin counterstain; primary magnification, ×40)

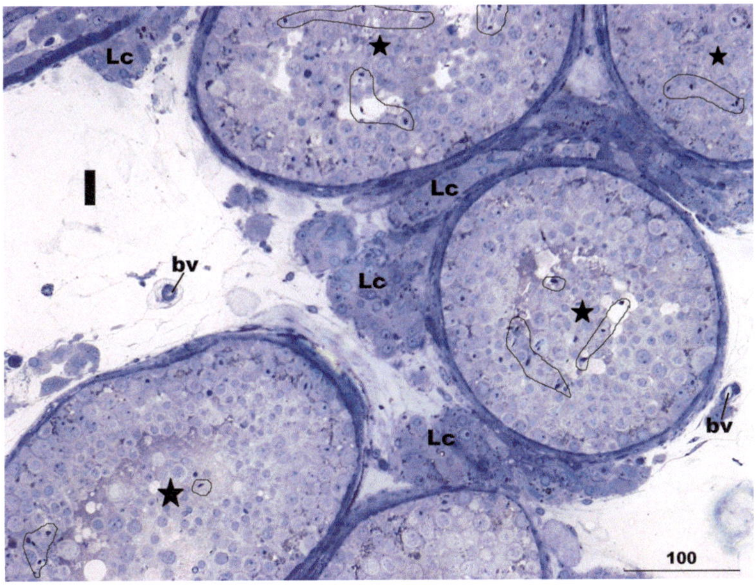

Fig. 8.11 Hypospermatogenesis in a testicular biopsy sample from a man with infertility. Within the seminiferous tubules (*stars*), no distinctive stages of spermatogenesis are visible. Spermatogenic cells mostly reach the round spermatid stage. Rare elongated spermatids (*encircled areas*) might be used for TESE/ICSI (testicular sperm extraction/intracytoplasmic injection of spermatozoon into the oocyte). The interstitial compartment (*I*) bears small blood vessels (*bv*) and Leydig cells (*Lc*), which are hyperplastic in some areas (Semithin section, toluidine blue; ×200. *Scale bar*, 100 μm)

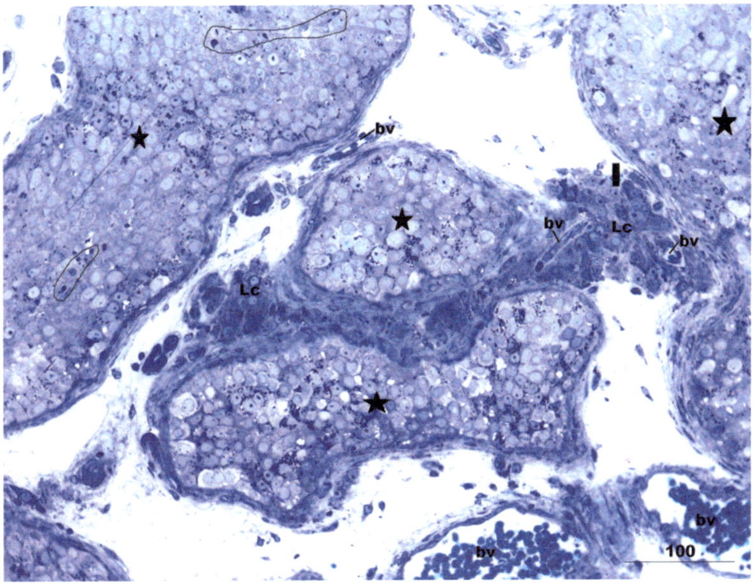

Fig. 8.12 Maturation arrest at the round spermatid stage. In some parts of the seminiferous tubules (*stars*), an abortive formation of elongated spermatids may be recorded (*encircled areas*). Within the loose connective tissue of the interstitium are clusters of Leydig cells (*Lc*) and medium-sized or small blood vessels (*bv*). *I* interstitial compartment (Semithin section, toluidine blue; ×200. *Scale bar*, 100 μm)

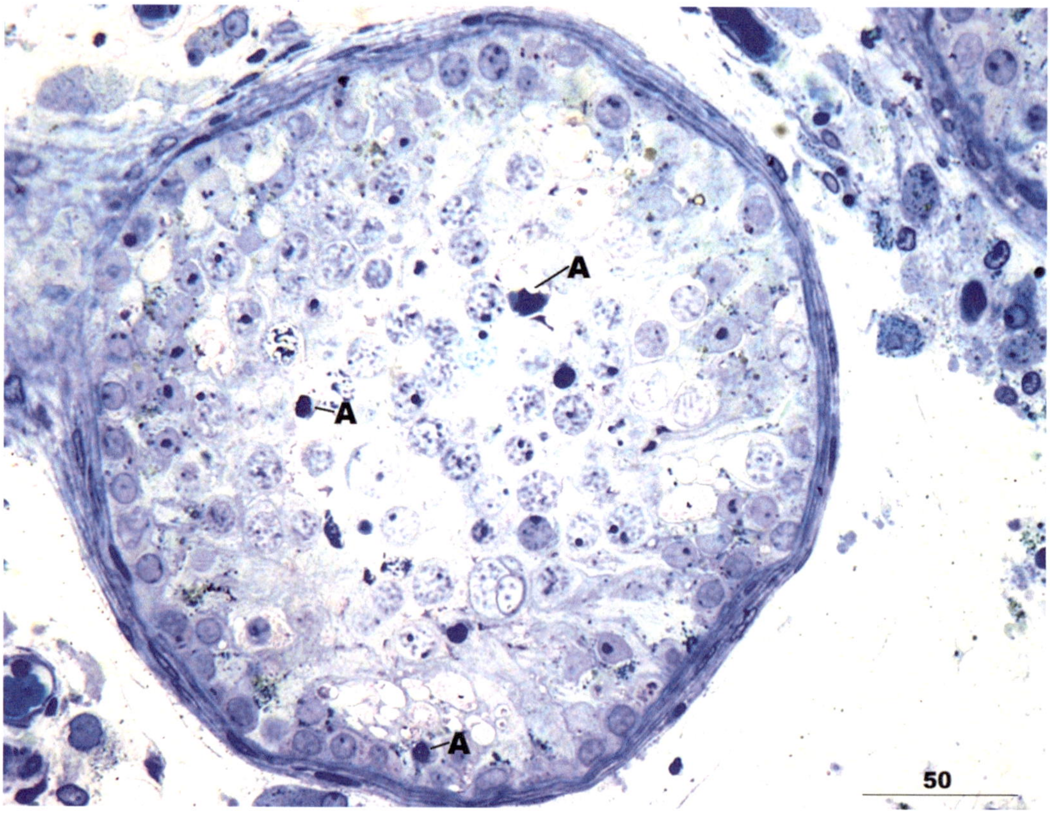

Fig. 8.13 Maturation arrest at the spermatocyte stage. Some cells undergo a process of apoptosis (*A*), or programmed cell death (Semithin section, toluidine blue; ×400. *Scale bar*, 50 μm)

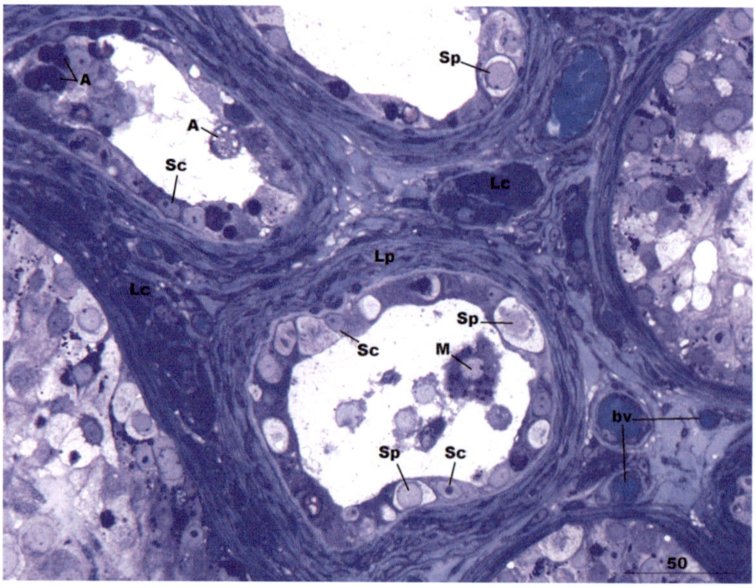

Fig. 8.14 "Mixed atrophy" of seminiferous tubules. The tubule in the center is lined with spermatogonia (*Sp*) and Sertoli cells (*Sc*) only. The Sertoli cells have a largely reduced cytoplasm. A significant number of remaining spermatogenic cells undergoes apoptosis (*A*). A presumptive macrophage (*M*) has invaded the seminiferous epithelium and is found in contact with desquamated spermatogenic cells within the tubular lumen. The lamina propria (*Lp*) of the tubule is enlarged. Leydig cells (*Lc*) of the interstitium are somewhat elongated and darkly stained. *bv* blood vessels (Semithin section, toluidine blue; ×400. *Scale bar*, 50 μm)

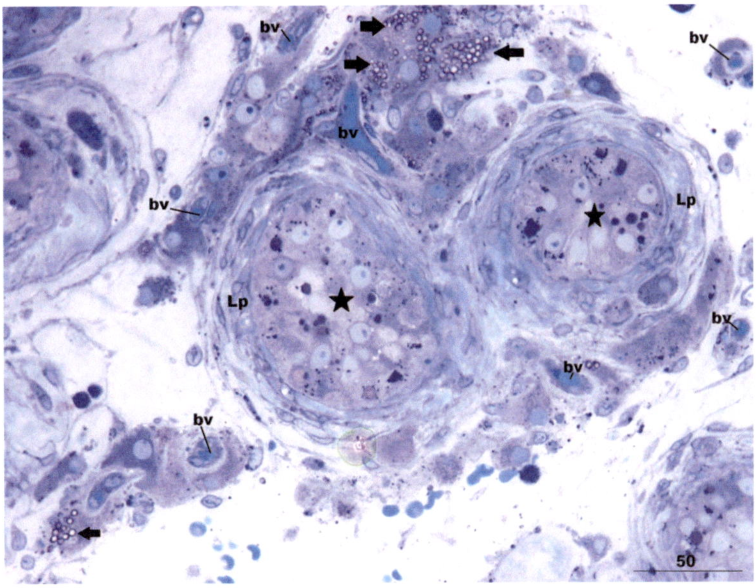

Fig. 8.15 Sertoli cell–only syndrome in a testicular biopsy specimen. The epithelium of the seminiferous tubules (*stars*) is lined with Sertoli cells exclusively. In their cytoplasm, the cells bear phagocytized material. Frequently, the tubules demonstrate thickening of the lamina propria (*Lp*). The cytoplasm of many Leydig cells is vacuolized (*arrows*). *bv* blood vessels (Semithin section, toluidine blue; ×400. *Scale bar*, 50 μm)

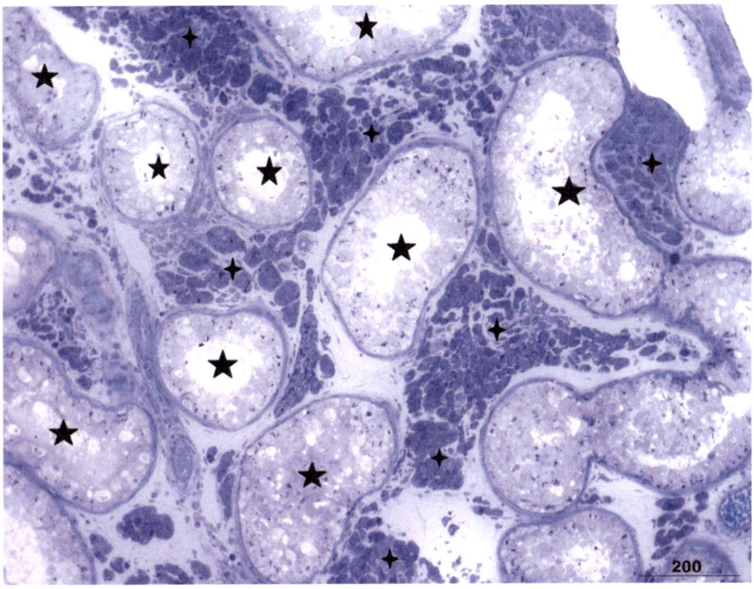

Fig. 8.16 Seminiferous tubules (*stars*) demonstrate maturation arrest at the primary spermatocyte stage. Between the tubules, the interstitial compartment contains many hypertrophic and hyperplastic Leydig cells (*crosslets*) (Semithin section, toluidine blue; ×100. *Scale bar*, 200 μm)

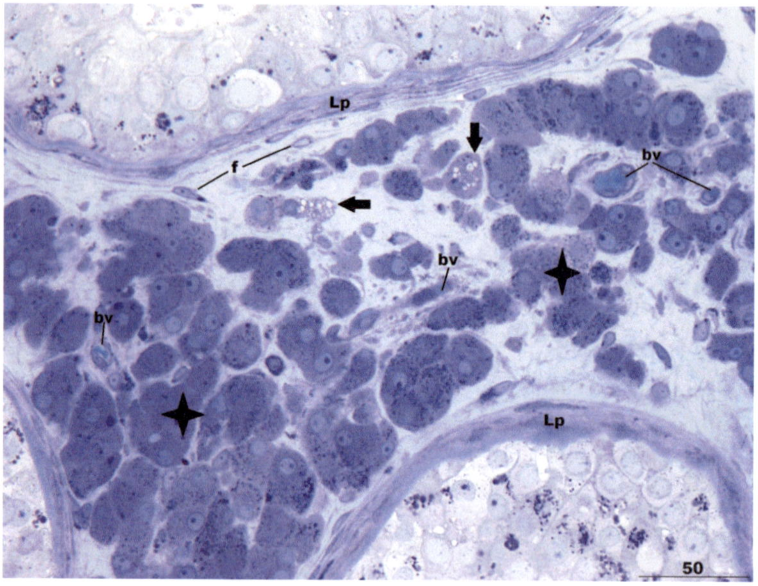

Fig. 8.17 Leydig cell hypertrophy and hyperplasia (*crosslets*). In some Leydig cells, an intensive vacuolization of the cytoplasm is visible (*arrows*). The lamina propria (*Lp*) of the seminiferous tubules shows a regular structure. *bv* blood vessels, *f* fibroblasts (Semithin section, toluidine blue; ×400. *Scale bar*, 50 μm)

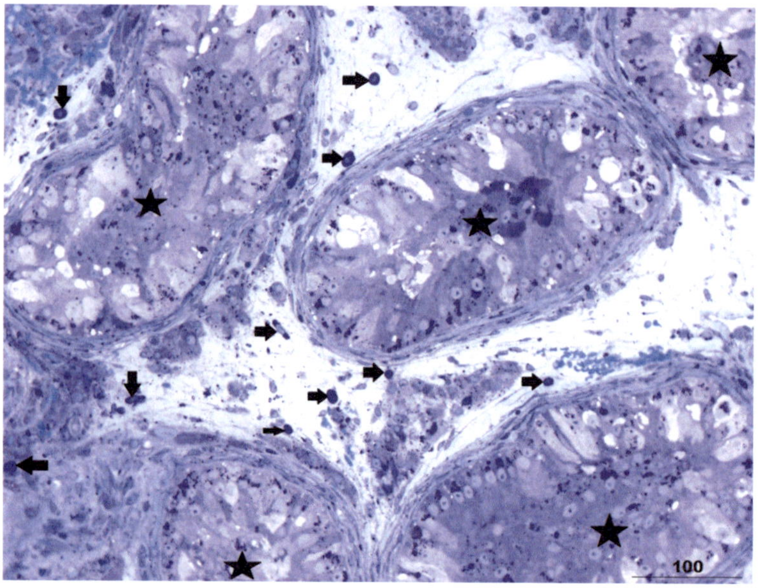

Fig. 8.18 Sertoli cell–only syndrome in a testicular biopsy specimen. The tubules (*stars*) are lined with Sertoli cells exclusively. In the loose connective tissue of the interstitium, there are many mast cells (*arrows*). These cells may frequently be encountered in the bioptic material of men with infertility (Semithin section, toluidine blue; ×200. *Scale bar*, 100 μm)

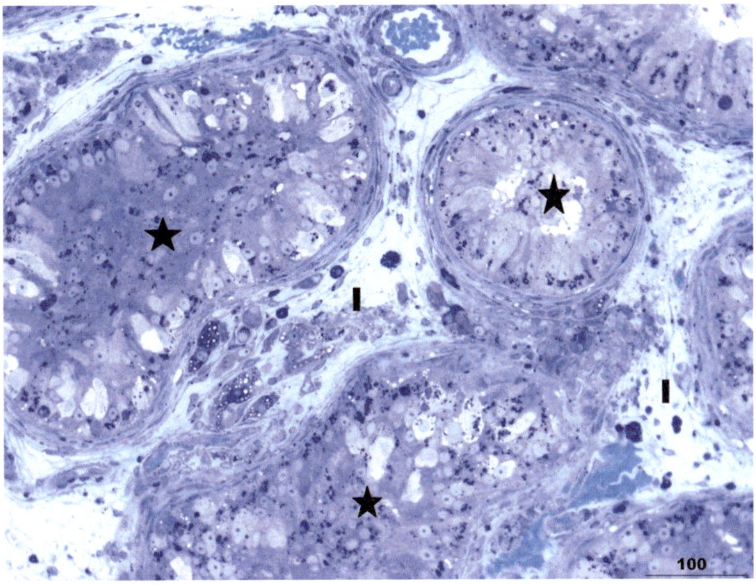

Fig. 8.19 Another example of Sertoli cell–only syndrome (*stars*) in a testicular biopsy specimen. The interstitium (*I*) bears many degenerating Leydig cells, mast cells, and macrophages (Semithin section, toluidine blue; ×200. *Scale bar*, 100 μm)

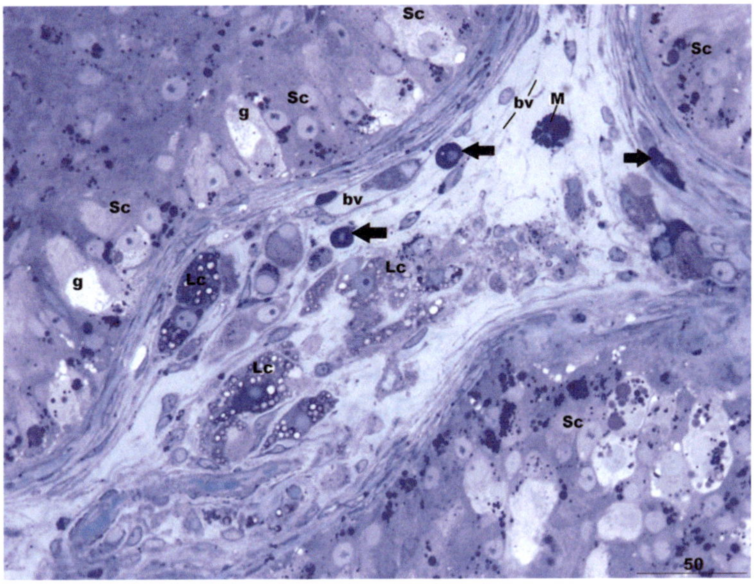

Fig. 8.20 Detail from Fig. 8.19 showing Sertoli cells only. Within the tubules, Sertoli cells (*Sc*) have a large amount of lipid droplets, glycogen (*g*), and phagocytized material. In the interstitial compartment, the Leydig cells (*Lc*) contain many vacuoles in their cytoplasm. An increased presence of mast cells (*arrows*) and macrophages (*M*) is notable. *bv* blood vessels (Semithin section, toluidine blue; ×400. *Scale bar*, 50 μm)

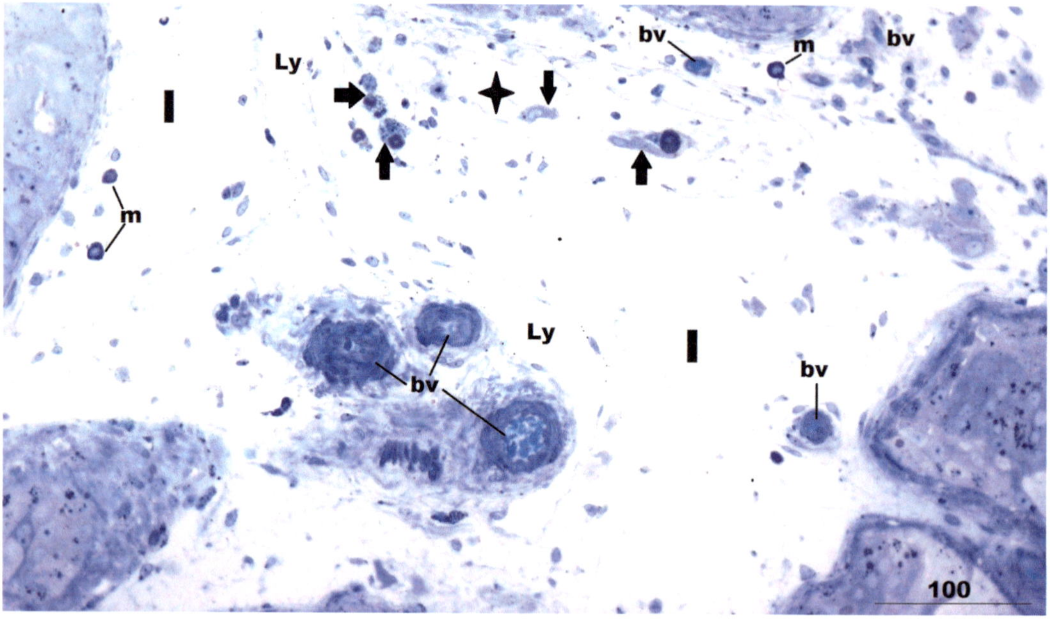

Fig. 8.21 Tubular sclerosis in a testicular biopsy specimen. In the case of tubular sclerosis, the seminiferous tubules are transformed into fibrous structures called *tubular shadows* (*crosslet*). The surrounding Leydig cells may be incorporated into the fibrous tissue (*arrows*). In addition, there is fibrosis of the interstitial compartment (*I*), which bears an increased number of mast cells (*m*). *bv* blood vessels, *Ly* lymph vessels (Semithin section, toluidine blue; ×200. *Scale bar*, 100 μm)

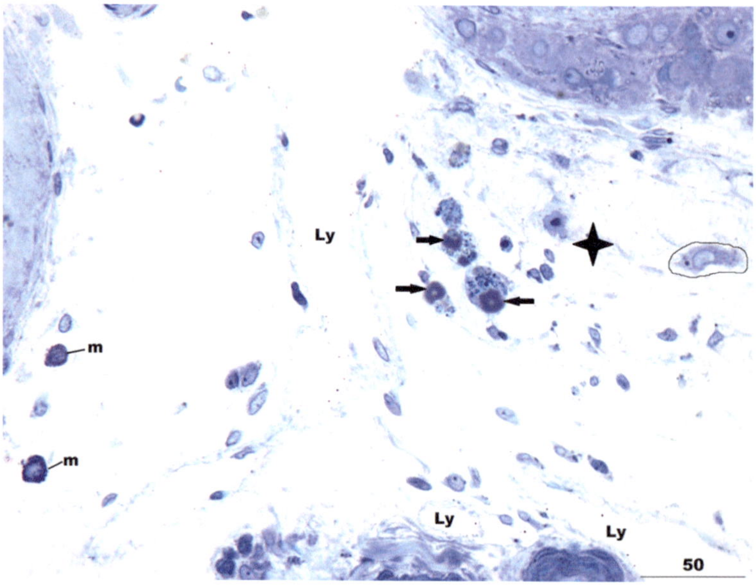

Fig. 8.22 Detailed view of the tubular sclerosis (*crosslet*) shown in Fig. 8.21. Most incorporated Leydig cells have large lipid droplets in their cytoplasm (*arrows*), in contrast to the same type of cell that maintains normal histological features (*encircled area*). *Ly* lymph vessels, *m* mast cells (Semithin section, toluidine blue; ×400. *Scale bar*, 50 μm)

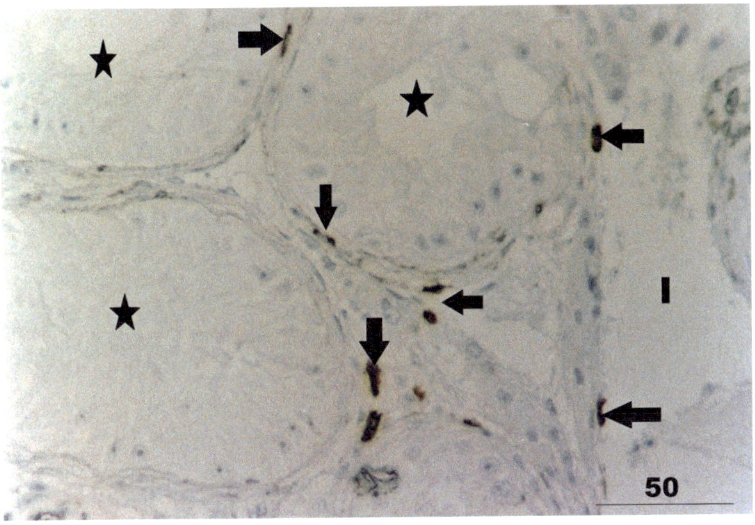

Fig. 8.23 Sertoli cell–only syndrome in a man with infertility. Seminiferous tubules (*stars*) are lined with Sertoli cells exclusively. Mast cell tryptase is expressed in abundant mast cells (*arrows*) that are found within the interstitial tissue (*I*) and close to the tubules (IHC [3,3′-diaminobenzidine and hematoxylin contrasting), ×400; *scale bar*, 50 μm)

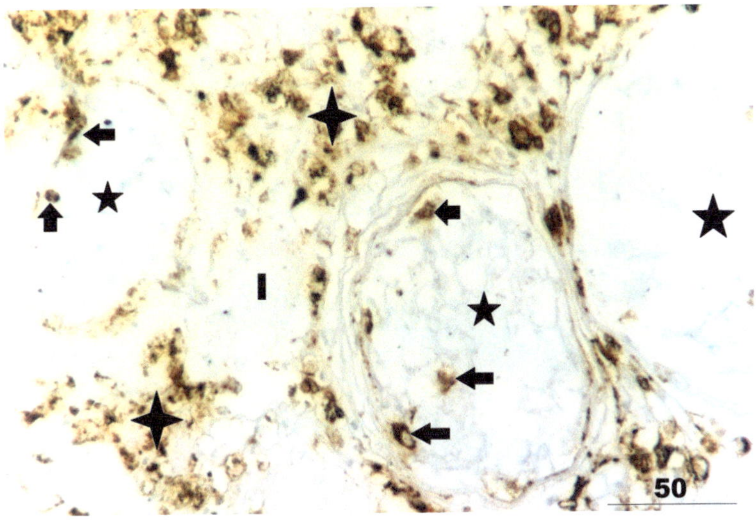

Fig. 8.24 Tubules with Sertoli cells only (*stars*); expression of KiM1P1 (macrophage marker). Macrophages (*crosslets*) infiltrate the interstitial testicular compartment (*I*). Some of these cells are found in the peritubular area and within the seminiferous tubules (*arrows*) (IHC [3,3′-diaminobenzidine and hematoxylin contrasting), ×400; *scale bar*, 50 μm)

Recommended Literature

Aggarwal N, Parwani AV. Spermatocytic seminoma. Arch Pathol Lab Med. 2009;133:1985–8.

Albers P, Albrecht W, Algaba F, Bokemeyer C, Cohn-Cedermark G, Fizazi K, et al. European Association of Urology. EAU guidelines on testicular cancer: 2011 update. Eur Urol. 2011;60:304–19.

Carpi A, Agarwal A, Sabanegh E, Todeschini G, Balercia G. Percutaneous biopsy of the testicle: a mini review with a proposal flow chart for non-obstructive azoospermia. Ann Med. 2011;43:83–9.

Cerilli LA, Kuang W, Rogers D. A practical approach to testicular biopsy interpretation for male infertility. Arch Pathol Lab Med. 2010;134:1197–204.

Dieckmann KP, Kulejewski M, Heinemann V, Loy V. Testicular biopsy for early cancer detection—objectives, technique and controversies. Int J Androl. 2011;34:7–13.

Heidenreich A. Contralateral testicular biopsy in testis cancer: current concepts and controversies. BJU Int. 2009;104(9 Pt B):1346–50.

Kleiman SE, Yogev L, Lehavi O, Hauser R, Botchan A, Paz G, et al. The likelihood of finding mature sperm cells in men with AZFb or AZFb-c deletions: six new cases and a review of the literature (1994–2010). Fertil Steril. 2011;95:2005–12.

Meissner A, Mamoulakis C, de la Rosette JJ, Pes MP. Clinical update on testicular microlithiasis. Curr Opin Urol. 2009;19:615–8.

Oosterhuis JW, Stoop H, Dohle G, Boellaard W, van Casteren N, Wolffenbuttel K, Looijenga LH. A pathologist's view on the testis biopsy. Int J Androl. 2011;34:14–9.

Schuppe HC, Meinhardt A, Allam JP, Bergmann M, Weidner W, Haidl G. Chronic orchitis: a neglected cause of male infertility? Andrologia. 2008;40:84–91.

Volkmann J, Müller D, Feuerstacke C, Kliesch S, Bergmann M, Mühlfeld C, Middendorff R. Disturbed spermatogenesis associated with thickened lamina propria of seminiferous tubules is not caused by dedifferentiation of myofibroblasts. Hum Reprod. 2011; 26:1450–61.

von Kopylow K, Kirchhoff C, Jezek D, Schulze W, Feig C, Primig M, et al. Screening for biomarkers of spermatogonia within the human testis: a whole genome approach. Hum Reprod. 2010;25:1104–12.

9 Inflammatory Conditions of the Testis

Hans-Christian Schuppe and Martin Bergmann

Infection and inflammation of the reproductive tract are accepted as important etiologic factors of male infertility [1–3]. With regard to their impact on reproductive function, orchitis and epididymo-orchitis are of particular concern. Whereas the clinical term *orchitis* is used mainly for acute, symptomatic disease due to local or systemic infection, a comprehensive classification also should include noninfectious disorders. The following etiologic factors of testicular inflammation must be considered [4–8] (Figs. 9.1–9.11):

- Microorganisms: local or systemic infection
- Chemical noxae
- Physical factors, trauma
- Preexisting testicular disorders of other origin
- Neoplastic processes (seminoma, carcinoma in situ [CIS, or testicular intraepithelial neoplasia])

Apart from etiologic factors, orchitis has been characterized by means of histomorphologic criteria such as type, location, and extent of inflammatory infiltrates, and according to its clinical course (acute vs. chronic) [6, 9]. Current classifications of infectious orchitis consider both etiologic and morphologic aspects, including spread of the disease. Notably, inflammatory conditions of the testis also may involve the epididymis, and vice versa ("epididymo-orchitis"). Retrograde ascent of urethral pathogens—that is, bacteria such as *Chlamydia trachomatis* or *Escherichia coli*—produces epididymo-orchitis secondary to epididymitis (also referred to as *bacterial* or *nonspecific orchitis*). Orchitis associated with systemic (most commonly viral) infections has been attributed to blood-borne dissemination of the microorganisms (also referred to as *viral orchitis*). The granulomatous type of infectious orchitis describes specific testicular involvement in patients with tuberculosis, lepra, brucellosis, or related infections (also referred to as *specific orchitis*). Chronic inflammatory lesions of unknown origin in elderly men have been classified as *nonspecific chronic epididymo-orchitis* and *idiopathic granulomatous orchitis* [3, 6]. Autoimmune orchitis, which can be induced in experimental animals as a vigorous organ-specific immune response by active immunization with homologous testis homogenates or adoptive transfer of germ cell–specific T lymphocytes, is not established as a clinical entity [9, 10].

Histopathology of acute bacterial epididymo-orchitis reveals edema and massive infiltration of both the interstitial compartment and seminiferous tubules, with neutrophils predominating [6]. Viral orchitis is characterized by multifocal perivascular as well as peri- and intratubular

H.-C. Schuppe, M.D. (✉)
Andrology Unit, Department of Urology, Pediatric Urology, and Andrology, Justus Liebig University, Giessen, Germany
e-mail: hans-christian.schuppe@derma.med.uni-giessen.de

M. Bergmann, Ph.D.
Institute of Veterinary Anatomy, Histology, and Embryology, Justus Liebig University, Giessen, Germany

D. Ježek (ed.), *Atlas on the Human Testis*,
DOI 10.1007/978-1-4471-2763-5_9, © Springer-Verlag London 2013

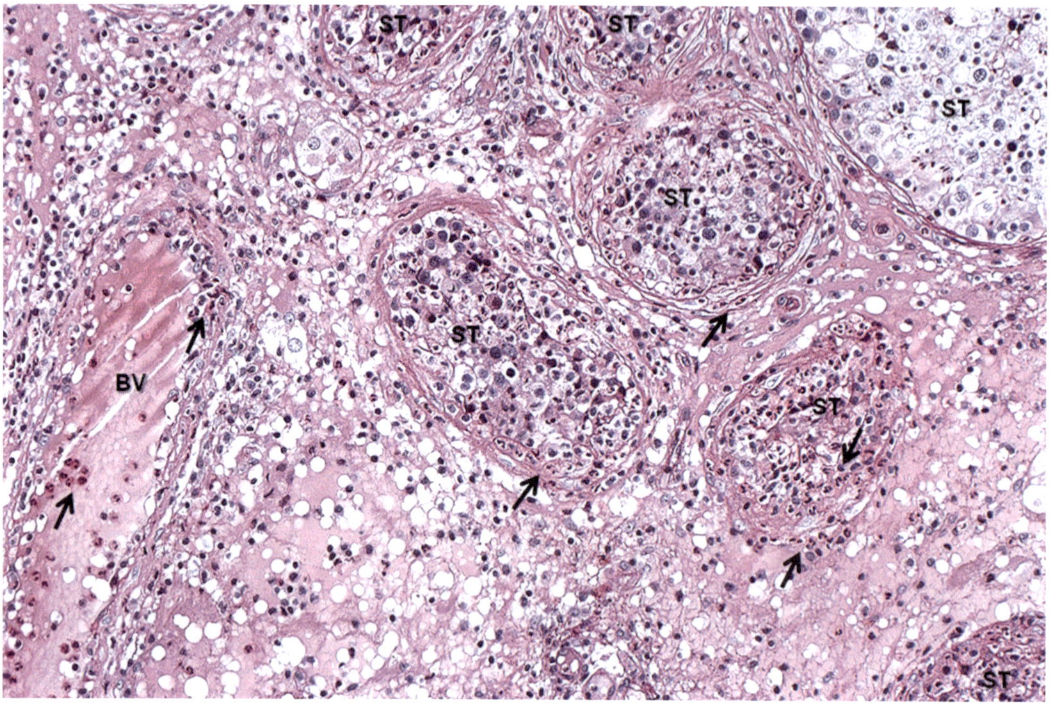

Fig. 9.1 Acute bacterial orchitis (epididymo-orchitis). The image shows massive infiltration of both the interstitium and seminiferous tubules (*ST*) with inflammatory cells, including numerous polymorphonuclear leukocytes (granulocytes). The lamina propria of the affected ST is thickened ("meshwork" pattern) and invaded by granulocytes, which also are found within the ST (*arrows*). The architecture of the seminiferous epithelium in the affected ST is largely disrupted, whereas adjacent tubules show hypospermatogenesis. Note the interstitial edema and enlarged venous blood vessel (*BV*) containing granulocytes (Periodic acid–Schiff [PAS] stain, ×10.) (From Köhn et al. [16], with permission)

Fig. 9.2 Subacute granulomatous orchitis. (**a**) Dense inflammatory infiltrate consisting predominantly of lymphocytes, involving the entire interstitial compartment as well as the lamina propria of the seminiferous tubules (*ST*; **c**). Focal necrosis (*asterisk*) is present, surrounded by epithelioid cells (macrophages) and residual tubular structures (**b**). Intact tubules reveal hypospermatogenesis. (Hematoxylin and eosin [H&E] stain, ×10.) (**b**) Residual structures of ST containing inflammatory cells, resulting in a "granulomatous" pattern. (H&E stain, ×40.) (**c**) Dense peritubular lymphocytic infiltrate (**d**) involving the lamina propria, showing the characteristic thickening ("meshwork" pattern). (H&E stain, ×40.) (**d**) CD3+ T lymphocytes (*red*) represent the vast majority of interstitial and peritubular immune cells, whereas only a few cells express CD45RA (*black*; B cells or naïve T cells). Some T cells are found within the ST. Note the intravascular lymphocytes (*arrow*). (Immunohistochemistry [IHC], ×40.) (**e**) Intra- and peritubular localization of CD68+ macrophages (*red*); interstitial mast cell containing tryptase (*black*; *arrow*) (IHC, ×40)

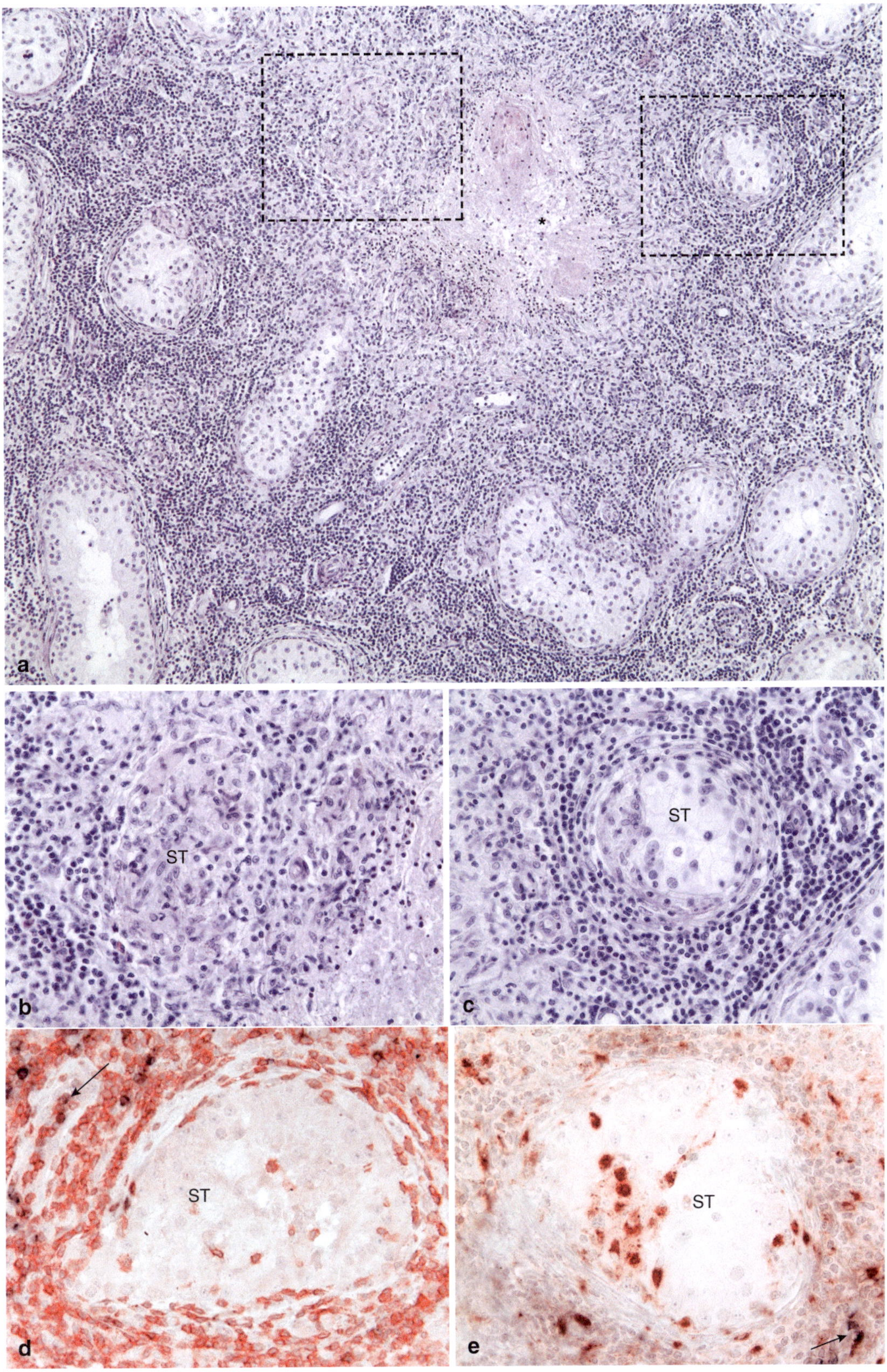
a
b
ST
c
ST
d
ST
e
ST

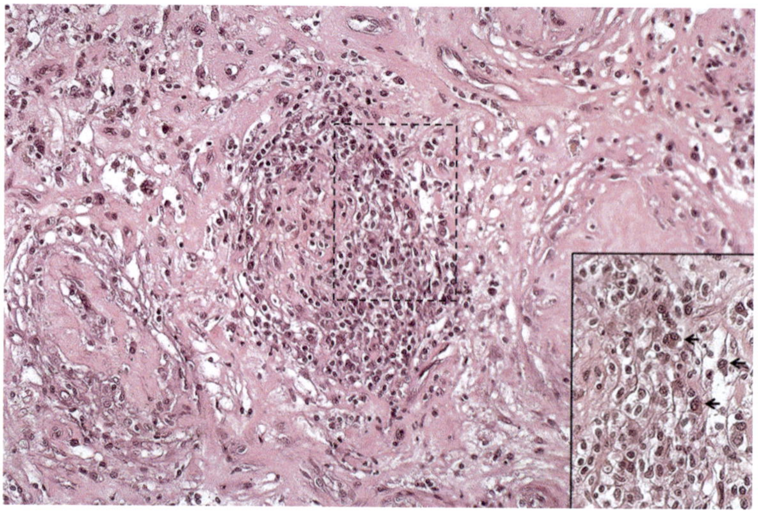

Fig. 9.3 Chronic granulomatous orchitis. The image shows a dense peritubular infiltrate with residual structures of a seminiferous tubule (granulomatous pattern); immune cells comprise lymphocytes, macrophages, and numerous plasma cells (*inset*; *arrows*). The focus of active inflammation is accompanied by complete hyalinization of the tubules ("tubular shadows") and interstitial fibrosis with sparsely distributed immune cells (H&E stain, ×10) (From Schuppe et al. [17], with permission)

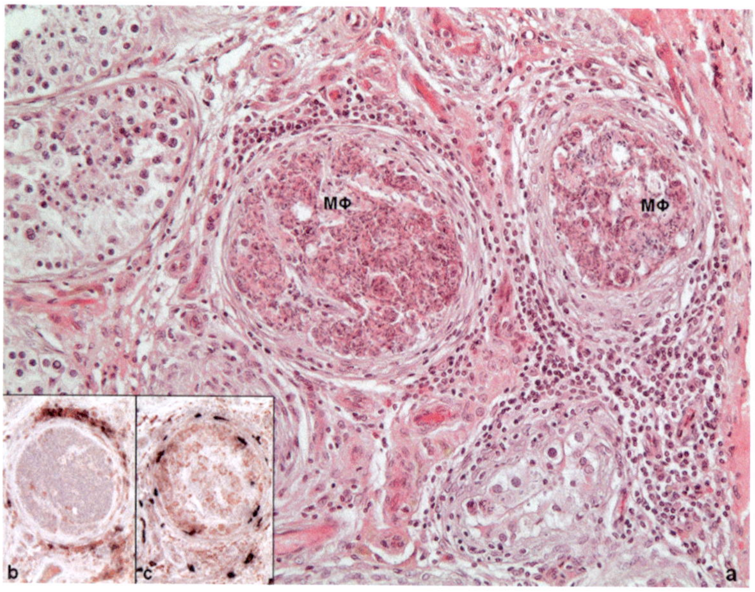

Fig. 9.4 Focal testicular inflammation associated with obstruction of the excurrent ducts. (**a**) Dense peritubular lymphocytic infiltrate (**b**), with some immune cells invading the thickened lamina propria. Two of the affected seminiferous tubules contain numerous macrophages (*MΦ*) phagocytosing sperm ("spermatophagy"; **c**), whereas only a few remaining spermatogonia can be identified. Adjacent tubules reveal hypospermatogenesis. (H&E stain, ×20.) (**b**) The interstitial, predominantly peritubular infiltrate consists of CD3+ T lymphocytes (*red*), with some cells (co-)expressing CD45RA+ (black; naïve T cells or B cells). Note the single T cells found within the seminiferous tubule. (IHC, ×40.) (**c**) Identification of intratubular CD68+ MΦ (*red*) phagocytosing sperm; some MΦ may be seen in the peritubular area. Note the increased number of mast cells containing tryptase (*black*) in a peritubular location, in the lamina propria, and inside the seminiferous tubule (IHC, ×40)

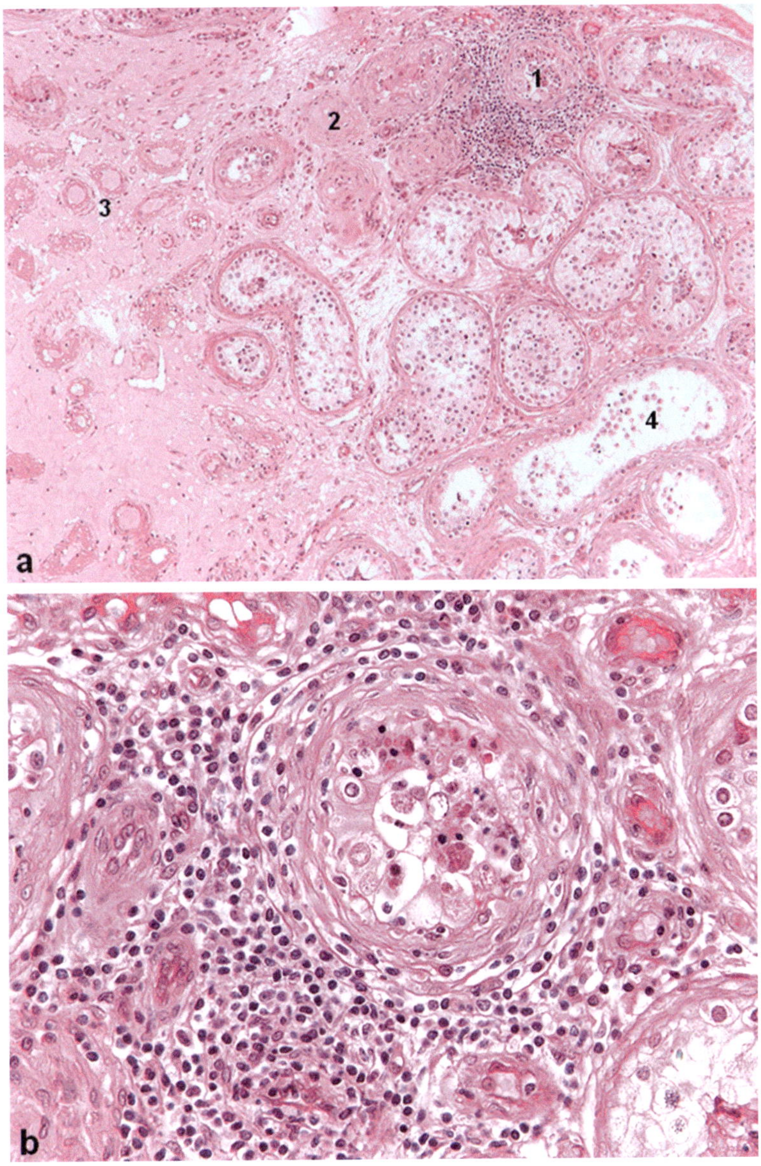

Fig. 9.5 Sequelae of mumps orchitis with persistent focal inflammation. (**a**) Dense peritubular lymphocytic infiltrate involving the lamina propria as well as adjacent blood vessels (*1*; **b**), tubular atrophy resulting in complete hyalinization ("tubular shadows"; *2*, *3*), and interstitial fibrosis (*3*). The adjacent seminiferous tubules show hypospermatogenesis; note the "flattened" epithelium with a complete loss of the adluminal compartment in some tubules (*4*; similar tubular damage has been described as "aspermatogenesis" in experimental autoimmune orchitis). (H&E stain, ×10.) (**b**) *Area 1* at higher magnification. Note the characteristic meshwork pattern of the affected lamina propria; the germinal epithelium is largely disrupted, with only a few germ cells remaining (H&E stain, ×40)

infiltrates with lymphocytes, plasma cells, neutrophils, and macrophages. Affected seminiferous tubules show degeneration of the germinal epithelium sparing a few spermatogonia and the Sertoli cells. Concomitant thickening of the lamina propria may result in complete hyalinization and fibrosis of the tubules with increased collagen production ("tubular shadows"). This pattern of tubular damage also has been described as "mixed atrophy" [11]. On the other hand, chronic inflammatory reactions following acute orchitis are characterized by multifocal lymphocytic infiltrates and tubular changes, which are typical morphologic features of experimental autoimmune orchitis [6, 8]. Similar lesions frequently are detected in testicular biopsy samples obtained from infertile but otherwise asymptomatic men [9, 12–14]. Moreover, despite the patchy distribution of the lesions, testicular volume and score counts for spermatogenesis may be reduced significantly [9].

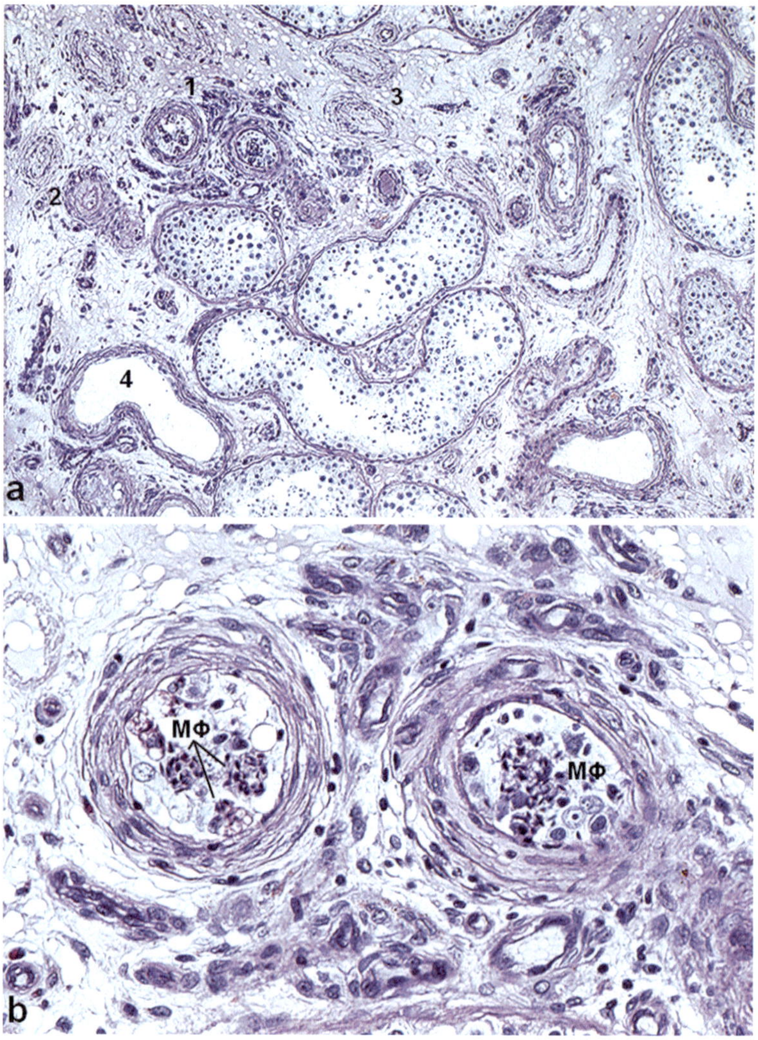

Fig. 9.6 Focal testicular inflammation associated with mixed atrophy of spermatogenesis (in an asymptomatic man with infertility). (**a**) Discrete inflammatory signs consisting of scattered interstitial and peritubular lymphocytes, in association with seminiferous tubules revealing the characteristic thickening and "meshwork" pattern of the lamina propria, disintegration of germinal epithelium, and intratubular macrophages (*MΦ*) phagocytosing sperm (*1*; **b**); progressive tubular atrophy resulting in tubular shadows (*2*, *3*); and seminiferous tubules with a complete loss of the adluminal compartment (*4*; "aspermatogenesis"). Adjacent tubules with a preserved architecture show hypospermatogenesis. (PAS stain, ×10.) (**b**) *Area 1* at higher magnification. Note the scattered lymphocytes in the perivascular and peritubular regions, as well as within the altered lamina propria, and intratubular MΦ phagocytosing sperm (PAS stain; ×20)

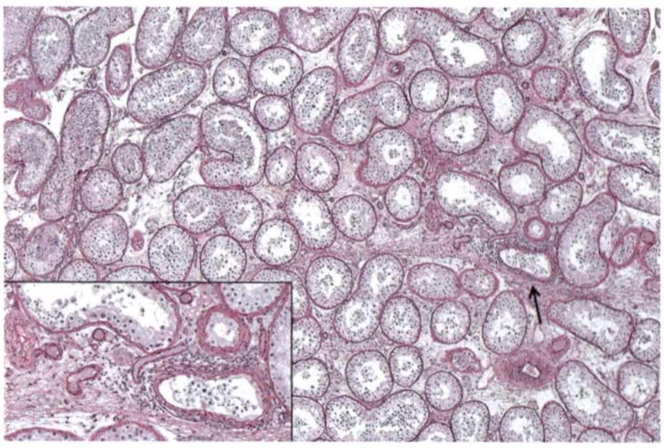

Fig. 9.7 Focal testicular inflammation (in an asymptomatic man with infertility); focal peritubular lymphocytic infiltration. Note the disintegration of the germinal epithelium in the affected tubules and the adjacent tubular shadows (*arrow*; *inset*). The remaining tissue shows hypospermatogenesis (PAS stain, ×4)

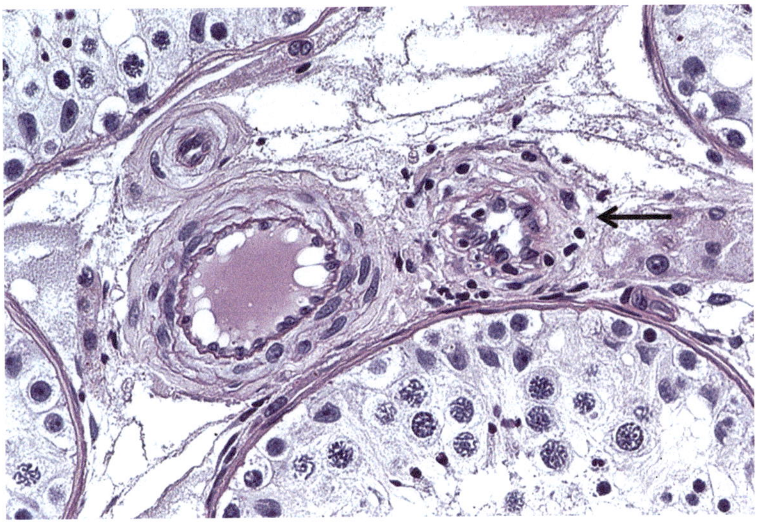

Fig. 9.8 Focal testicular inflammation (in an asymptomatic man with infertility); discrete perivascular lymphocytic infiltration involving the vessel wall (*arrow*). Note the hypospermatogenesis in the adjacent seminiferous tubules (PAS stain, ×40)

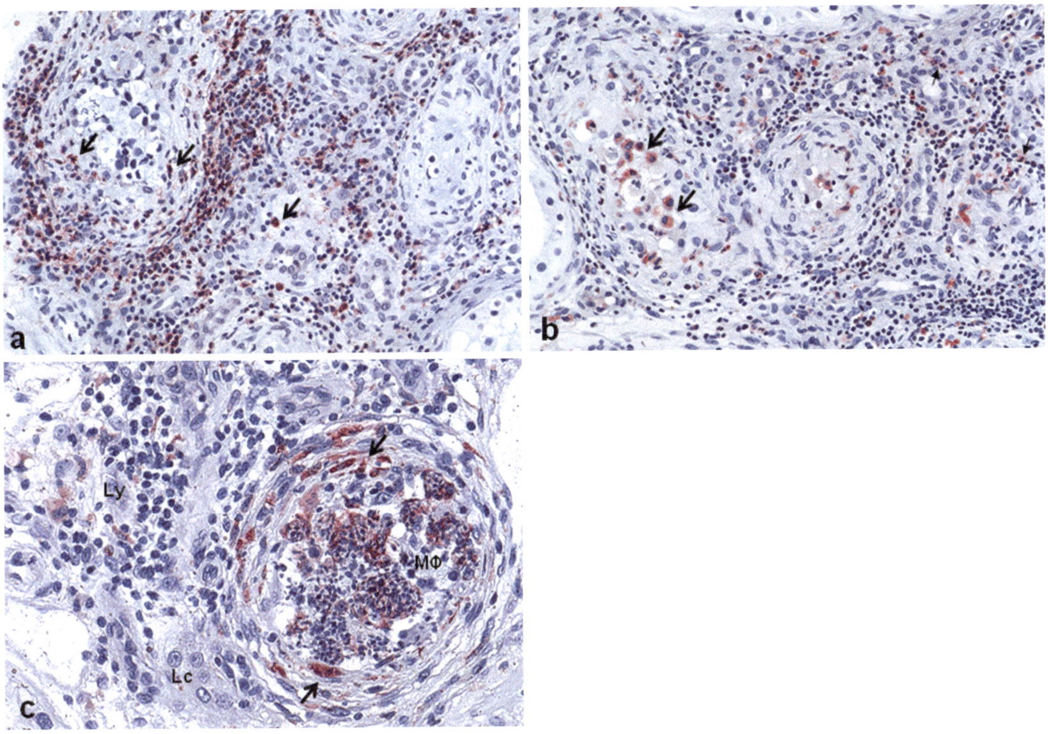

Fig. 9.9 Immunopathology of testicular inflammatory reactions. (**a**) Dense interstitial and peritubular infiltration of immune cells, predominantly CD45RO+lymphocytes (activated T cells). Note the involvement of the altered lamina propria and intratubular lymphocytes (*arrows*). (IHC, ×20.) (**b**) The lymphocytic infiltration is accompanied by CD68+ macrophages, which are most prominent within the affected seminiferous tubules (*arrows*). (Single granulocytes also react with the anti-CD68 antibody used; IHC, ×20.) (From Schuppe et al. [17], with permission.) (**c**) CD68^{+} macrophages (*MΦ*) in the altered lamina propria (*arrows*) and in the tubular lumen; intratubular MΦ phagocytosing sperm. Note the dense peritubular infiltration of lymphocytes (*Ly*) (IHC, ×40) *Lc* Leydig cells

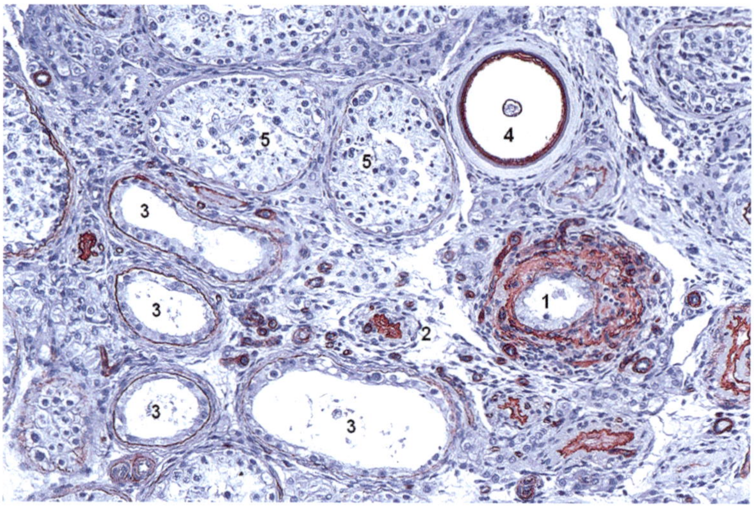

Fig. 9.10 Focal testicular inflammation and concomitant tubular damage. The seminiferous tubule affected by focal lymphocytic infiltration contains only a few remaining cells of the basal compartment of the germinal epithelium. Alteration of the lamina propria (thickening, "meshwork" pattern) is reflected by increased amounts of type IV collagen and vascularization (*1*). Note the adjacent "tubular shadow" containing type IV collagen (*2*); the seminiferous tubules with a complete loss of the adluminal compartment (*3*; "aspermatogenesis"), some of which show prominent staining for type IV collagen at the basement membrane; and a concentric spherical concrement (*4*). The remaining intact tubules show hypospermatogenesis (*5*) (IHC, ×20)

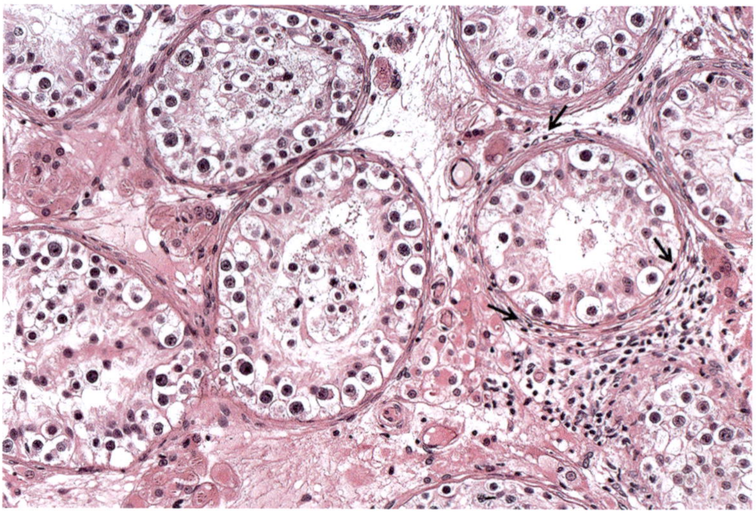

Fig. 9.11 Focal inflammatory reaction associated with carcinoma in situ. The image shows sparse peritubular lymphocytic infiltrate involving the lamina propria (*arrows*), which reveals moderate thickening. Note the CIS cells in the basal compartment of all seminiferous tubules (H&E stain, ×20) (From Köhn et al. [16]; with permission)

The hallmark of chronic inflammatory reactions in the human testis following acute bacterial or viral orchitis, as well as those observed among infertile men (i.e., infiltration of activated inflammatory T lymphocytes), indicates a profound disturbance of local immunoregulation and thus testicular immune privilege [15]. Infiltrating inflammatory cells obviously overcome the immunosuppressive influence in the testis. The predominantly peritubular location of lymphocytes and characteristic morphologic changes in the seminiferous tubules resembling experimental autoimmune orchitis support the concept that a T cell–mediated autoimmune response is induced [8].

Acknowledgments The authors thank Dr. F. Birke and Dr. B. Katzer (Fulda, Germany) and Prof. G. Haidl (University of Bonn, Germany) for the kind donation of tissue specimens (Figs. 9.2 and 9.3), and Th. Haggeney (Giessen) for preparing the immunohistochemistry. The work of M. Bergmann and H.-C. Schuppe is supported and funded by the LOEWE focus group "MIBIE" (Male Infertility during Infection and Inflammation), an excellence initiative of the German state government of Hessen.

References

1. Dohle GR, Colpi GM, Hargreave TB, Papp GK, Jungwirth A, Weidner W, The EAU Working Group on Male Infertility. EAU guidelines on male infertility. Eur Urol. 2005;48:703–11.
2. Rusz A, Pilatz A, Wagenlehner F, Linn T, Diemer T, Schuppe H-C, et al. Influence of urogenital infections and inflammation on semen quality and male fertility. World J Urol. 2012;30:23–30.
3. Weidner W, Krause W, Ludwig M. Relevance of male accessory gland infection for subsequent fertility with special focus on prostatitis. Hum Reprod Update. 1999;5:421–32.
4. Bols B, Jensen L, Jensen A, Braendstrup O. Immunopathology of in situ seminoma. Int J Exp Pathol. 2000;81:211–7.
5. Jahnukainen K, Jörgensen N, Pöllänen P, Giwercman A, Skakkebaek NE. Incidence of testicular mononuclear cell infiltrates in normal human males and in patients with germ cell neoplasia. Int J Androl. 1995;18:313–20.
6. Mikuz G, Damjanov I. Inflammation of the testis, epididymis, peritesticular membranes, and scrotum. Pathol Annu. 1982;17:101–28.
7. Nistal M, Riestra ML, Paniagua R. Focal orchitis in undescended testes. Discussion of pathogenic mechanisms of tubular atrophy. Arch Pathol Lab Med. 2002;126:64–9.
8. Schuppe H-C, Meinhardt A. Immune privilege and inflammation of the testis. In: Markert U, editor. Immunology of gametes and embryo implantation. Basel: Karger; 2005. p. 1–14 (Chemical immunology and allergy; vol 88).
9. Schuppe H-C, Meinhardt A, Allam JP, Bergmann M, Weidner W, Haidl G. Chronic orchitis – a neglected cause of male infertility? Andrologia. 2008;40: 84–91.
10. Suominen JJ. Sympathetic auto-immune orchitis. Andrologia. 1995;27:213–6.
11. Sigg C, Hedinger C. Quantitative and ultrastructural study of germinal epithelium in testicular biopsies with "mixed atrophy". Andrologia. 1981;13:412–24.
12. El Demiry MI, Hargreave TB, Busuttil A, Elton R, James K, Chisholm GD. Immunocompetent cells in human testis in health and disease. Fertil Steril. 1987;48:470–9.
13. Hofmann N, Kuwert E. Die chronische, nicht-Erreger-bedingte Orchitis. Z Hautkr. 1979;54:173–80.
14. Suominen J, Söderström KO. Lymphocyte infiltration in human testicular biopsies. Int J Androl. 1982;5: 461–6.
15. Hedger MP. Immunophysiology and pathology of inflammation in the testis and epididymis. J Androl. 2011;32:625–40.
16. Köhn FM, Schuppe HC, Jung A, Schulze W. Andrological relevance of testicular biopsies [in German]. J Dtsch Dermatol Ges. 2005;3:532–57.
17. Schuppe HC, Pilatz A, Diemer T. Erkrankungen des Hodens: Entzündungen. In: Krause W, Weidner W, Diemer T, Sperling H, editors. Andrologie. 4th ed. Stuttgart: Georg Thieme; 2011. p. 266–72.

Cryptorchidism (Undescended Testis)

10

Ljerka Banek, Viviana Kozina, Iva Marolt, and Tomislav Banek

The undescended testis may be present at birth (congenital cryptorchidism) or, more frequently, may ascend during childhood (ascending testis, acquired cryptorchidism). The acquired undescended testis frequently is confused with the retractile testis (a variant of the normal organ). If a retractile or an undescended testis is still present in adulthood, it may be the cause of infertility and even a place for the development of a testicular carcinoma.

The aim of this chapter is to present a histologic review of acquired undescended testes located in the high scrotal position or in the inguinal canal. Overall, 50 cryptorchid patients (aged 17–25 years) were analyzed. The biopsy samples from these patients are part of the tissue collection archive of the Department of Histology and Embryology, School of Medicine, University of Zagreb.

In the anamnesis, these patients stated that since childhood, they had one gliding testis or a testis located in the groin. After an extensive clinical examination and ultrasound measurement of the organs, a urologist confirmed the diagnosis of unilateral undescended testis in all patients. In 40 patients, the dystopic testis was sliding from the inguinal canal to the high scrotal position. In ten patients, the undescended testis was fixed in the inguinal canal. After informed consent, the patients underwent biopsy and orchiopexy of the undescended testis. The biopsy specimens were fixed in Gendre's solution, embedded in paraffin, and cut into sections 7 μm thick. The sections were stained with hematoxylin and eosin (H&E) and according to Masson.

L. Banek, M.D., Ph.D. (✉) • V. Kozina, MEd
I. Marolt, DMD • T. Banek, M.D., Ph.D.
Department of Histology and Embryology,
School of Medicine, University of Zagreb,
Šalata 3 10000 Zagreb, Croatia
e-mail: ljbanek@mef.hr

D. Ježek (ed.), *Atlas on the Human Testis*,
DOI 10.1007/978-1-4471-2763-5_10, © Springer-Verlag London 2013

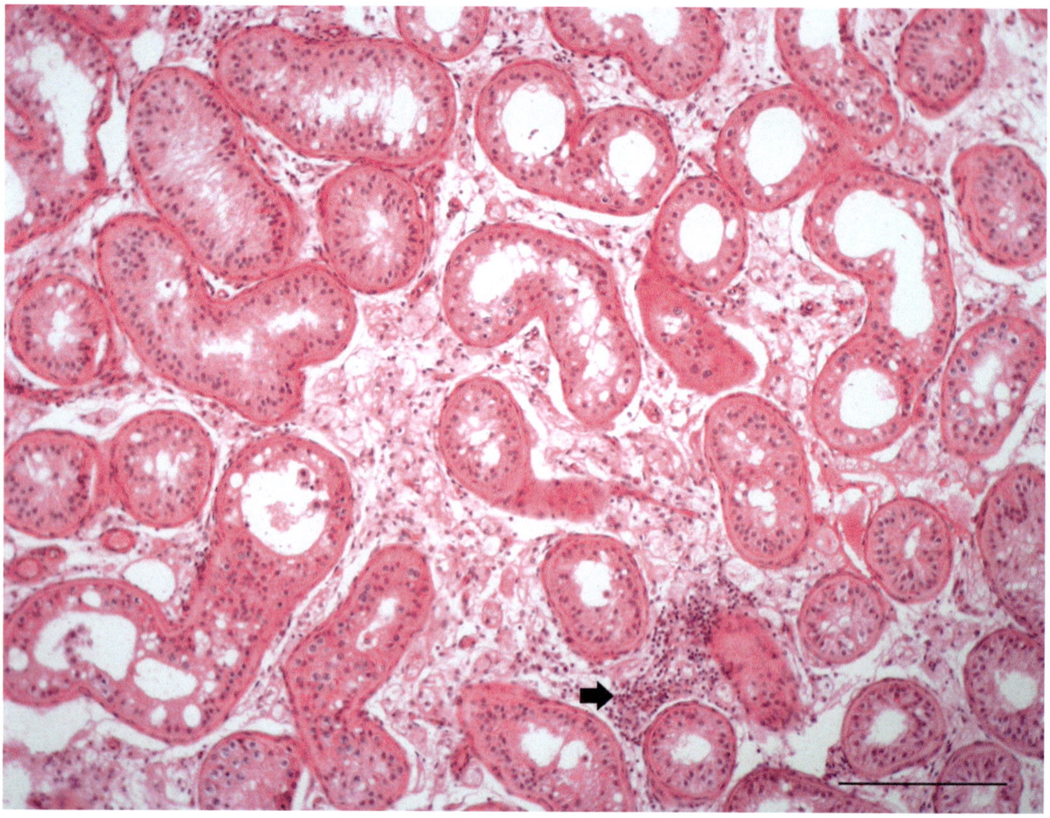

Fig. 10.1 High scrotal unilateral cryptorchid gliding testis. Shown is a biopsy specimen from a 17-year-old patient. In the testicular interstitial tissue, an infiltrate of inflammatory cells is present (*arrow*) (H&E stain, ×100; *scale bar*, 200 μm)

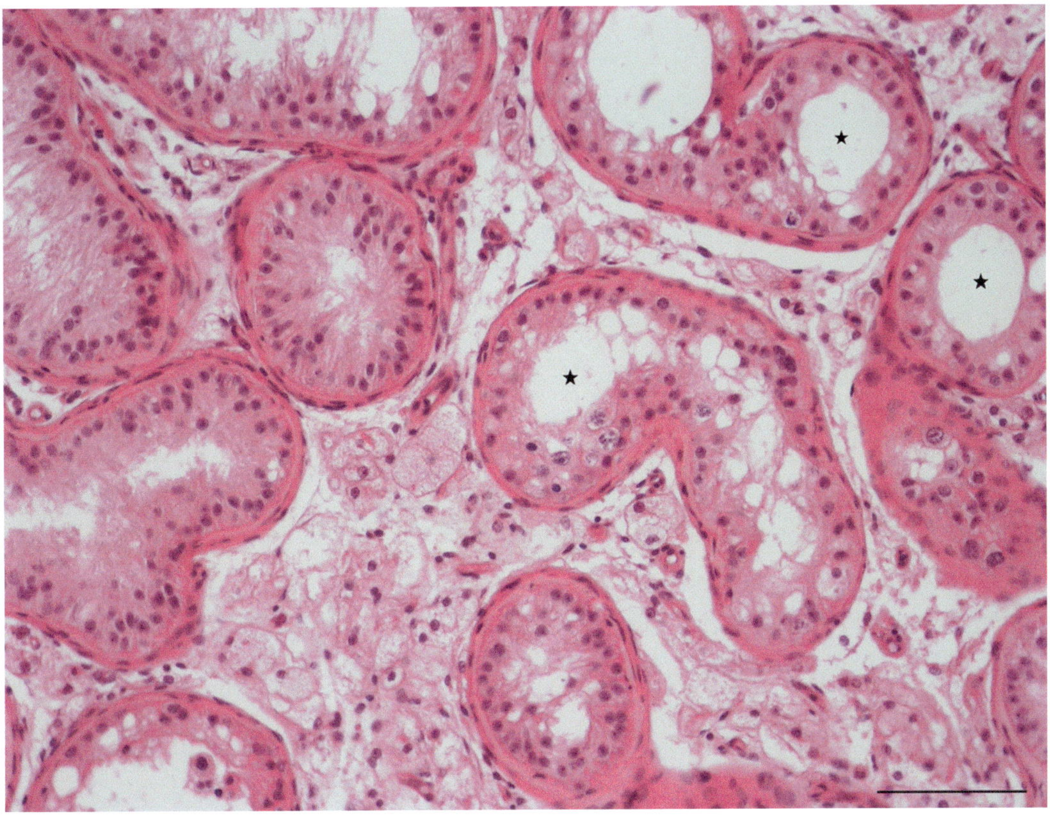

Fig. 10.2 Detail from Fig. 10.1. The biopsy specimen shows a maturation arrest of spermatogenesis (*asterisks*). The other tubules are lined by Sertoli cells only (H&E stain, ×200; *scale bar*, 100 μm)

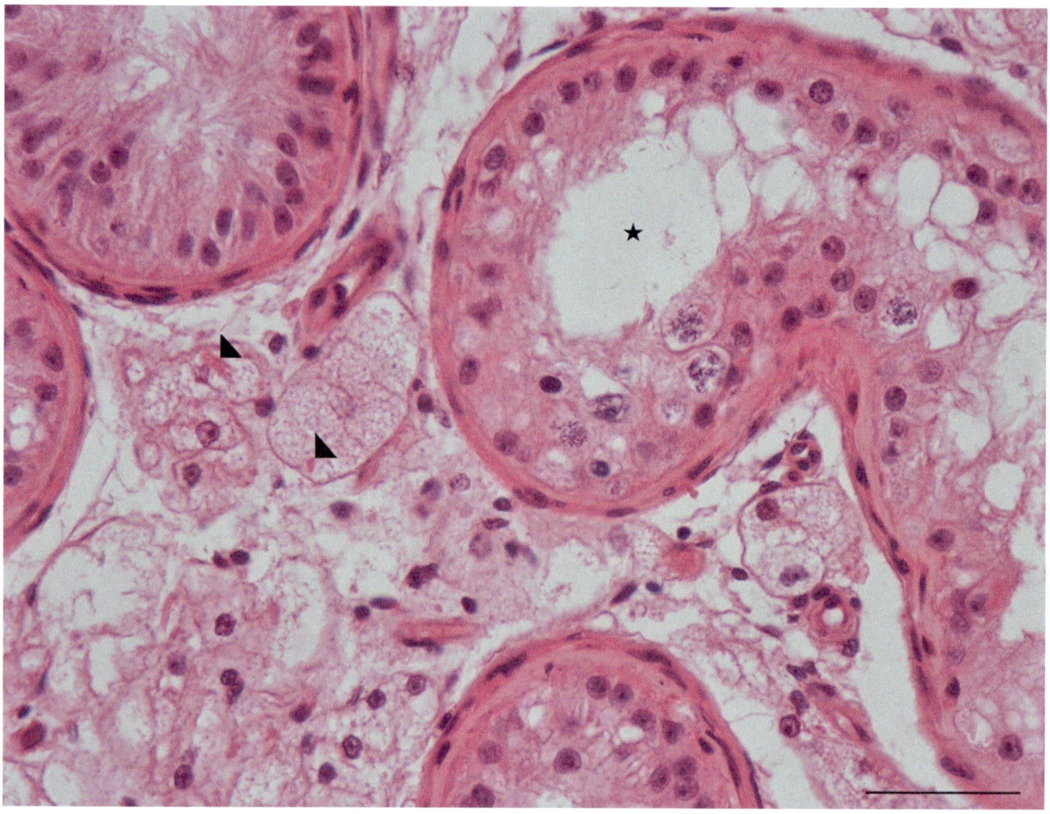

Fig. 10.3 High scrotal unilateral cryptorchid testis. The seminiferous tubules are lined by Sertoli cells and degenerated spermatogonia with some maturation to the primary spermatocyte stage (*asterisk*). Neighboring tubules contain only mature Sertoli cells. Leydig cells show marked cytoplasmic vacuolization (Reinke's crystals, *arrowheads*) (H&E stain, ×400; *scale bar*, 50 μm)

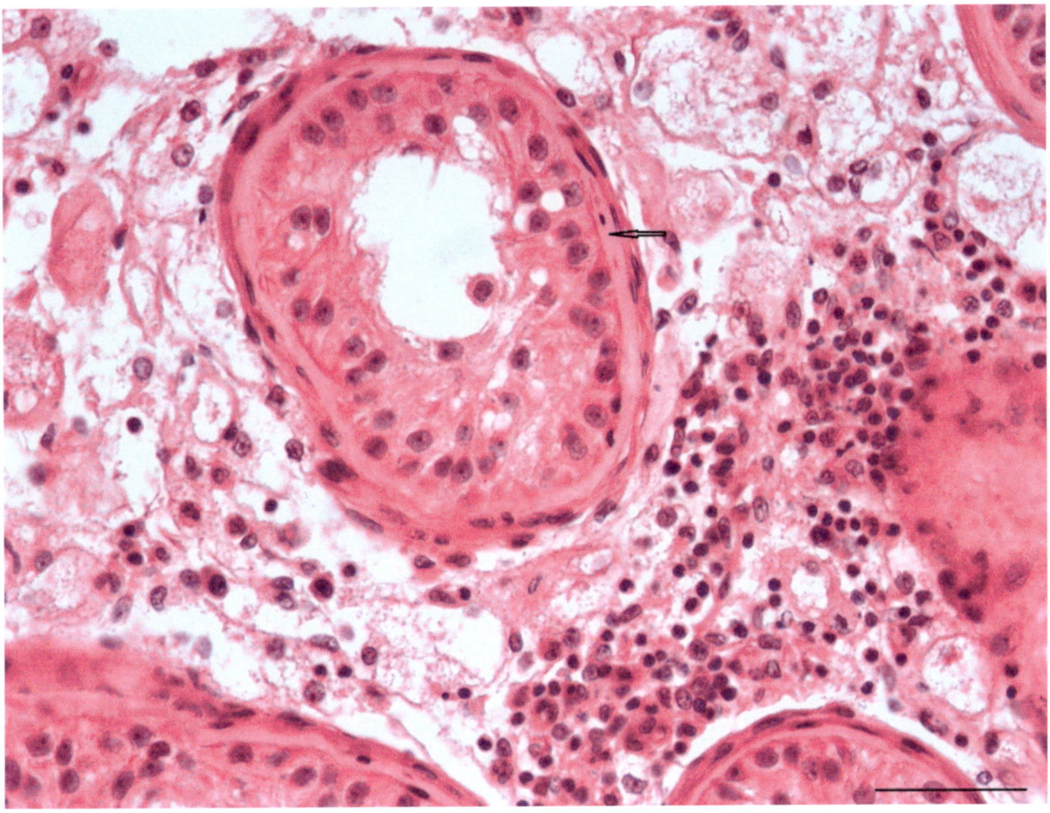

Fig. 10.4 Detail from a high scrotal unilateral cryptorchid gliding testis. The testicular interstitium is infiltrated by macrophages and lymphocytes. The tubules are lined by Sertoli cells exclusively. Note some thickening and hyalinization of the lamina propria of the tubules (*arrow*) (H&E stain, ×400; *scale bar*, 50 μm)

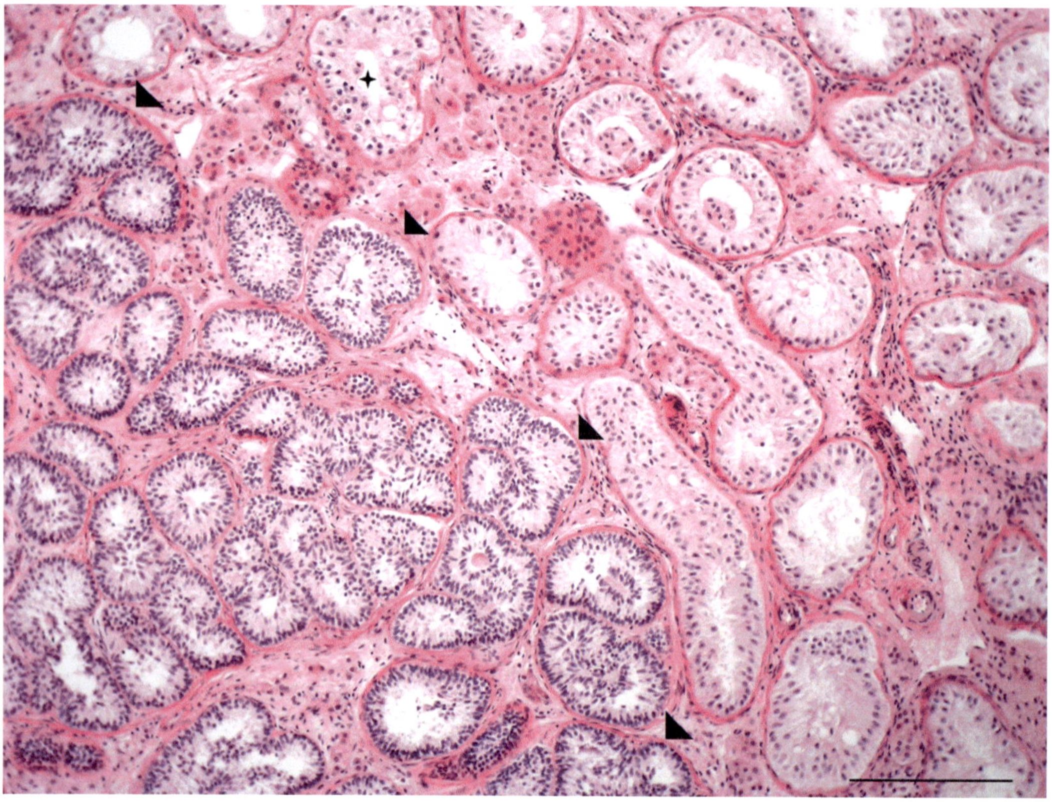

Fig. 10.5 High scrotal unilateral cryptorchid gliding testis obtained from a 20-year-old man. The biopsy specimen shows seminiferous tubules lined by mature Sertoli cells, which may support spermatogenesis (*crosslet*). Also present are small tubules lined by immature Sertoli cells (*arrowheads*) (H&E stain, ×100; *scale bar*, 200 μm)

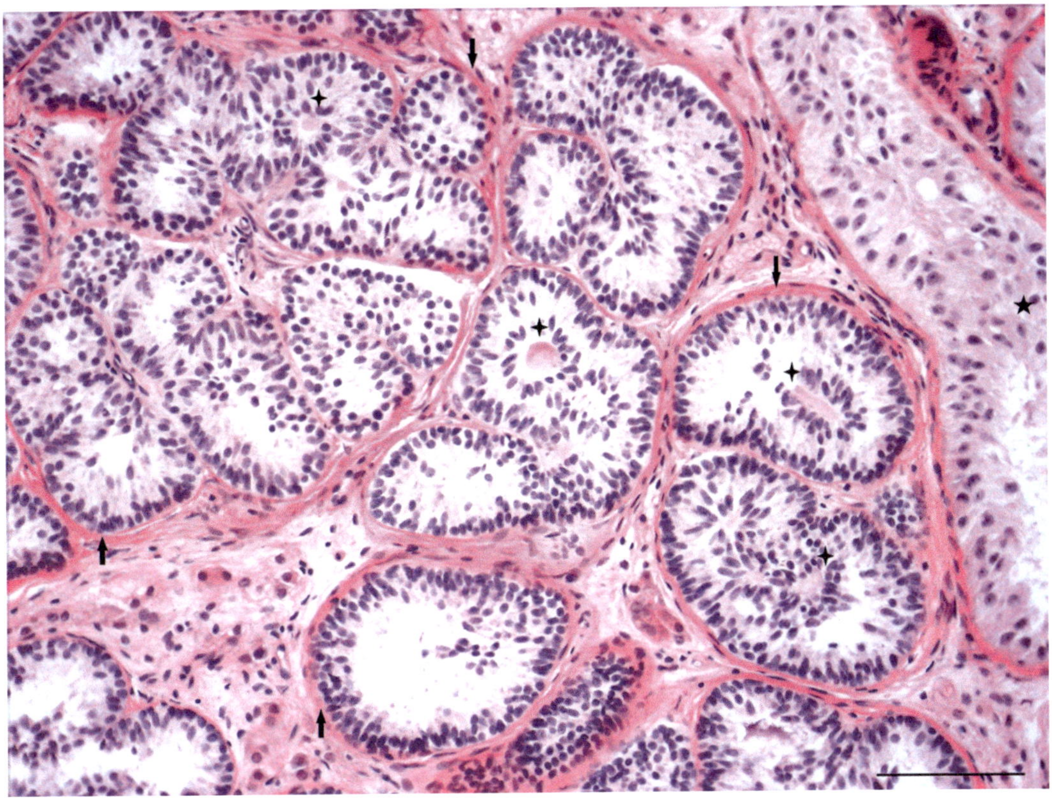

Fig. 10.6 Detail from Fig. 10.5. Sertoli cell nodules consist of small tubules lined by immature Sertoli cells and are surrounded by a thickened basement membrane (*arrows*). Hyaline deposits (*crosslets*) are visible in Sertoli cell nodules of an undescended testis. In contrast, some tubules are lined by mature Sertoli cells (*asterisk*) (H&E stain, ×200; *scale bar*, 100 μm)

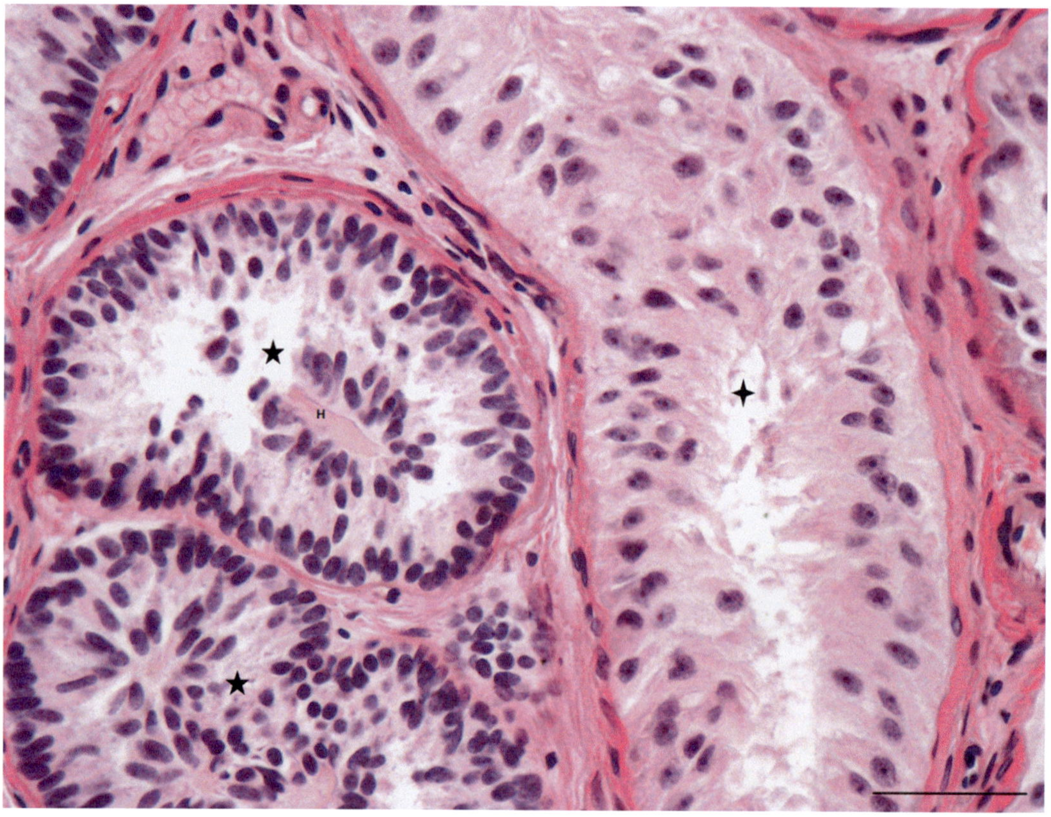

Fig. 10.7 Detail from Figs. 10.5 and 10.6: high scrotal unilateral cryptorchid gliding testis from a 20-year-old man. This sample allows comparison of immature (*stars*) with mature (*crosslet*) Sertoli cells. The immature cells within nodules are smaller; they contain minor nuclei without a clear nucleolar complex and have scanty cytoplasm. Hyaline deposits (*H*) in the middle of the nodules may originate from an increased production of basement membrane material. In contrast, mature Sertoli cells have a well-defined nucleolar complex and plentiful cytoplasm (H&E stain, ×400; *scale bar*, 50 μm)

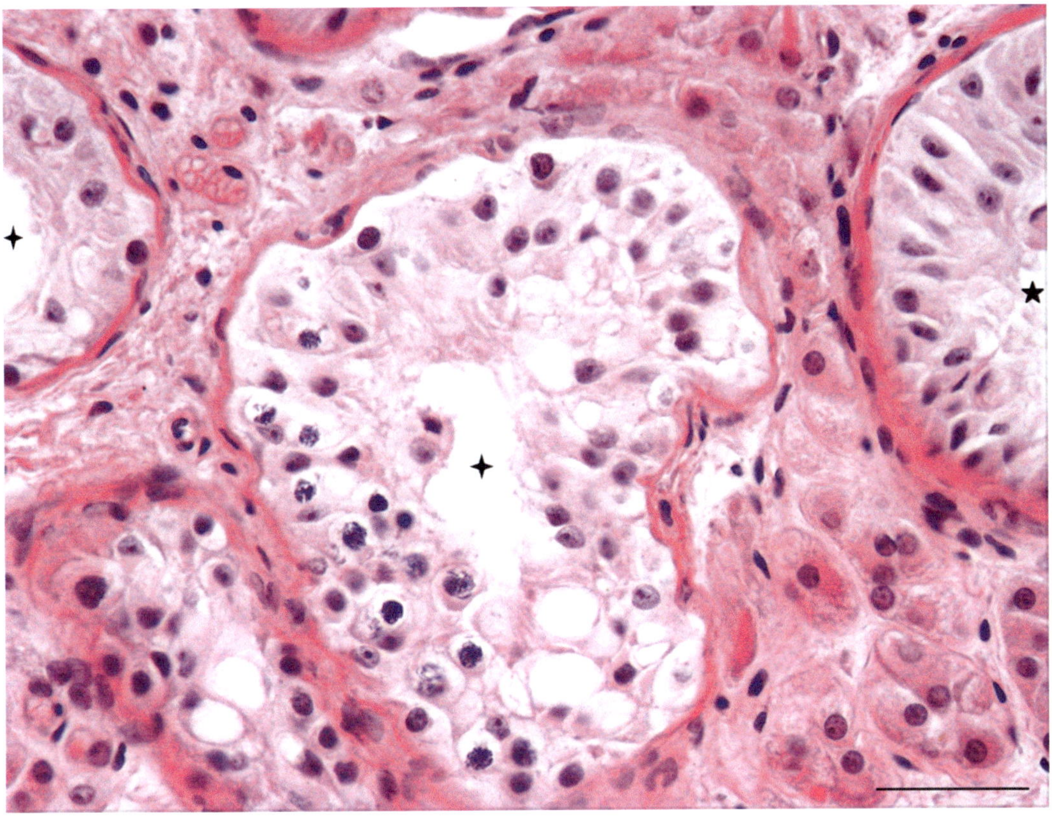

Fig. 10.8 Detail from the specimen shown in Fig. 10.7. Seminiferous tubules lined by mature Sertoli cells contain degenerated spermatogonia and spermatocytes (*crosslets*). Also present is a tubule lined by Sertoli cells only (*star*) (H&E stain, ×400; *scale bar*, 50 μm)

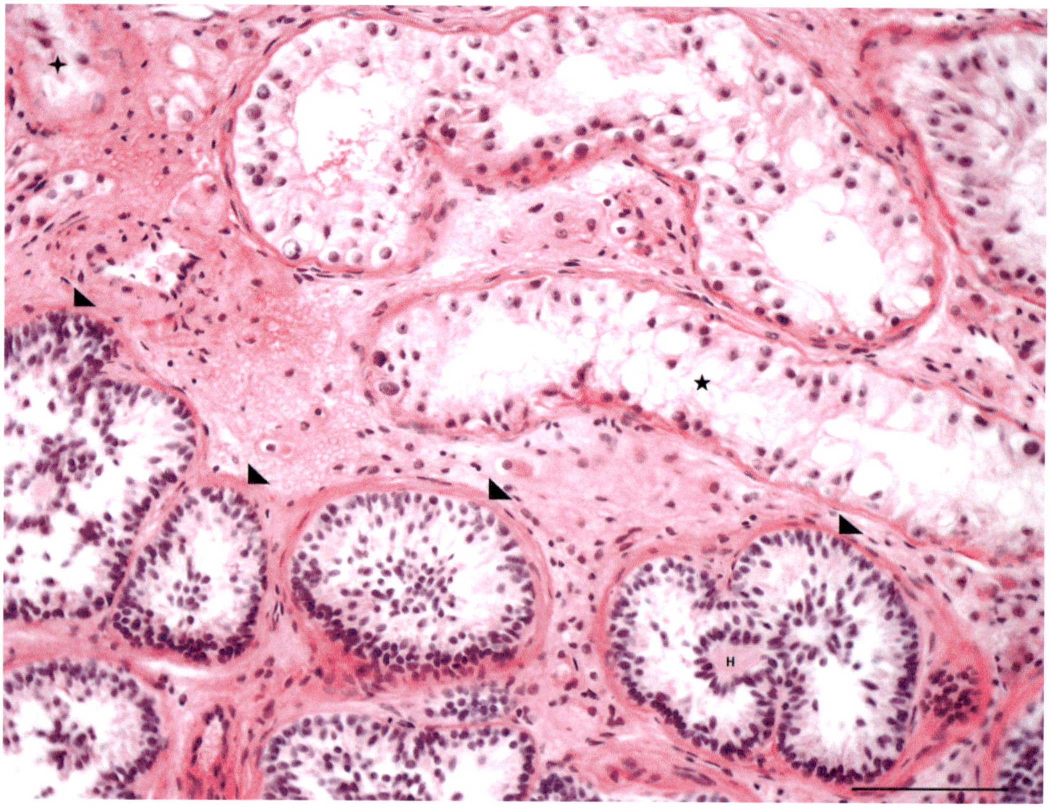

Fig. 10.9 High scrotal unilateral cryptorchid gliding testis in a 20-year-old man. The seminiferous tubules are lined by mature Sertoli cells and spermatogonia (*star*). In contrast, the Sertoli cell nodules are lined by immature Sertoli cells (*arrowheads*). The nodules sometimes contain hyaline deposits (*H*). Tubular sclerosis also is present (*crosslet*) (H&E stain, ×200; *scale bar*, 100 μm)

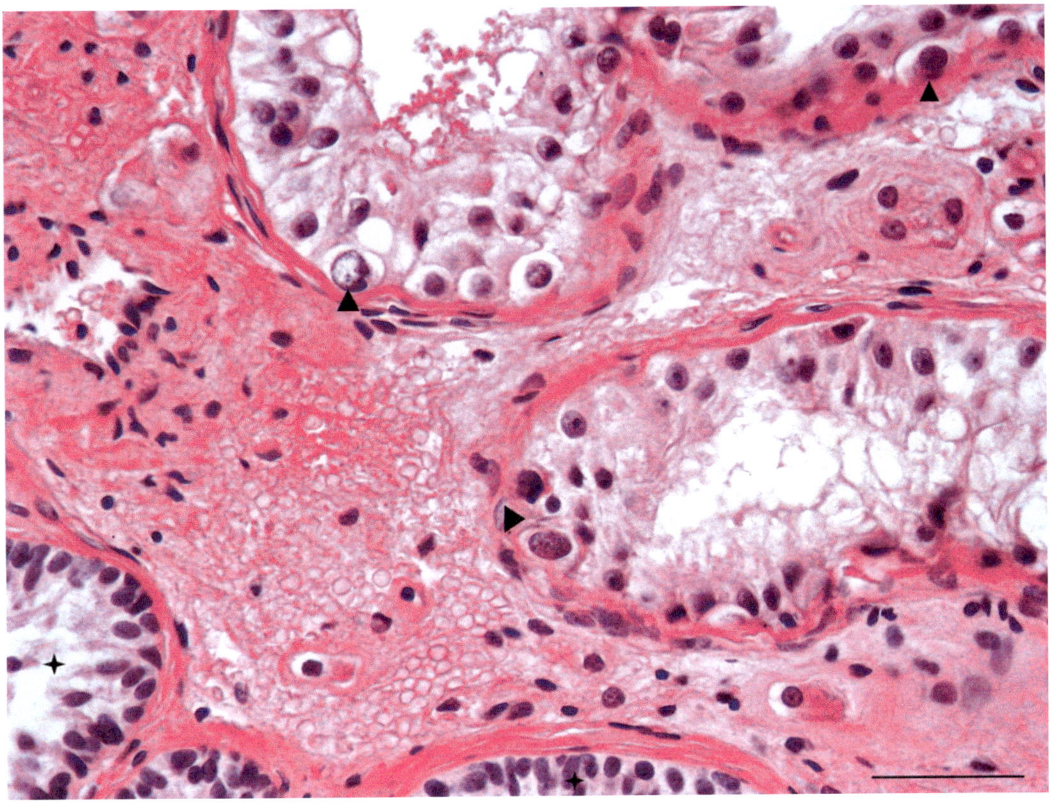

Fig. 10.10 Detail from Fig. 10.9. The seminiferous tubules are lined by mature Sertoli cells and spermatogonia. Some of the latter are of giant size (*arrowheads*). The *crosslets* mark Sertoli cell nodules (H&E stain, ×400; *scale bar*, 50 μm)

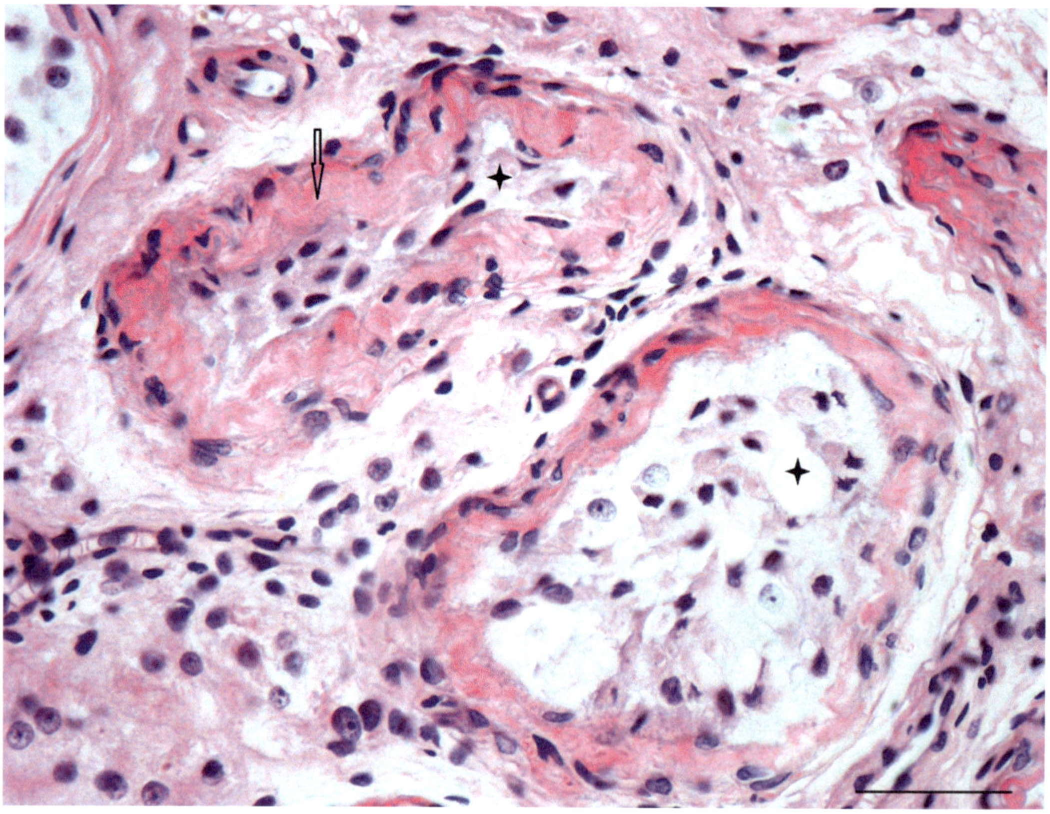

Fig. 10.11 Part of a biopsy sample from a patient with a high scrotal unilateral cryptorchid gliding testis. Tubular sclerosis is present. The reduced diameter of the tubules is significant. The basement membrane and tunica propria demonstrate thickening and hyalinization (*arrow*). In addition, the basement membrane is undulated. Sertoli cells show atrophy (*crosslets*) (H&E stain, ×400; *scale bar*, 50 μm)

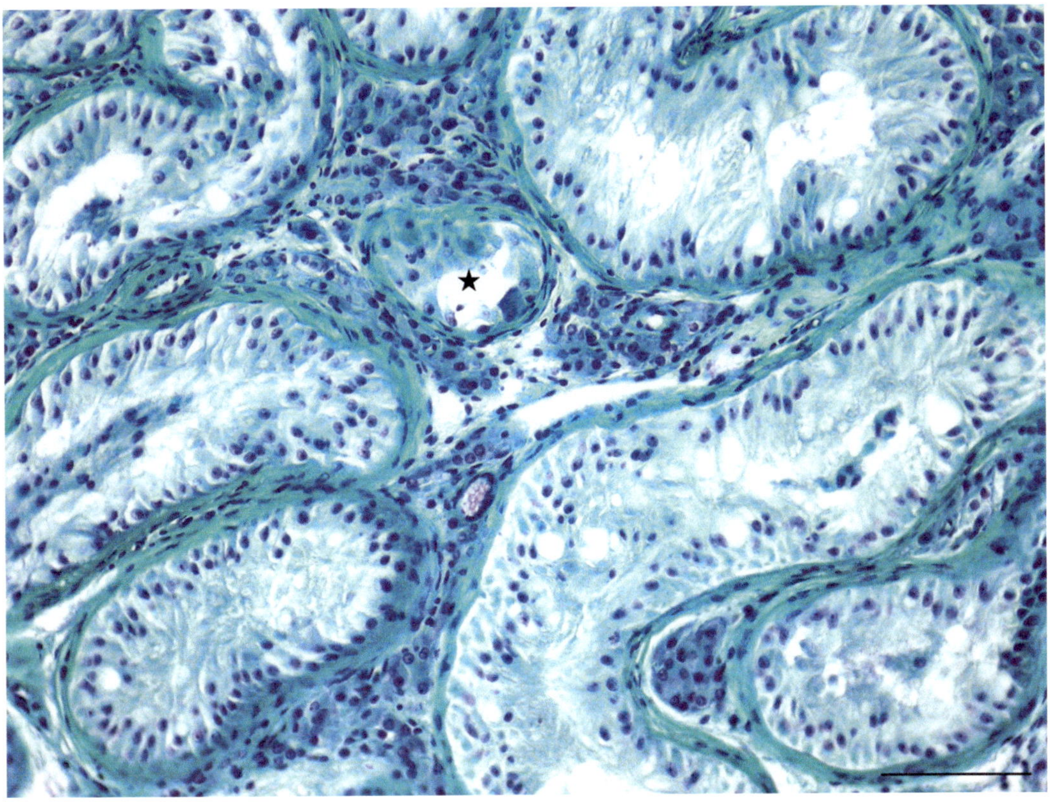

Fig. 10.12 Biopsy sample of a high scrotal unilateral cryptorchid gliding testis obtained from a 19-year-old man. The tubules are lined by mature Sertoli cells only. Note that one tubule is considerably smaller in diameter than the others (*star*). Leydig cells are somewhat prominent (Masson stain, ×200; *scale bar*, 100 μm)

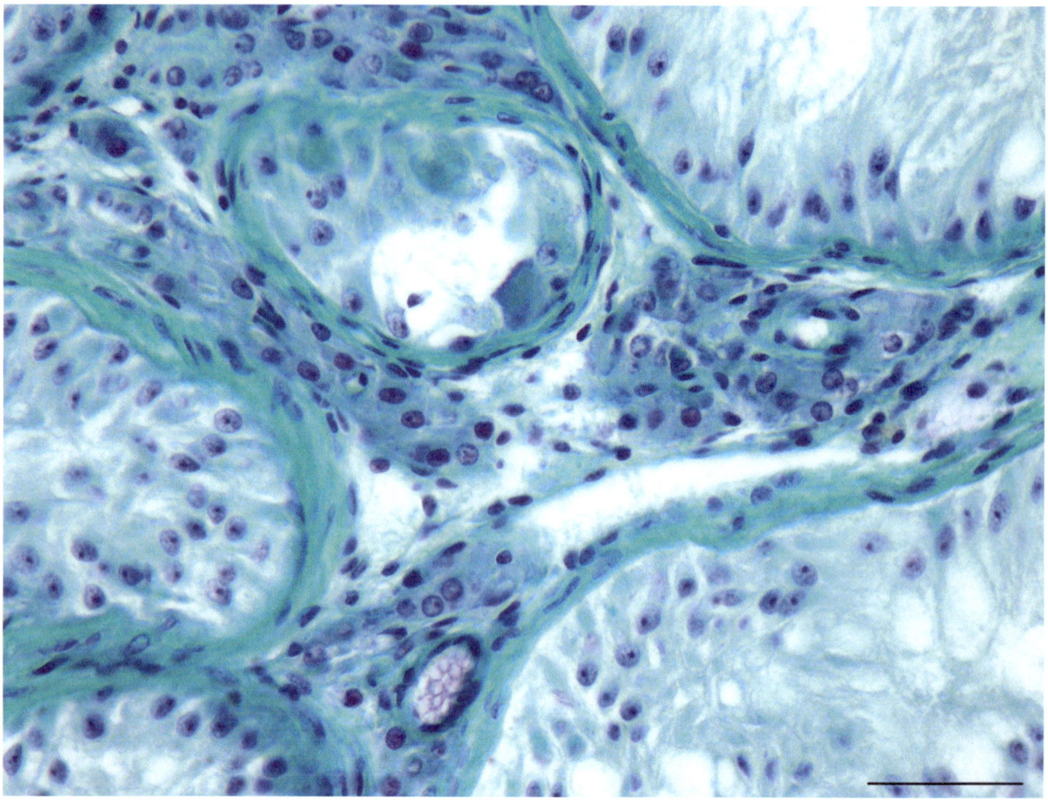

Fig. 10.13 Detail of the tubule shown in Fig. 10.12. Some Sertoli cells lining this tubule have a condensed cytoplasm (Masson stain, ×400; *scale bar*, 50 μm)

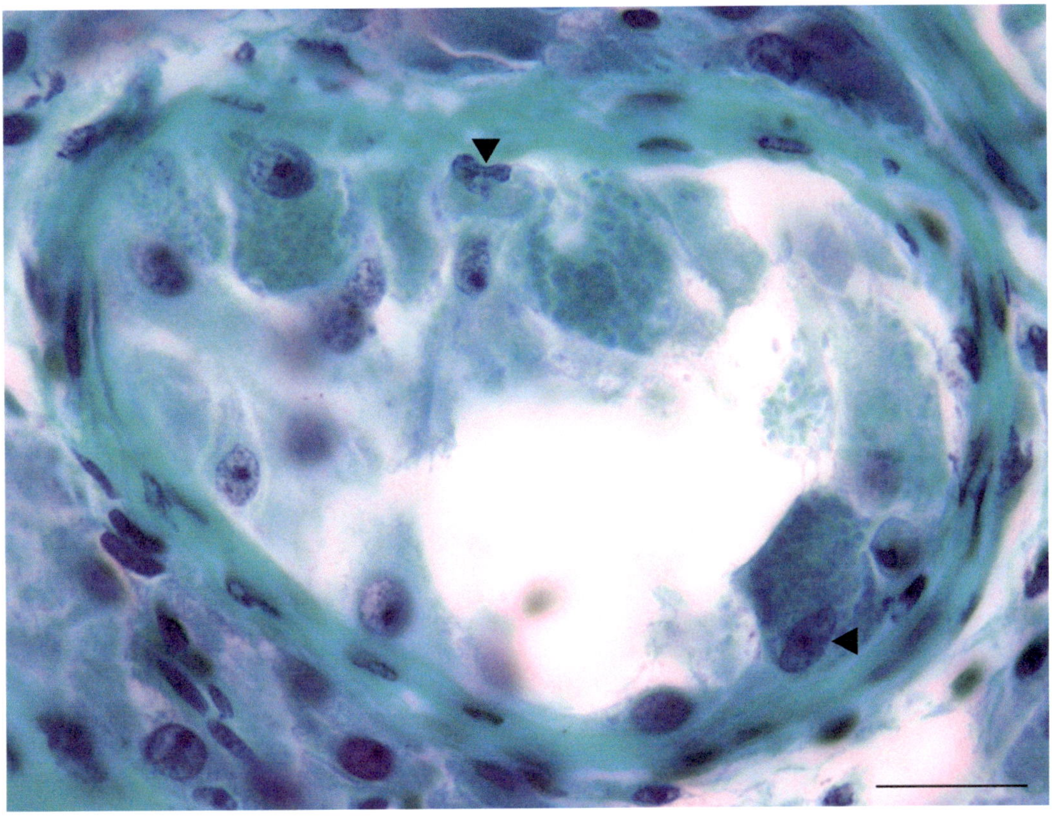

Fig. 10.14 Detail from Figs. 10.12 and 10.13 showing a high scrotal unilateral cryptorchid gliding testis in a 19-year-old man. The cytoplasm of the Sertoli cells is granulated. The nuclei of these cells may be wrinkled (*arrowheads*) (Masson stain, ×1000; *scale bar*, 20 μm)

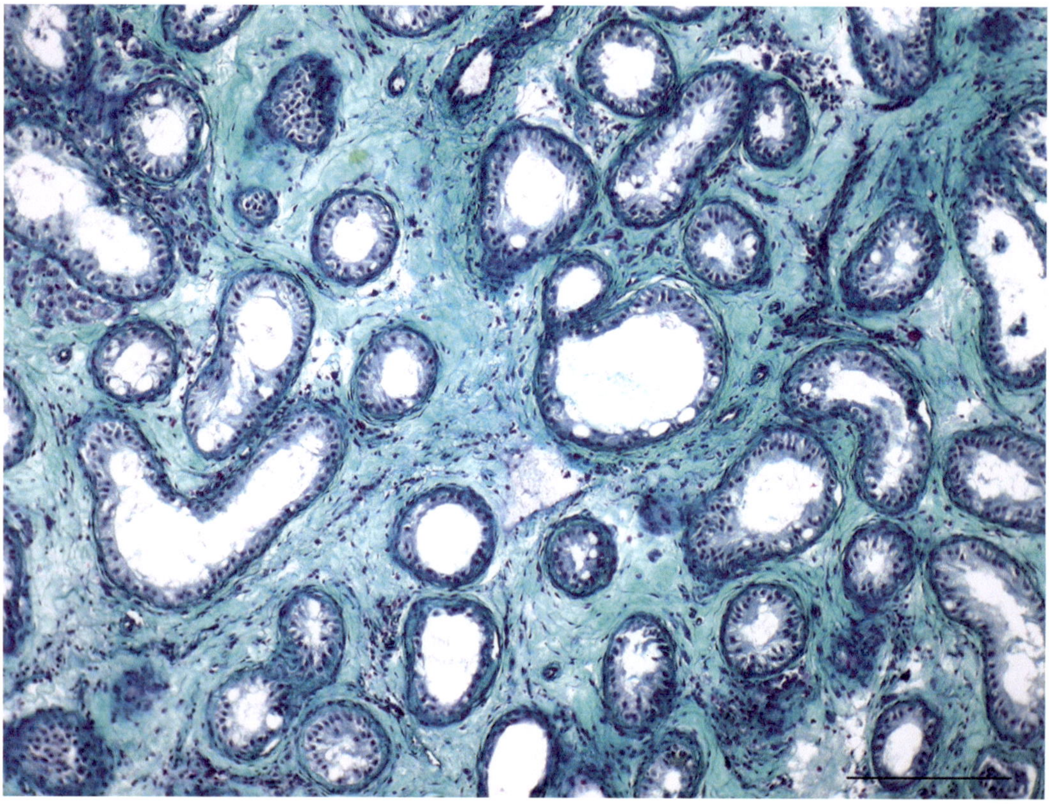

Fig. 10.15 Biopsy specimen from a 21-year-old man with a high scrotal unilateral cryptorchid gliding testis. The tubular diameter is considerably smaller than in normal testicular parenchyma. Note the absence of spermatogenic cells. The epithelium consists of Sertoli cells only. Many cells have cytoplasmic vacuoles. There is significant fibrosis of the testis interstitium (Masson stain, ×100; *scale bar*, 200 μm)

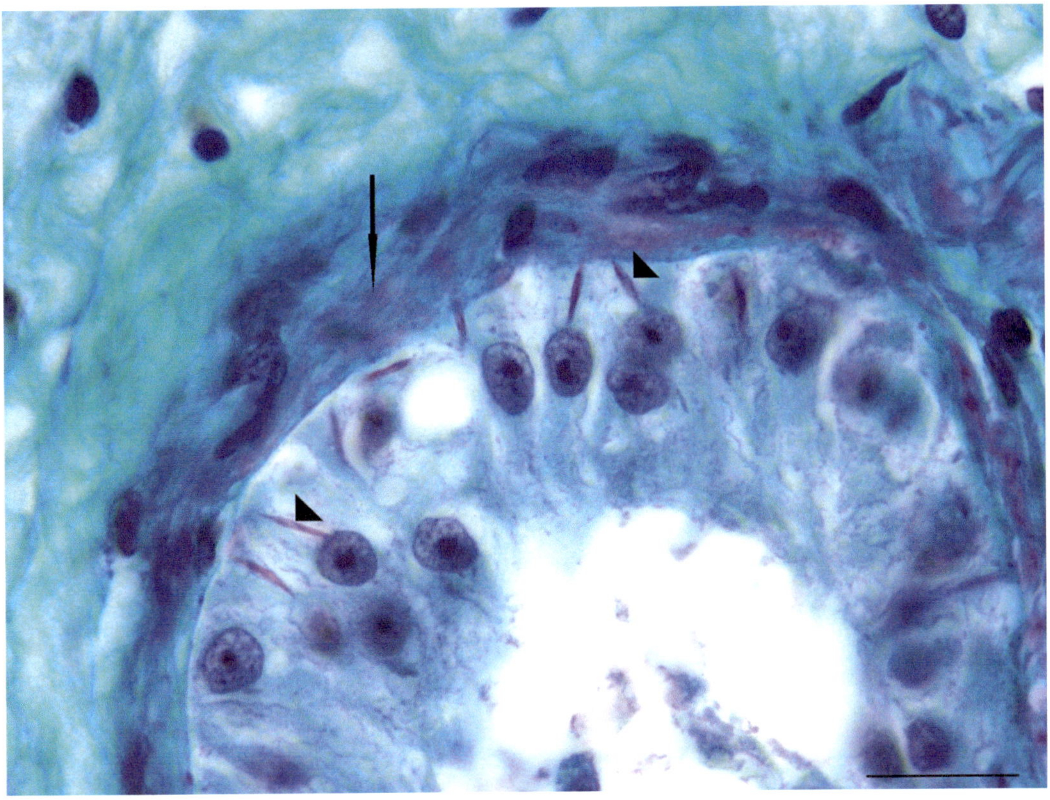

Fig. 10.16 Detail from Fig. 10.15. The tubule is lined by mature Sertoli cells only. In the Sertoli cell cytoplasm, many slender spindle-shaped Charcot-Böttcher crystals are present (*arrowheads*). The tunica propria is thickened (*arrow*) (Masson stain, ×1000; *scale bar*, 20 μm)

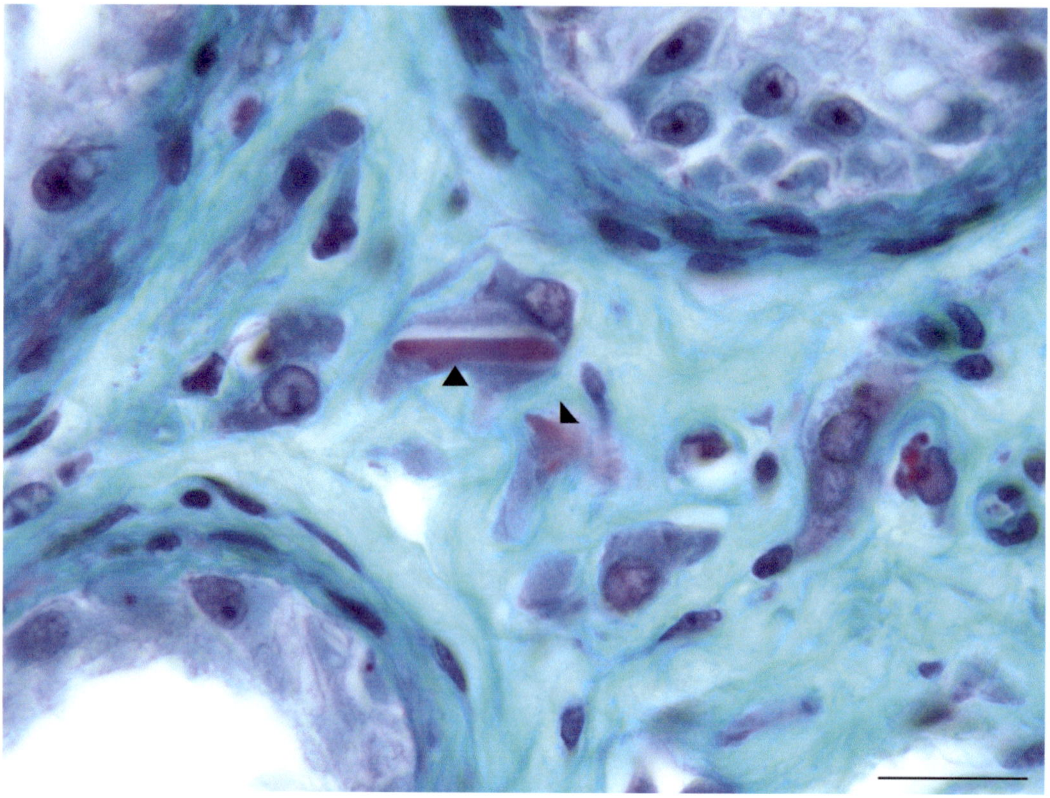

Fig. 10.17 Another detail from Fig. 10.15 showing interstitial fibrosis. The Leydig cells are wrinkled and have a scanty cytoplasm, sometimes with Reinke's crystals (*arrowheads*). The tubules demonstrate thickening of the tunica propria. The Sertoli cells have a condensed cytoplasm (Masson stain, ×1000; *scale bar*, 20 μm)

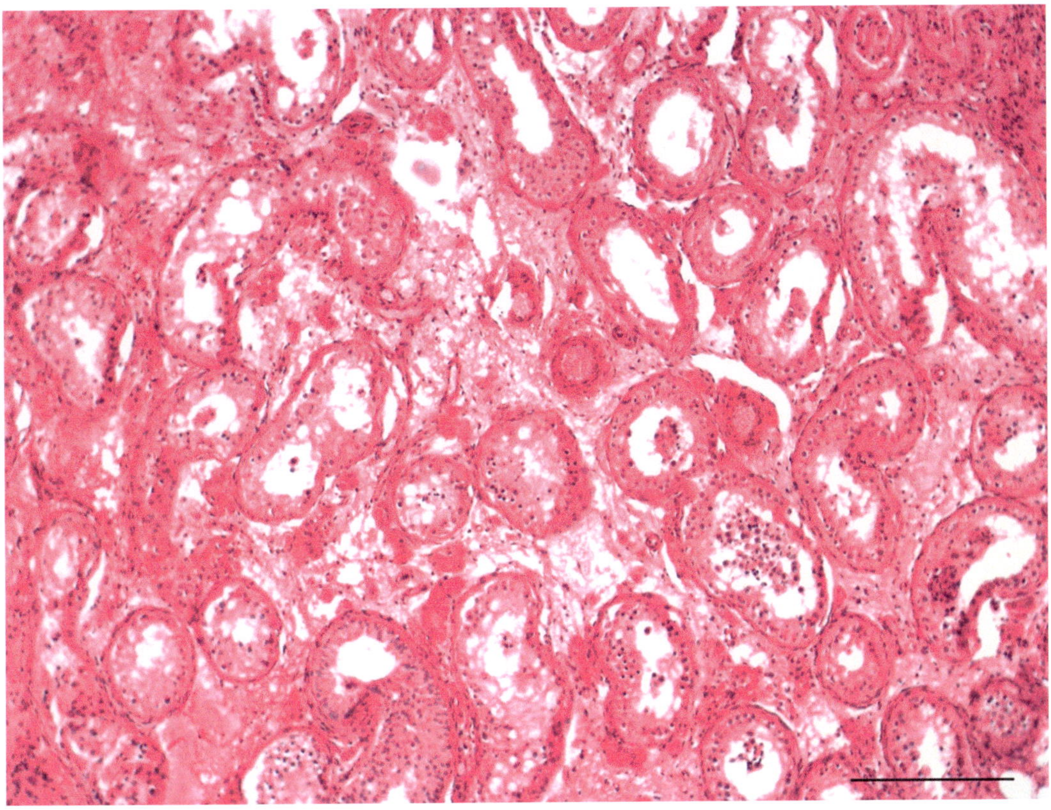

Fig. 10.18 Testicular biopsy sample from a testis fixed in the inguinal canal. The tubules have a moderately narrowed diameter. The seminiferous epithelium consists of mature Sertoli and mostly degenerated spermatogenic cells (H&E stain, ×100; *scale bar*, 200 μm)

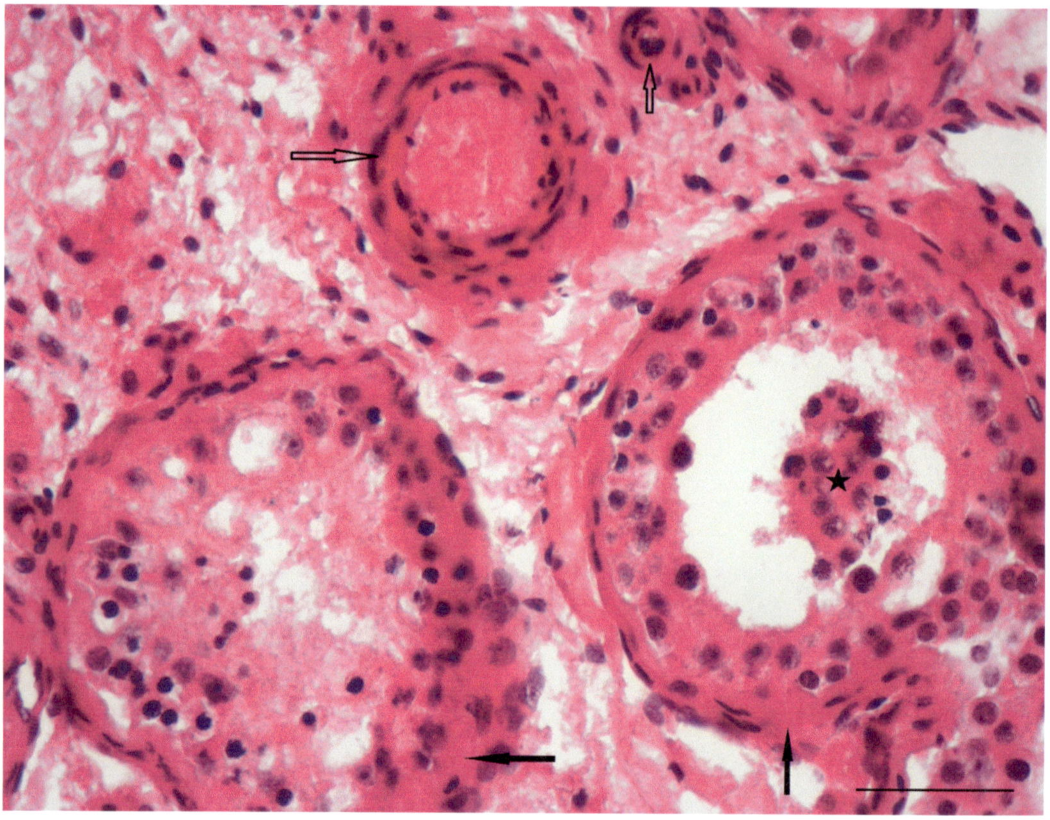

Fig. 10.19 Spermatogenetic "stop". The mature Sertoli cells contain degenerated spermatogonia and spermatocytes in their cytoplasm. The lumen of seminiferous tubules is filled with desquamated spermatogenic and Sertoli cells. The basement membrane of the tubules is thickened and undulated (*black arrows*). Blood vessels in the interstitium have a thickened wall as well (*transparent arrows*) (H&E stain, ×400; *scale bar*, 50 μm)

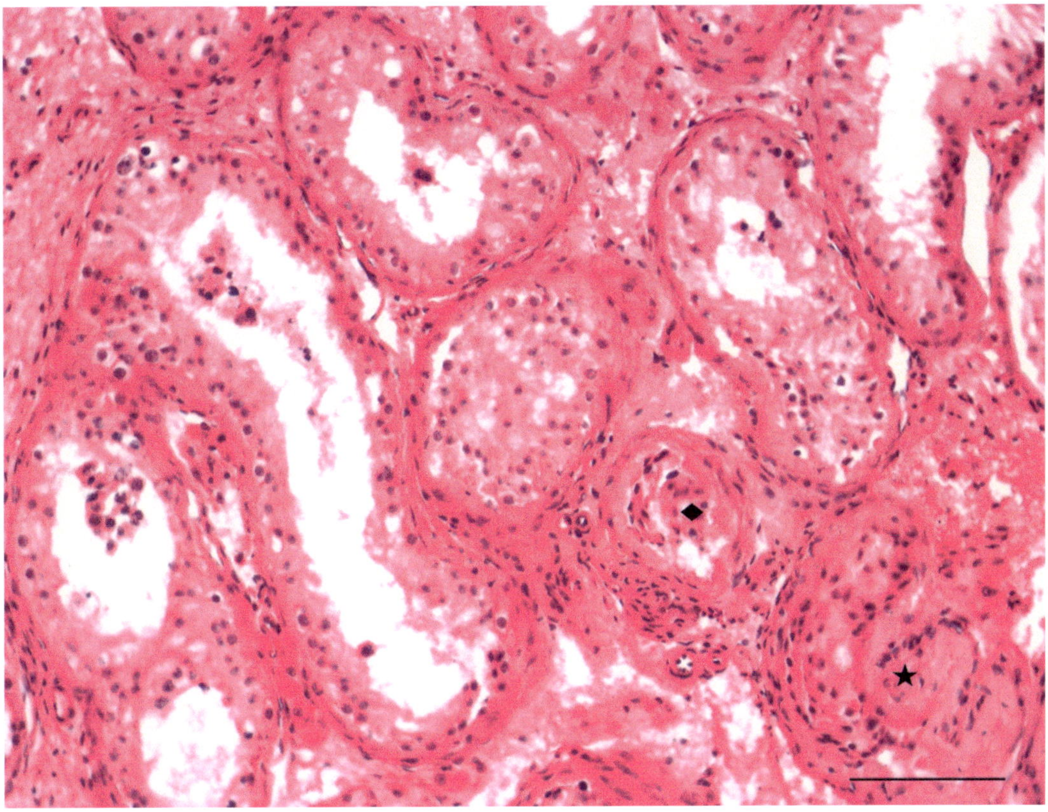

Fig. 10.20 Detail from the specimen depicted in Fig. 10.19. Although some tubules contain spermatogenic cells, others are completely fibrosed (*asterisk, diamond*). Seminiferous tubules with a significantly reduced lumen and a thickened basement membrane are lined only with atrophic Sertoli cells (H&E stain, ×200; *scale bar*, 100 μm)

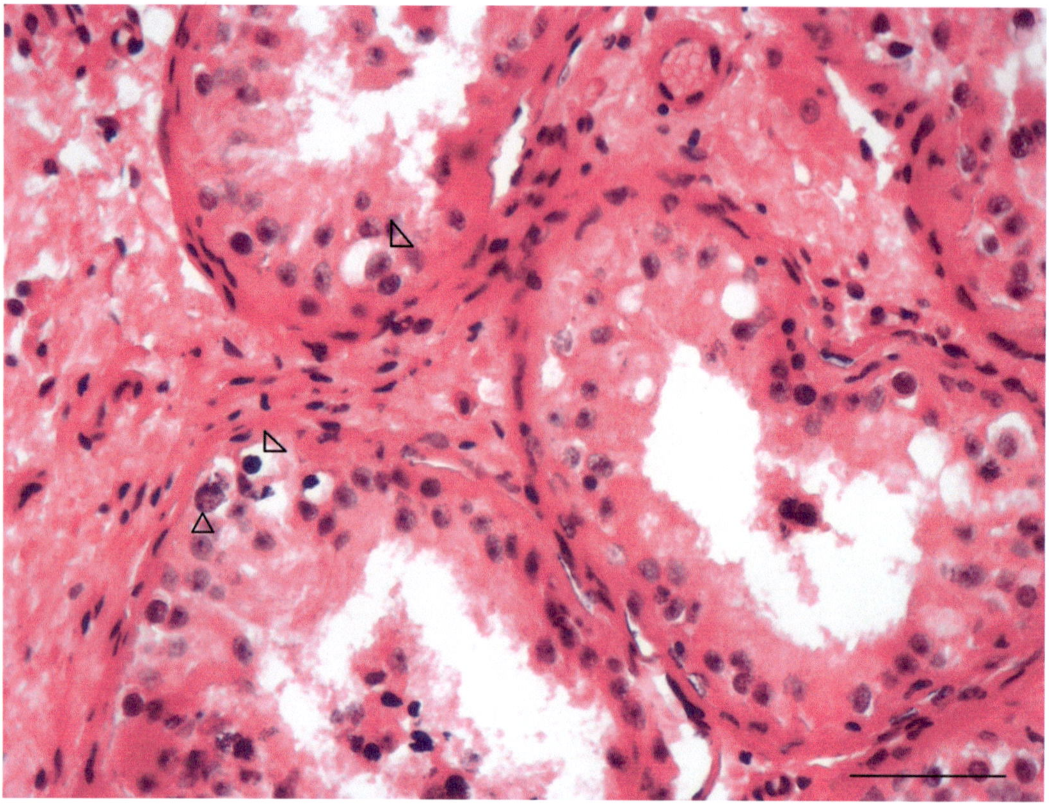

Fig. 10.21 Part of a testicular biopsy sample (testis retained in the inguinal canal). Spermatogenic and Sertoli cells (*arrowheads*), although present within the seminiferous tubules, show various degenerative changes and desquamation (H&E stain, ×400; *scale bar*, 50 μm)

Recommended Literature

Govender D, Sing Y, Chetty R. Sertoli cell nodules in the undescended testis. A histochemical, immunohistochemical and ultrastructural study of hyaline deposits. J Clin Pathol. 2004;57:802–6.

Hack WW, Meijer RW, Bos SD, Haasnoot K. A new clinical classification for undescended testis. Scand J Urol Nephrol. 2003a;37:43–7.

Hack WW, Meijer RW, van der Voort-Doedens LM, Bos SD, De Kok ME. Previous testicular position in boys referred for an undescended testis: futher explanation of the late orchidopexy enigma? BJU Int. 2003b; 92:293–6.

John Radcliffe Hospital Cryptorchidism Study Group. Cryptorchidism: a prospective study of 7500 consecutive male births 1984–8. Arch Dis Child. 1992;67: 892–9.

La Scala GC, Ein SH. Retractile testes: an outcome analysis on 150 patients. J Pediatr Surg. 2004;39:1014–7.

Rusnack SL, Wu H, Huff DS, Snyder HM, Zderic SA, Carr MC, Canning DA. The ascending testis and the testis undescended since birth share the same histopathology. J Urol. 2002;168:2590–1.

Sharpe MR, McKinnell C, Kivlin C, Fisher S. Proliferation and functional maturation of Sertoli cells and their relevance to disorders of testis function in adulthood. Reproduction. 2003;125:769–84.

11 Klinefelter's Syndrome

Manuel Nistal and Pilar González-Peramato

Klinefelter's syndrome is characterized by the presence of an abnormal number of X chromosomes and primary testicular insufficiency. Testicular lesions involve both the seminiferous tubules and testicular interstitium and are (a) atrophy of the seminiferous tubules that, through a hyalinization process, may lead to the complete disappearance of the tubules, with only ghost tubules remaining and (b) an apparent increase in the number of Leydig cells, which form characteristic nodules that tend to coalesce and delete the testicular structure.

This syndrome is the most common genetic cause of male infertility. The estimated incidence of this disorder varies according to the population studied: 1:500 male newborns and infants, 1 % of institutionalized adults with mental retardation, 3.4 % of patients undergoing consultation for infertility, 11 % of azoospermic males, and 0.7 % of oligozoospermic patients [1–3].

M. Nistal, M.D., Ph.D. (✉)
Department of Pathology, University Hospital La Paz, Madrid, Spain

Department of Anatomy, Histology, and Neuroscience, University Autonoma of Madrid, Madrid, Spain
e-mail: mnistal.hulp@salud.madrid.org

P. González-Peramato, M.D., Ph.D.
Department of Pathology, University Hospital La Paz, Madrid, Spain

Department of Pathology, University Autonoma of Madrid, Madrid, Spain

The most frequent karyotype in this syndrome is 47,XXY (80 %), followed by 46,XY/47,XXY (10 %). Although it should be taken into account that more than 30 different karyotypes have been reported, the following karyotypes comprise most of the remaining cases (10 %): 46,XY/48,XXXY; 46,XX/47,XXY; 47,XXY/46,XX/46,XY; 46,XY/45,XO/47,XXY; 46XX/47XXY/48XXXY;48,XXXY/49,XXXXY; and 47,X,Xi(Xq)Y, 46XX [4–7]. The classic form of Klinefelter's syndrome results from a meiotic X chromosome nondisjunction in the meiotic cells of either parent. Mosaicisms are the result of a mitotic nondisjunction in the zygote after fertilization.

The most frequent clinical symptoms in adult patients are [8]:

1. Eunuchoid phenotype. The elevated stature is the result of disproportionate growth of the lower extremities. The span/height ratio is less than 1.
2. Variable degrees of virilization, from the absence of secondary sex characteristics to normal masculinization.
3. Gynecomastia, which is present in more than half the patients and is bilateral in most cases.
4. Mental retardation in many patients, although 80 % of them have an IQ allowing them to participate in social activities. The behavior of these patients has been described as immature, lacking initiative or enterprise, and emotionally unstable, with poor concentration capacity.
5. Several associated disorders, including chronic bronchitis; cervical rib kyphosis, scoliosis, or

D. Ježek (ed.), *Atlas on the Human Testis*,
DOI 10.1007/978-1-4471-2763-5_11, © Springer-Verlag London 2013

pectus excavatum; and a high incidence of hypothalamic, hypophyseal, thyroid, and pancreatic dysfunctions.

6. Supernumerary X chromosome material in men, which is associated with a reduction in left temporal lobe gray matter, giving rise to verbal and language deficits.
7. Normally developed external genitalia in most patients. Testicular size usually is reduced. The longest testicular axis usually measures less than 2.5 cm, although testicular length is normal in other Klinefelter's syndrome variants showing karyotypic mosaicisms. Testicular consistency is firm in 50 % of cases and soft in 20 %. The incidence of cryptorchidism is low in 47,XXY patients, although it often is higher in variants showing more than two X chromosomes.

The phenotypic variation of Klinefelter's syndrome is wide and may be caused not only by inactivation of the supranumeric X chromosome but also by polymorphism in the androgen receptor [9].

Histology

Infancy

Testicular lesions occur early in Klinefelter's syndrome. Early germ cell loss takes place once the stage of testis differentiation has begun [10]. Most aborted human fetuses with 47,XXY show germ cell loss as early as the 18th to 22th gestational weeks. Germ cell loss is associated with an irregular distribution of germ cells: tubules with germ cells alternate with other tubules devoid of them. Neither the Leydig cells nor the remaining testicular structures show lesions. Testosterone levels in cord blood are low in some cases and normal in others. Most newborns with 47,XXY have a tubular fertility index lower than 30 % and a decreased number of germ cells per tubular section. Sertoli and Leydig cell lesions have not been reported. The two postnatal waves of germ cell proliferation (at "minipuberty" and at about 4 years of age) occurring in the normal testis have not been observed in patients with 47,XXY. On the contrary, germ cells decrease in number from the second year of life onward in cases with associated cryptorchidism (Fig. 11.1) [11].

During infancy, tubular diameter remains low and static, with a subsequent small testicular volume. This occurs despite the observation that levels of testosterone, insulin-like peptide 3, follicle-stimulating hormone, luteinizing hormone, antimüllerian hormone (AMH), and inhibin B do not differ from those in controls [12]. Moreover, a testosterone peak has been observed in the first months of life, coinciding with minipuberty [13]. If one takes into account that testicular volume is determined by the grade of growth in length and tortuousness of the seminiferous tubules, as well as by the tubular diameter and that all these parameters depend on Sertoli cell trophism, one may assume that, in infancy, these testes have at least two impaired cell types: Sertoli cells and germ cells.

Early Adolescence

Normally in early adolescence, a new wave of germ cell proliferation coinciding with hypothalamic-pituitary-testicular axis activation may be expected. However, in patients with Klinefelter's syndrome, if germ cell proliferation takes place at all, it occurs only in some tubules in which a few spermatocytes may be observed; otherwise, the germ cells completely disappear [14–16]. In half the cases, as puberty progresses, residual spermatogonia disappear, and there is an increase in tubular hyalinization and Leydig cells (Fig. 11.2).

Adulthood

In adulthood, the most frequent histologic picture is that of an apparent nodular Leydig cell hyperplasia in which it is difficult to see even ghost tubules (Figs. 11.3 and 11.4). The hyperplasia is apparent only because quantitative and qualitative studies have shown that the actual number of Leydig cells does not increase [17]. Eosinophilic inclusions larger than Reinke's

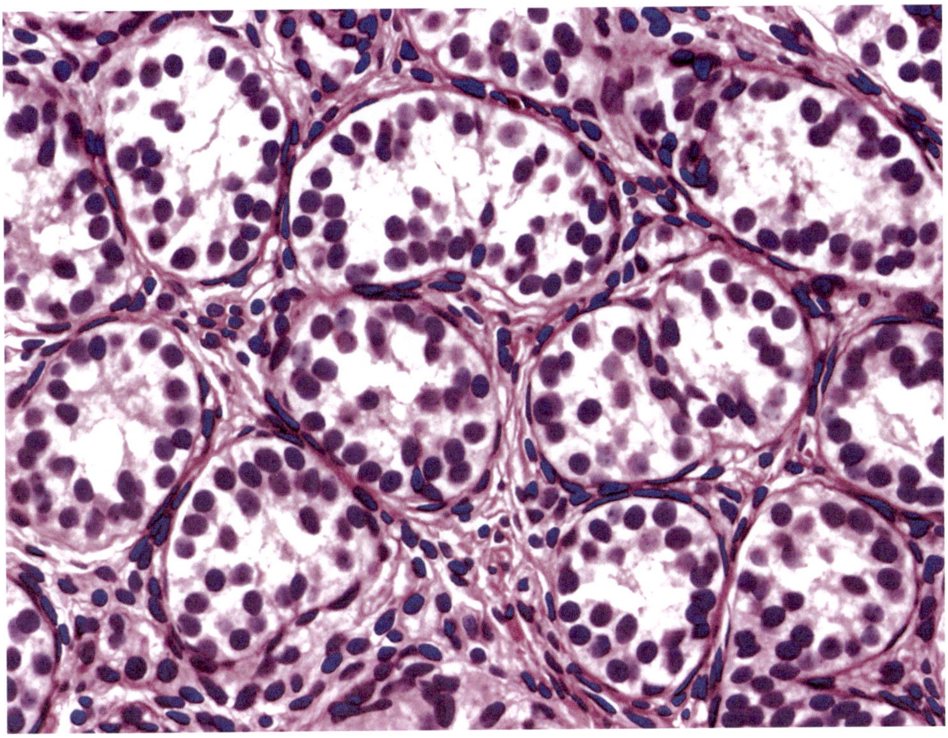

Fig. 11.1 Klinefelter's syndrome in a 3-year-old boy. The seminiferous tubules show Sertoli cells arranged in a pseudostratified pattern. Germ cells are absent. In the interstitium, only undifferentiated Leydig cells can be seen

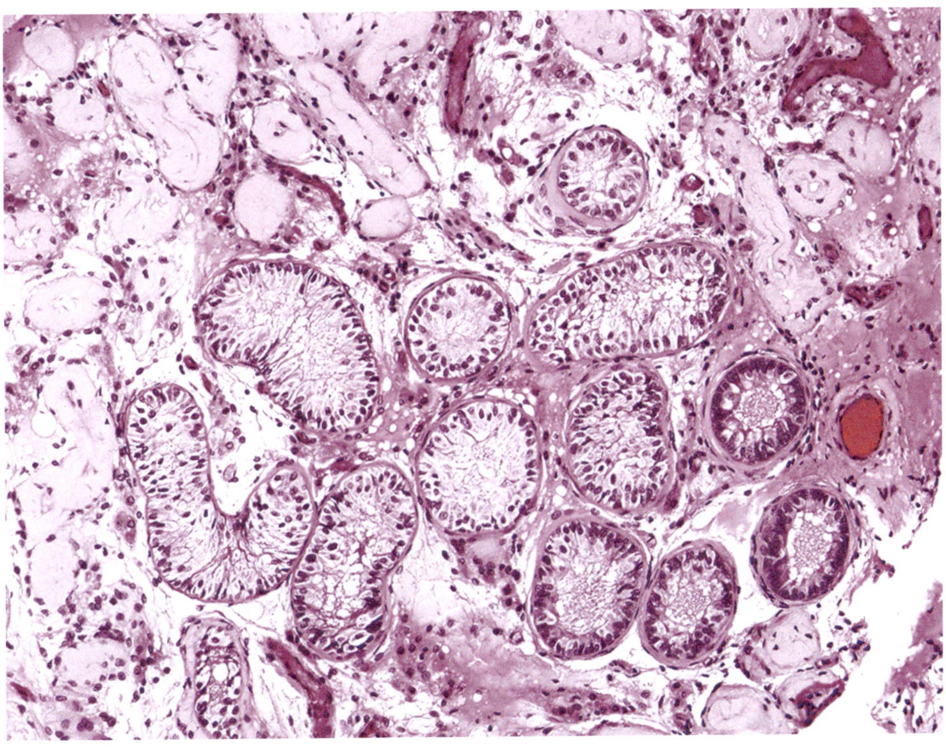

Fig. 11.2 Klinefelter's syndrome in a 16-year-old pubertal patient. SCO tubules are adjacent to completely hyalinized tubules. In the interstitium, isolated groups of Leydig cells can be seen

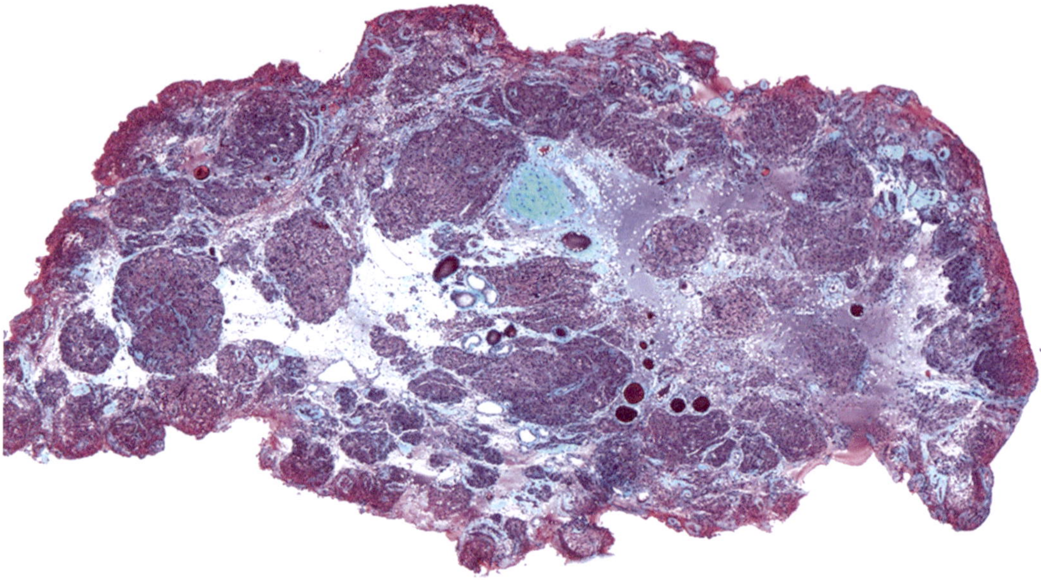

Fig. 11.3 Klinefelter's syndrome in an adult patient. This low-magnification image shows a testicular biopsy sample from a patient under consultation for infertility. Nodular hyperplasia of Leydig cells is highlighted

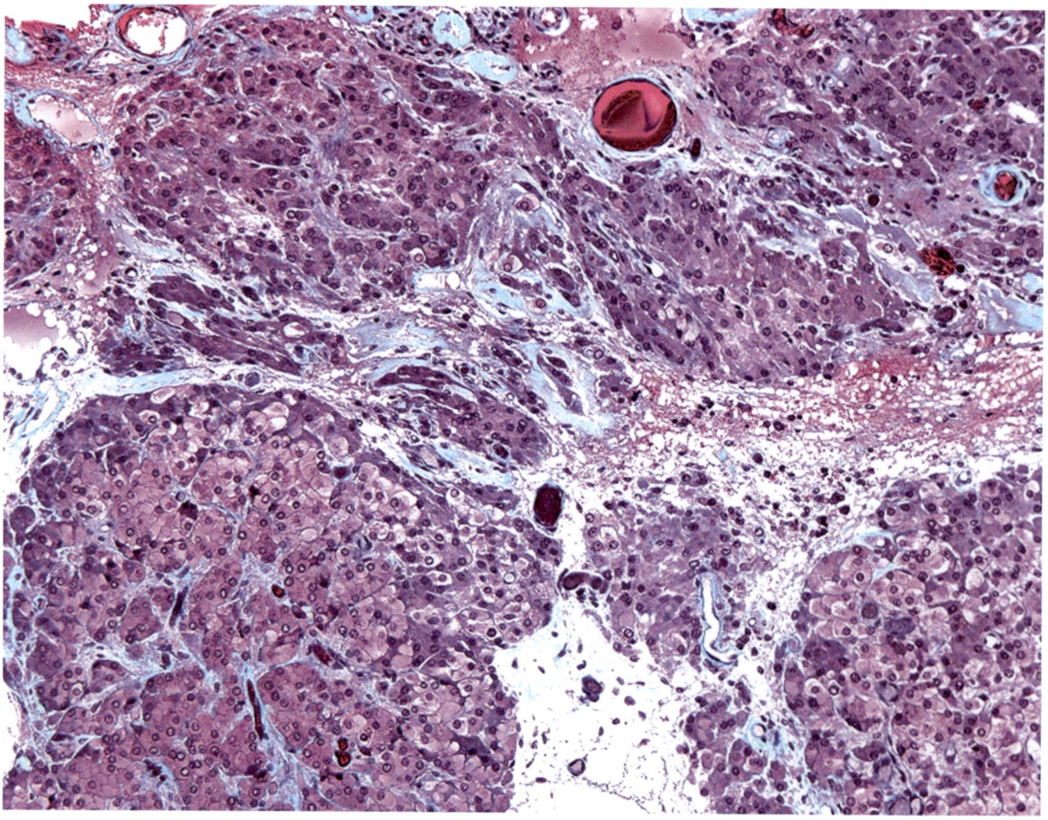

Fig. 11.4 Adult patient with Klinefelter's syndrome. Leydig cell nodules show marked hyperplasia and hypertrophy. Among the nodules, hyalinized tubules can be observed

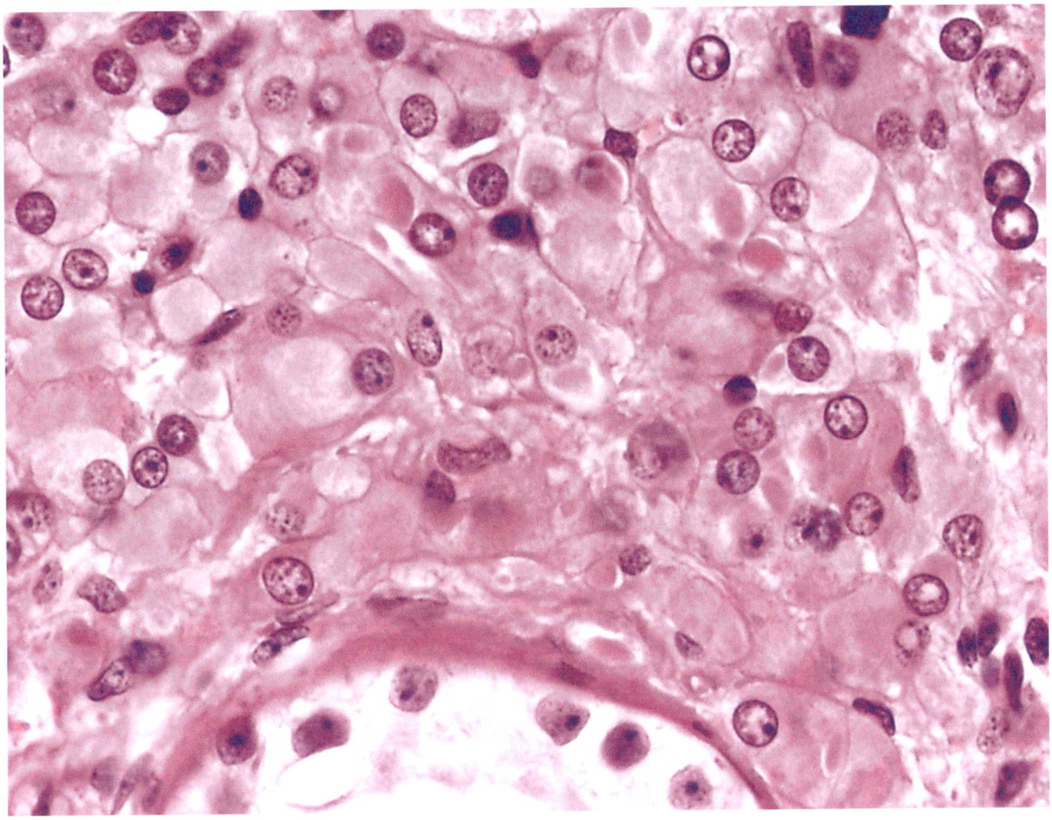

Fig. 11.5 Hypertrophied Leydig cells showing in their cytoplasm frequent eosinophilic inclusions surrounded by a clear halo and scarce Reinke crystals

crystals may be observed in the cytoplasm of Leydig cells (Figs. 11.5 and 11.6). These crystalloid inclusions also are present in intratubular Leydig cells in hyalinized tubules (Fig. 11.7). There is a wide variety of ultrastructural Leydig cells: from fibroblast-like cells to cells with hypertrophied cytoplasm and cells with signs of exhaustion [18, 19]. In 75 % of cases, isolated groups of Sertoli cell-only (SCO) tubules (Fig. 11.8), or even some foci with spermatogenesis, persist. Prepubertal Sertoli cells cannot turn into adult Sertoli cells at puberty, and those that persist focally have dysgenetic or involutive features [20, 21]. This finding correlates with the low synthesis of inhibin B. Two types of SCO tubules may be distinguished: Type A tubules have Sertoli cells with a dysgenetic morphology including a pseudostratified distribution; elongated hyperchromatic nuclei with small, peripherally placed nucleoli; the absence of annulate lamellae; and weak immunoreaction to AMH. Type B tubules have a central lumen, and their Sertoli cells show an adult, mature pattern or signs of involution (many nuclear folds). In some testes, both tubule types are present. Peritubular cells have little myoid differentiation and cannot produce elastic fibers (Fig. 11.9) [22]. Leydig cells, despite their hyperplasic appearance, produce normal testosterone levels in only some cases [23].

Patients with 47,XXY classically have been considered infertile, and their spermiogram results show azoospermia as a rule, although some spermatozoa have been found in 8.4 % of patients [24], and 50 % of patients show focal spermatogenesis. This finding has led to new expectations regarding the potential fertility of these patients through testicular sperm extraction therapies combined with intracytoplasmic sperm injection [25].

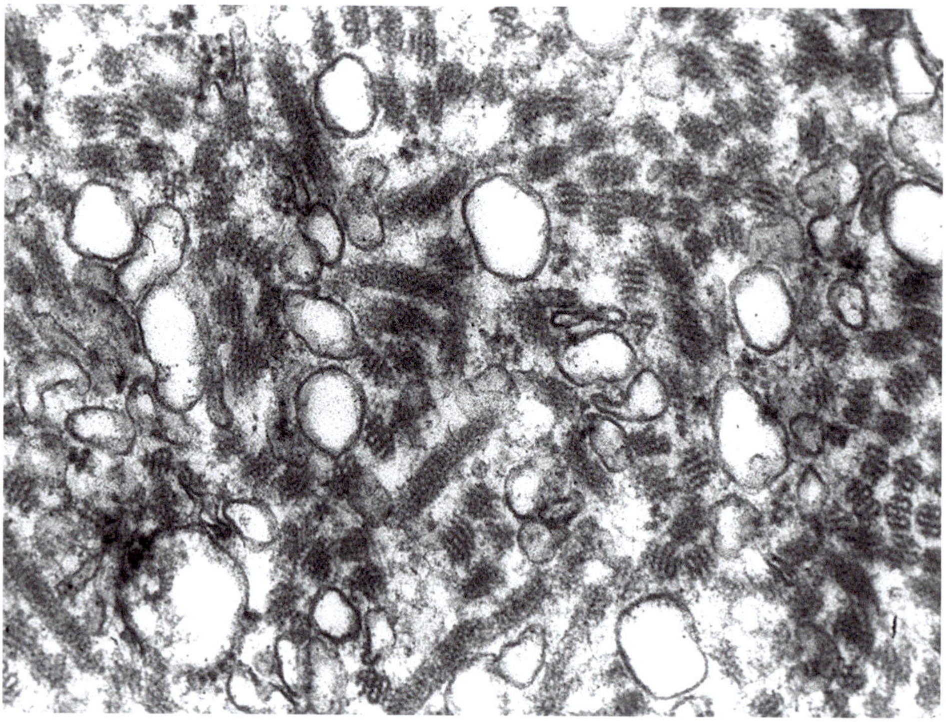

Fig. 11.6 Ultrastructure of a paracrystalline inclusion with several parallel lamellae resulting from several infoldings of a single lamella among smooth endoplasmic reticulum tubules

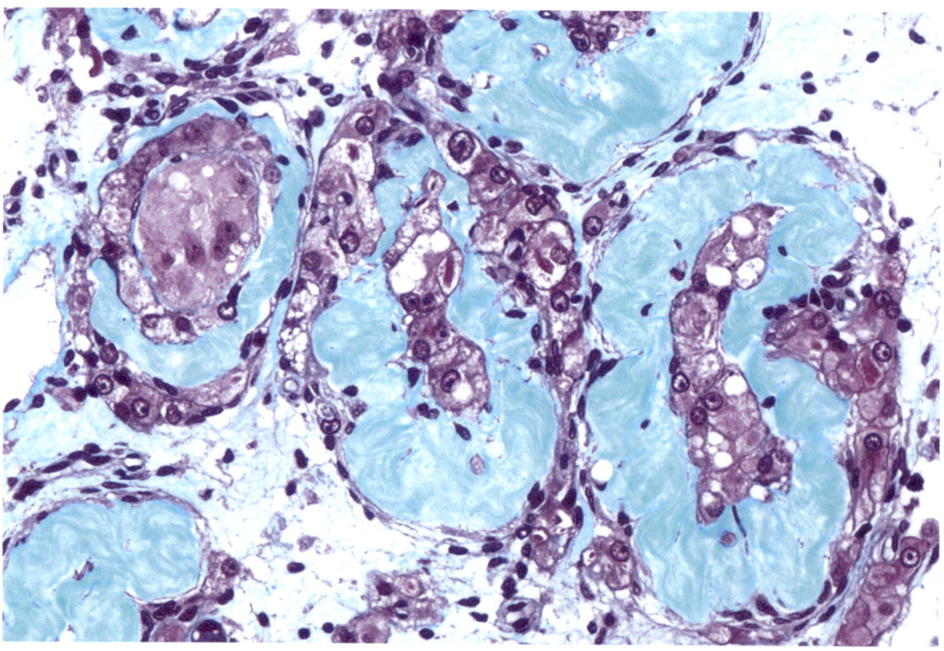

Fig.11.7 Infiltrative activity of Leydig cells inside a seminiferous tubule. Shown are several stages of Leydig cell internalization that end in the cells being completely surrounded by the basal membrane of the atrophic tubule. Note the presence of one Reinke crystal in a Leydig cell inside the tubule in the center of the figure

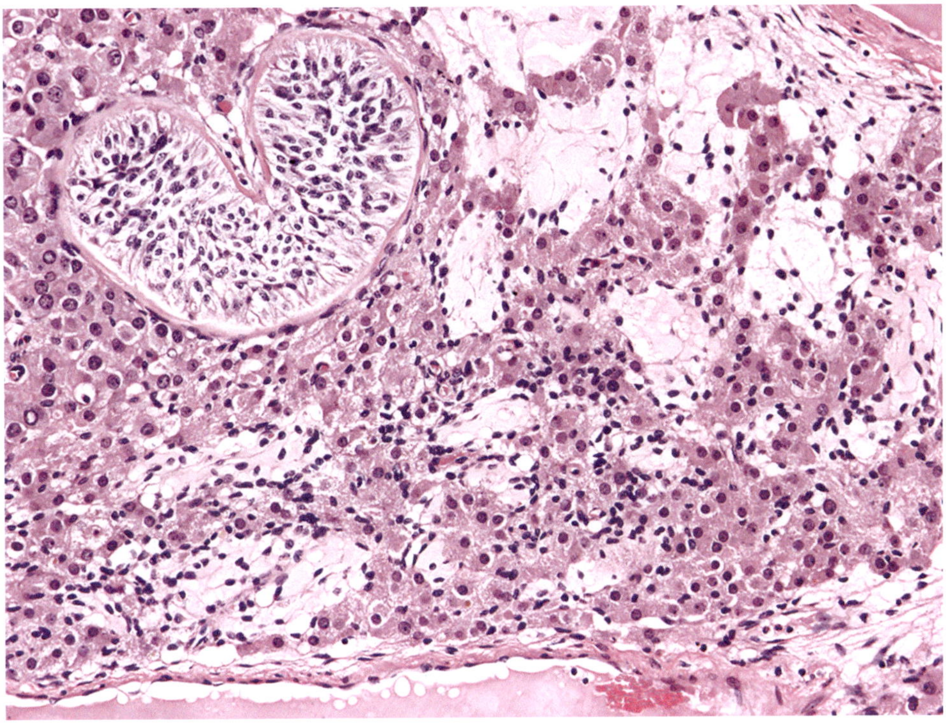

Fig. 11.8 Testicular biopsy specimen from an infertile patient with 47,XXY karyotype. Shown are tubules lined only by Sertoli cells with immature features surrounded by numerous Leydig cells and hyalinized tubules

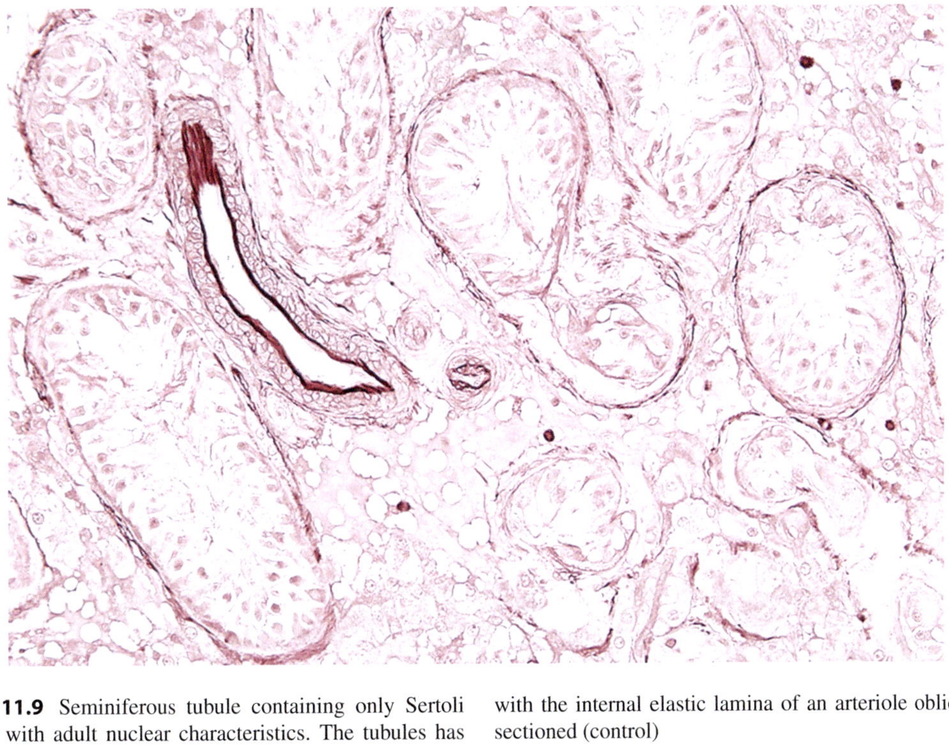

Fig. 11.9 Seminiferous tubule containing only Sertoli cells with adult nuclear characteristics. The tubules has markedly decreased elastic fibers in their walls compared with the internal elastic lamina of an arteriole obliquely sectioned (control)

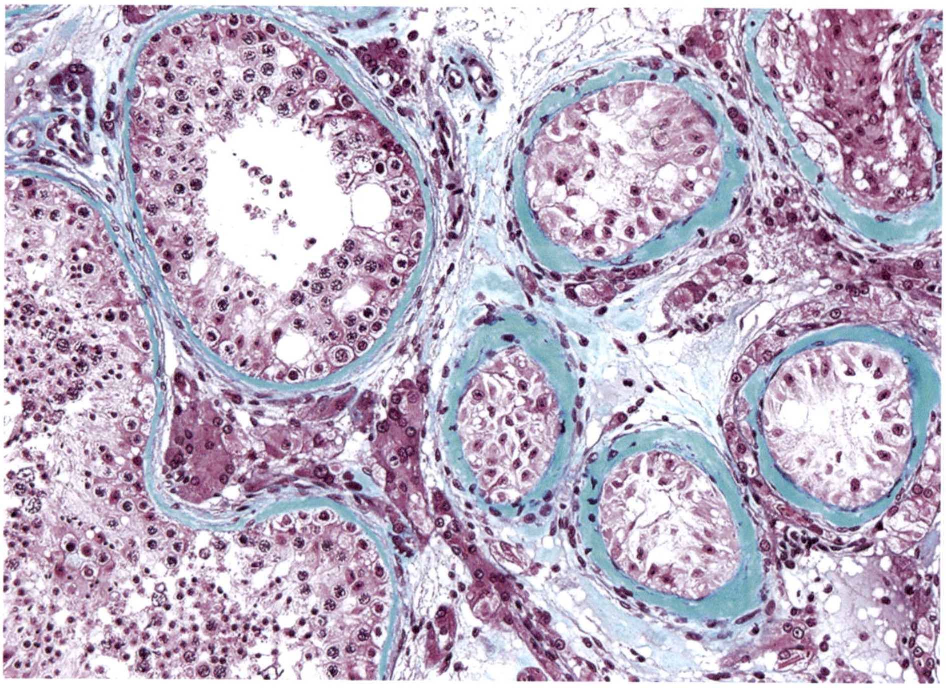

Fig. 11.10 A patient with 46,XX/47,XXY. Mixed atrophy with a mosaic pattern is present. Seminiferous tubules with spermatogenesis adjacent to SCO tubules have markedly thickened walls

Variants of Klinefelter's Syndrome

46,XX/47,XXY

Klinefelter's syndrome 46,XX/47,XXY is the most frequent variant. The clinical symptoms of 47,XXY appear attenuated; gynecomastia is present in only 33 % of patients and azoospermia in 50 % of cases. The testes are larger, and spermatogenesis is more conservative because of the presence of spermatogonia with a normal chromosomal constitution (Fig. 11.10).

46,XX

Patients with 46,XX Klinefelter syndrome and normal external genitalia have most of the same clinical symptoms as those with the 47,XXY variant but have harmonious body proportions, normal or small stature, and normal intelligence. The testicular parenchyma shows an SCO pattern with hyalinized tubules and the absence of spermatogenesis (Fig. 11.11).

48,XXYY

The incidence of Klinefelter's syndrome 48,XXYY is estimated at approximately 0.04 in 1000 live births. Aggressive and antisocial behavior, together with a greater degree of mental retardation and a higher frequency of congenital malformations than observed in men with 47,XXY Klinefelter's syndrome, has been attributed to patients with 48,XXYY [26]. A peculiar finding is the presence of concentric smooth endoplasmic reticulum cisternae in the Leydig cell cytoplasm (Figs. 11.12 and 11.13) [18].

48,XXXY and 49,XXXYY

In both 48,XXXY and 49,XXXYY Klinefelter's syndrome, cryptorchidism frequently is present,

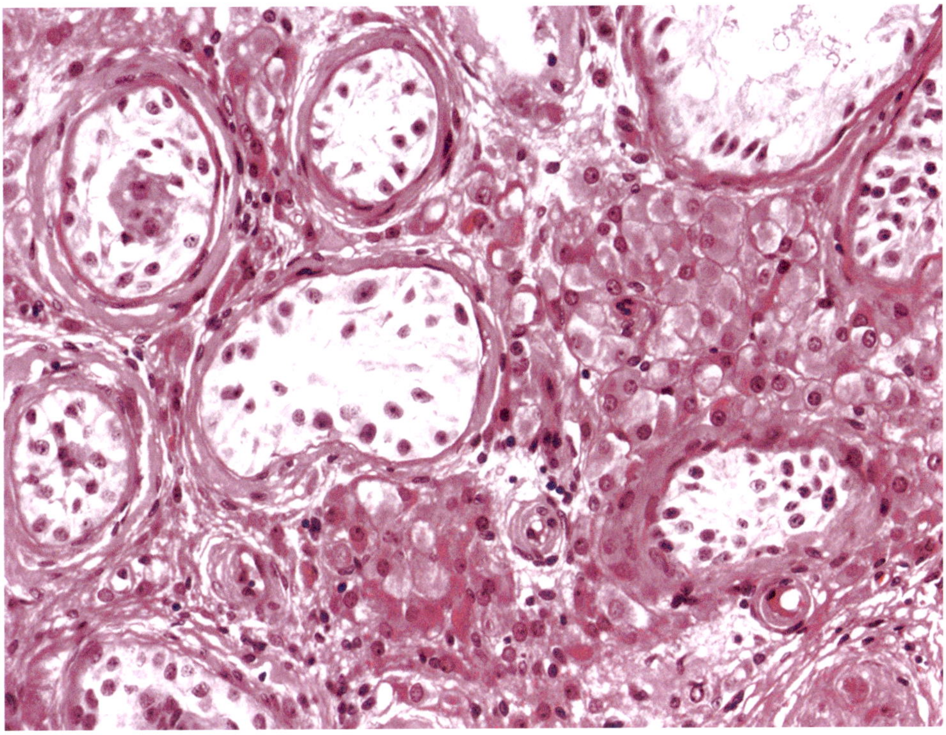

Fig. 11.11 Infertile male with 46,XX. SCO seminiferous tubules are lined with adult Sertoli cells. Germ cells are absent, and the walls are thickened. Diffuse Leydig cell hyperplasia is seen in the interstitium

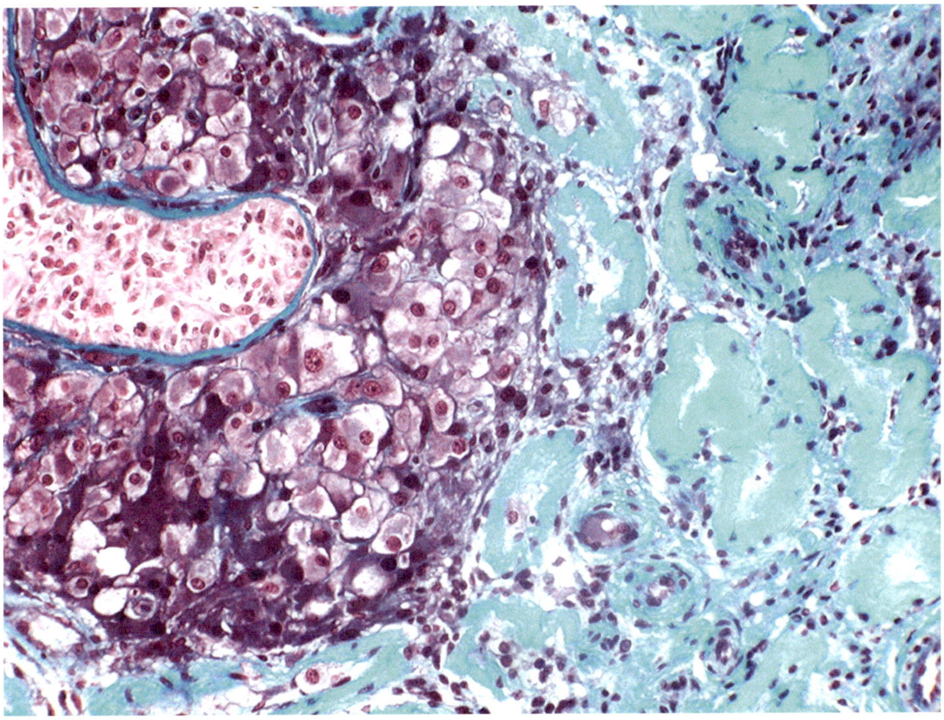

Fig. 11.12 Patient with 48,XXYY. Seminiferous tubules with immature Sertoli cells are surrounded by hypertrophic Leydig cells. The remaining parenchyma consists of hyalinized seminiferous tubules

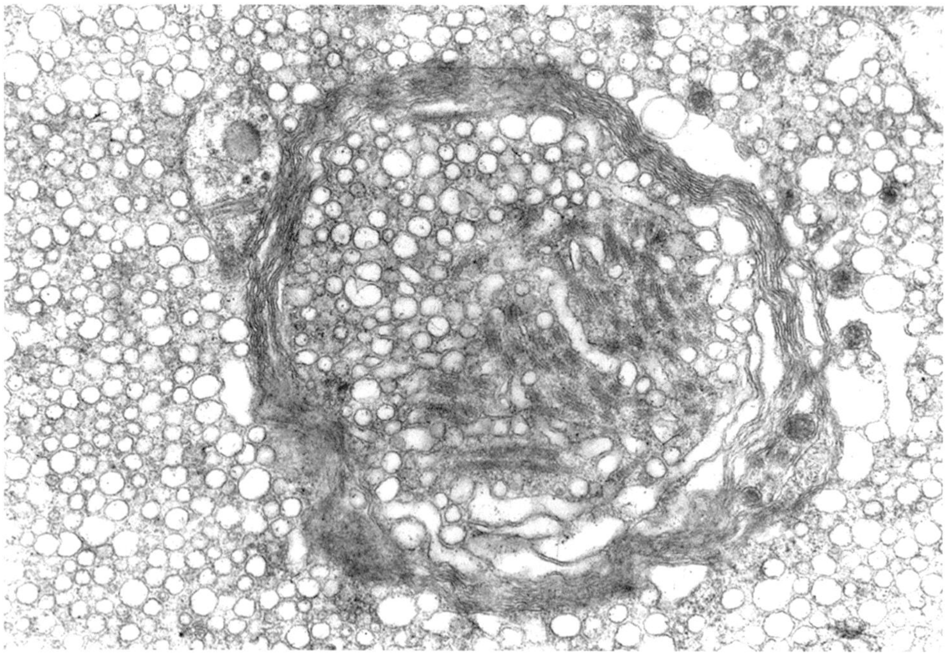

Fig. 11.13 Patient with 48,XXYY. The ultrastructure of Leydig cell cytoplasm shows an inclusion formed by concentric lamellae surrounding paracrystalline inclusions and smooth endoplasmic reticulum cisternae

as are multiple congenital malformations and psychological features [27].

Sometimes, 47,XXY Klinefelter's syndrome and its variants begin clinically with a central precocious puberty, which is caused by a mediastinal germ cell tumor secreting human chorionic gonadotropin in most cases [28] and is idiopathic in most of the remaining cases [29], although a hamartoma of the third ventricle has been found in isolated cases.

Although Klinefelter's syndrome is the typical example of chromosomal anomaly with hypergonadotropic hypogonadism, the association of Klinefelter's syndrome with pituitary disorders is not exceptional [30]. Some patients with the 47,XXY karyotype have *hypogonadotropic* hypogonadism. Testicular biopsy samples from these patients show diffuse tubular hyalinization and Leydig cell depletion or even absence (Figs. 11.14 and 11.15) [31].

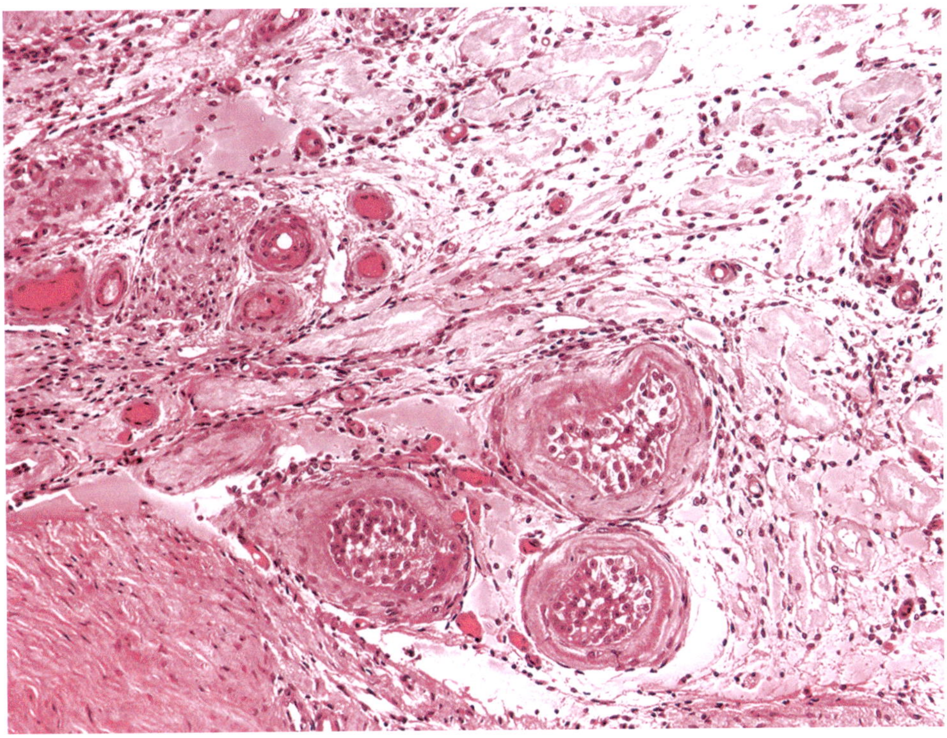

Fig. 11.14 Adult patient with 47,XXY, small testes, and hypogonadotropic hypogonadism. Note the tubular hyalinization, isolated SCO tubules of spherical nuclei, and absence of Leydig cells in the interstitium

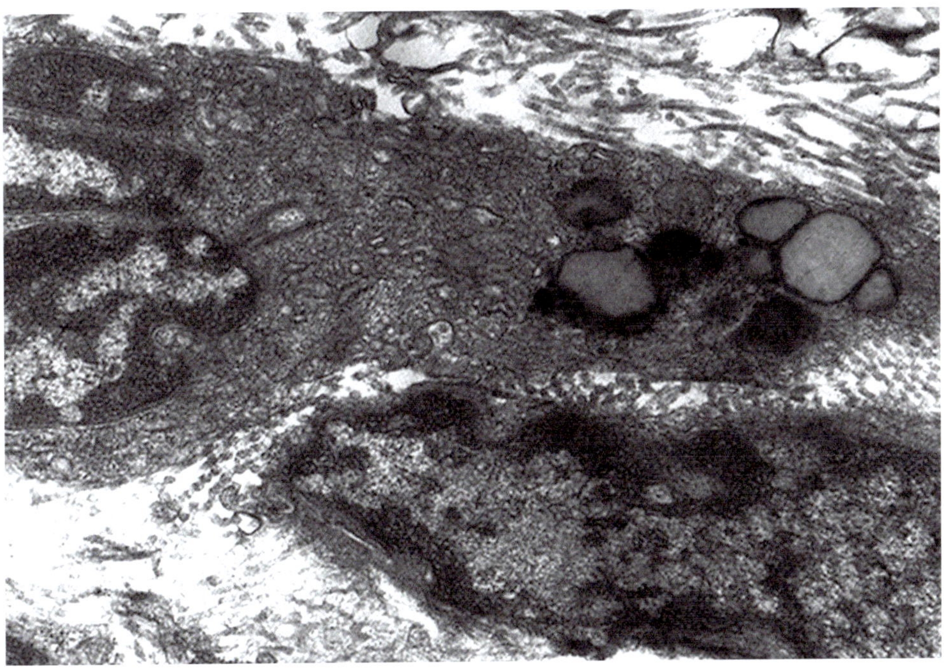

Fig. 11.15 Adult patient with 47,XXY and hypogonadotropic hypogonadism. Shown are fibroblast-like cells with scarce smooth endoplasmic reticulum cisternae and abundant lipids and lysosomes. Abundant collagen fibers may be observed among the cells

References

1. Foresta C, Galeazzi C, Bettella A, Marin P, Rossato M, Garolla A, et al. Analysis of meiosis in intratesticular germ cells from subjects affected by classic Klinefelter's syndrome. J Clin Endocrinol Metab. 1999;84:3807–10.
2. De Braekeleer M, Dao T-N. Cytogenetics studies in male infertility: a review. Hum Reprod. 1991;6:245–50.
3. Yoshida A, Kazukiyo M, Shirai M. Chromosome abnormalities and male infertility. Assist Reprod Rev. 1996;6:93–9.
4. Hatakayama S. The male reproductive system. In: Kovacs K, Asa SL, editors. Functional endocrine pathology. Oxford: Blackwell; 1991. p. 585–607.
5. Griffin JE, Wilson JD. Chapter 37: Disorders of sexual differentiation. In: Walsh PC, Retik AB, Stamey TA, Vaughan Jr ED, editors. Campbell's urology, vol. 2. 6th ed. Philadelphia: WB Saunders Co.; 1992. p. 1509–42.
6. Arps S, Koske-Westphal T, Meinecke P, Meschede D, Nieschlag E, Harprecht W, et al. Isochromosome Xq in Klinefelter syndrome: report of 7 new cases. Am J Med Genet. 1996;64:580–2.
7. Filiadis IF, Syrrou MB, Bai MC, Georgiou IA, Pagoulatos GN, Gianakopoulos X. Infertility and multiple urogenital abnormalities in a male with mosaic 46, XY/45, X0/47, XXY karyotype and mixed phenotype. Urol Int. 1998;61:111–4.
8. Smyth CM, Bremner WJ. Klinefelter syndrome. Arch Intern Med. 1998;158:1309–14.
9. Bojesen A, Hertz JM, Gravholt CH. Genotype and phenotype in Klinefelter syndrome – impact of androgen receptor polymorphism and skewed X inactivation. Int J Androl. 2011;34:e642–8.
10. Hunt PA, Worthman C, Levinson H, Stallings J, LeMaire R, Mroz K, et al. Germ cell loss in the XXY male mouse: altered X chromosome dosage affects prenatal development. Mol Reprod Dev. 1998;49:101–11.
11. Müller J, Skakkebaek NE, Ratcliffe SG. Quantified testicular histology in boys with sex chromosome abnormalities. Int J Androl. 1995;18:57–62.
12. Cabrol S, Ross JL, Fennoy I, Bouvattier C, Roger M, Lahlou N. Assessment of Leydig and Sertoli cell functions in infants with nonmosaic Klinefelter syndrome: insulin-like peptide 3 levels are normal and positively correlated with LH levels. J Clin Endocrinol Metab. 2011;96:E746–53.
13. Lahlou N, Fennoy I, Carel JC, Roger M. Inhibin B and anti-Müllerian hormone, but not testosterone levels, are normal in infants with nonmosaic Klinefelter syndrome. J Clin Endocrinol Metab. 2004;89:1864–8.
14. Aksglaede L, Wikstrom AM, Rajpert-De Meyts E, Dunkel L, Skakkebaek NE, Juul A. Natural history of seminiferous tubule degeneration in Klinefelter syndrome. Hum Reprod Update. 2006;12:39–48.
15. Wikström AM, Hoei-Hansen CE, Dunkel L, Rajpert-De Meyts E. Immunoexpression of androgen receptor and nine markers of maturation in the testes of adolescent boys with Klinefelter syndrome: evidence for degeneration of germ cells at the onset of meiosis. J Clin Endocrinol Metab. 2007;92:714–9.
16. Nistal M, González-Peramato P. Pathological evolution of testicular lesions in Klinefelter's syndrome. Rev Horm Crecim. 2009;12:1–10.
17. Nistal M, Santamaría L, Paniagua R. Quantitative and ultrastructural study of Leydig cells in Klinefelter's syndrome. J Pathol. 1985;146:323–31.
18. Nistal M, Paniagua R, Lopez-Pajares I. Ultrastructure of Leydig cells in Klinefelter's syndrome with 48, XXYY. Virchows Arch B Cell Pathol. 1978;28:39–46.
19. Paniagua R, Nistal M, Bravo MP. Leydig cell types in primary testicular disorders. Hum Pathol. 1984;15:181–90.
20. Skakkebaek E. Two types of tubules containing only Sertoli cells in adults with Klinefelter syndrome. Nature. 1969;223:643–5.
21. Nistal M, Paniagua R, Abaurrea MA, Santamaria L. Hyperplasia and the immature appearance of Sertoli cells in primary testicular disorders. Hum Pathol. 1982;13:3–12.
22. Martin R, Santamaria L, Nistal M, Fraile B, Paniagua R. The peritubular myofibroblasts in the testes from normal men and men with Klinefelter's syndrome. A quantitative, ultrastructural, and immunohistochemical study. J Pathol. 1992;168:59–66.
23. Regadera J, Codesal J, Paniagua R, Gonzalez-Peramato P, Nistal M. Immunohistochemical and quantitative study of interstitial and intratubular Leydig cells in normal men, cryptorchidism, and Klinefelter's syndrome. J Pathol. 1991;164:299–306.
24. Lanfranco F, Kamischke A, Zitzmann M, Nieschlag E. Klinefelter's syndrome. Lancet. 2004;364:273–83.
25. Van Saen D, Tournaye H, Goossens E. Presence of spermatogonia in 47, XXY men with no spermatozoa recovered after testicular sperm extraction. Fertil Steril. 2012;97:319–23.
26. Garry MB. Two cases of 48, XXYY: with discussion on the behaviour of prepubertal and postpubertal patients. N Z Med J. 1980;92:49–51.
27. Tartaglia N, Ayari N, Howell S, D'Epagnier C, Zeitler P. 48, XXYY, 48, XXXY and 49, XXXXY syndromes: not just variants of Klinefelter syndrome. Acta Paediatr. 2011;100:851–60.
28. Volkl TM, Langer T, Aigner T, Greess H, Beck JD, Rauch AM, Dorr HG. Klinefelter syndrome and mediastinal germ cell tumors. Am J Med Genet A. 2006;140:471–81.
29. Bertelloni S, Baroncelli GI, Battini R, Saggese G. Central precocious puberty in Klinefelter syndrome: a case report with longitudinal follow-up of growth pattern. Am J Med Genet. 1996;65:52–5.
30. Bahíllo-Curieses MP, Fournier-Carrera M, Morán-López J, Martínez-Sopena MJ. Klinefelter syndrome and short stature: an unusual combination. Endocrine. 2011;39:294–5.
31. Sabbaghian M, Meybodi AM, Rahimian M, Sadighi Gilani MA. Occurrence of 47,X,i(X)(q10),Y Klinefelter variant with hypogonadotropic hypogonadism. Fertil Steril. 2011;96:e115–7.

12 Testicular Dysgenesis Syndrome and Carcinoma In Situ Testis

Ewa Rajpert-De Meyts, Kristian Almstrup, and Niels E. Skakkebæk

It has been well established that some disorders of sexual development (DSDs) are associated with dysgenesis of the gonad, often resulting in a "streak" gonad [1]. If a testis has developed, the signs of dysgenesis may be very variable and include poorly developed seminiferous tubules with undifferentiated Sertoli cells, few or a total lack of germ cells, and a diminished number or complete depletion of Leydig cells. The fetal dysfunction of Leydig cells may result in the inhibition of descent of the testis (cryptorchidism) and a lack of normal development of the male genitals, including the penis (causing, e.g., hypospadias or micropenis). The severe DSD phenotypes are clearly caused by genetic defects. In the most severe case—so-called 46,XY pure gonadal dysgenesis—the genetically male individual is sex reversed. In milder cases—for example, 45,X/46,XY or some genetic mutations—the individual may have an "intersex" phenotype ranging from almost normal male to almost normal female. For a detailed description of DSD features in the testis, the reader is referred to Chap. 11. Individuals with severe gonadal dysgenesis have a significantly increased risk of germ cell tumor [1–5]. During preventive gonadectomies, usually performed in phenotypic females, a precursor lesion often is found: either a gonadoblastoma or a pattern similar to the adult carcinoma in situ (CIS) testis, although a combination of both also may appear in some cases [6–9].

During the past several decades, it has been noted that mild histological signs of dysgenesis also may be present in testes of normally virilized adult men with reproductive disorders such as infertility and oligozoospermia, a history of cryptorchidism, and testicular cancer, suggesting that these disorders are of fetal origin and, in reality, may be a mild form of DSD. The growing incidence trends of these disorders likely are etiologically linked to environmental or lifestyle factors, perhaps in combination with some genetic predisposition [10, 11]. This has been supported by convincing epidemiological evidence that germ cell cancer and some cases of poor semen quality may be of fetal origin, as the risk factors often are associated with the pre- or perinatal period [12–15]. Taken together, the evidence led us to propose that a number of male reproductive disorders, both those manifested at birth (cryptorchidism, hypospadias) and those manifested in young adulthood (testicular cancer, reduced spermatogenesis), are components of one syndrome: testicular dysgenesis syndrome (TDS) [10]. The histological signs, illustrated in this chapter, include poorly developed seminiferous tubules, often with immature Sertoli cells or a Sertoli cell—only (SCO) pattern; intratubular microcalcifications (microliths); large clumps (micronodules) of morphologically heterogeneous Leydig cells; and the presence of CIS,

E.R.-De Meyts, M.D., D.M.Sc. (✉)
K. Almstrup, M.Sc., Ph.D. • N.E. Skakkebæk, M.D., D.M.Sc.
Department of Growth and Reproduction, Copenhagen University, Rigshospitalet, Blegdamsvej 9, DK-2100 Copenhagen, Denmark
e-mail: erm@rh.dk

D. Ježek (ed.), *Atlas on the Human Testis*,
DOI 10.1007/978-1-4471-2763-5_12, © Springer-Verlag London 2013

the precursor of testicular germ cell tumors (TGCTs) [16–22]. The evidence that CIS testis is a dysgenetic feature is strongly supported by gene expression studies, which have documented the expression of embryonic genes in CIS cells, which now are commonly considered developmentally arrested and transformed gonocytes.

From a clinical point of view, it is important to mention that the recognition of CIS in a testicular biopsy sample is not always easy for the untrained eye and requires proper preparation of the tissue in a fixative, such as Bouin's or GR, or a buffered solution of formalin [23–25]. The recognition of CIS is greatly facilitated by immunohistochemical (IHC) staining for one of the CIS markers, such as placental-like alkaline phosphatase (PLAP) [26], OCT4 [27], AP2γ [28], M2A/D2-40/PDPN [29, 30], or LIN28 [31], as recently reviewed [32, 33]. In this chapter, several features of CIS are shown in detail, whereas Chap. 15 is dedicated to invasive testicular tumors. For details concerning the number of biopsies and the surgical aspects, the reader is referred to review articles [34, 35].

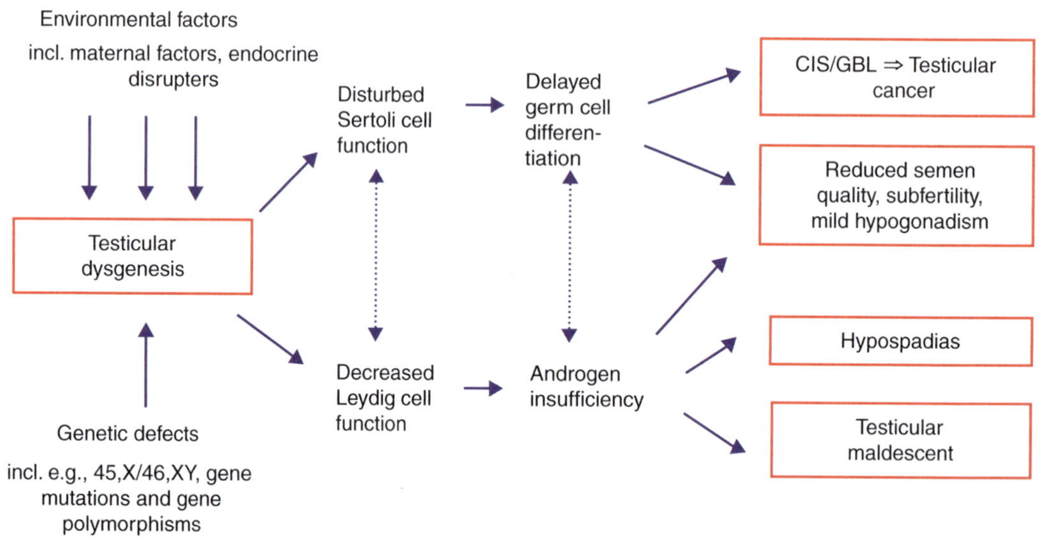

Fig. 12.1 Pathogenesis of reproductive disorders grouped within testicular dysgenesis syndrome (TDS) TDS. According to the TDS hypothesis, various environmental and lifestyle factors acting on a genetically predisposed individual during pre- or perinatal development (which are relatively common), or known genetic aberrations (e.g., mosaicism for sex chromosome aneuploidy; 45,X/46,XY; *SRY* mutations, which are rare), cause testicular dysgenesis, which is associated with an impairment of development and function of somatic cells (Sertoli and Leydig cells). This leads to a delay in germ cell differentiation and hormone deficiency, causing genital malformations (hypospadias or cryptorchidism) early in life or other disorders, such as testicular cancer and reduced semen quality, later in life. It is important to remember that not all cases of these phenotypes belong to TDS; only cases linked to pre- or perinatal testicular dysgenesis are part of TDS. *CIS* carcinoma in situ, *GBL* gonadoblastoma (Adapted from Skakkebæk et al. [11], with permission from Oxford University Press)

Subfertile/Infertile Patients with Signs of TDS (Without Malignancy)

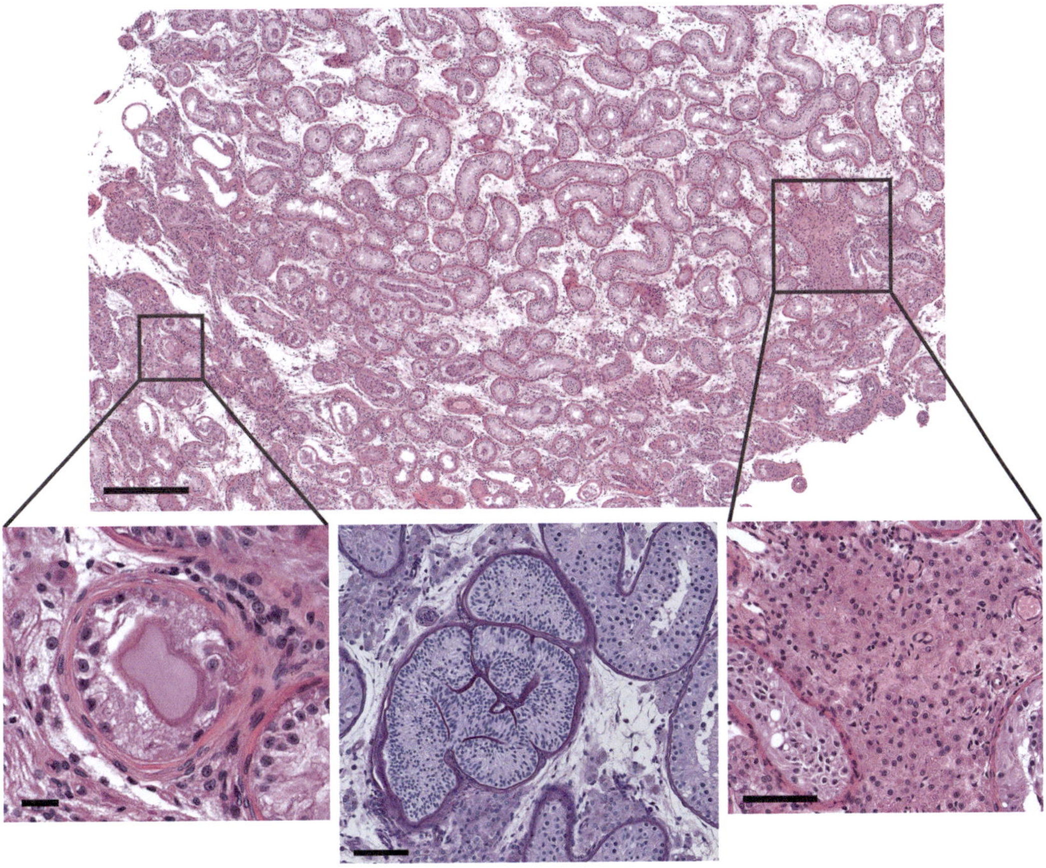

Fig. 12.2 Azoospermia and microlithiasis in a 23-year-old man with infertility. The biopsy specimen shows very poor spermatogenesis. Most tubules display the SCO pattern; the rest contain ongoing spermatogenesis but with spermatogenic maturation arrest. Only a few tubules contain abnormal round spermatids; no late spermatids were detected. The tissue shows several signs of TDS: severely undifferentiated and distorted tubules (shown on periodic acid-Schiff [PAS] staining in the middle bottom image), microliths (*left box*, shown in detail below), and micronodules of heterogeneous Leydig cells (*right box*, shown in detail below) (*Magnification bar*: *large image*, 500 μm; *lower left*, 25 μm; *lower middle and right*, 100 μm)

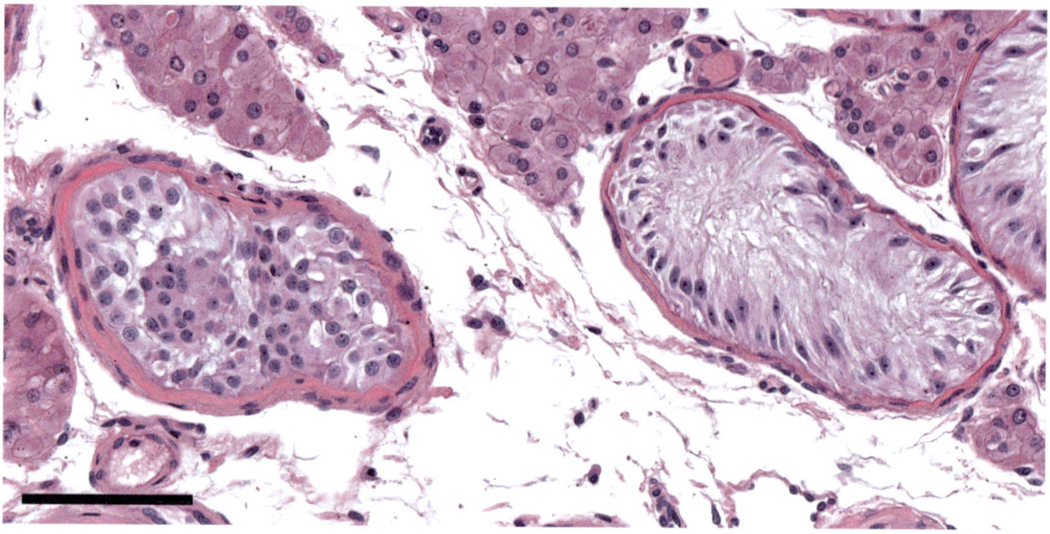

Fig. 12.3 Biopsy specimen from the left testis of a 22-year-old man diagnosed at birth with sex chromosome aneuploidy 45,X[8]/46,XdefY[2]. The patient was karyotyped because of very severe hypospadias. A bilateral biopsy performed in infancy revealed the presence of a rudimentary ovary and uterus on the right side and a small testis on the left. A rebiopsy in young adulthood showed a mixture of SCO pattern tubules and tubules with spermatogenesis arrested at the spermatocyte stage. A representative section from the rebiopsy is shown here. Two SCO tubules are visible: a tubule with only partly differentiated Sertoli cells *(left)* and a tubule with fully mature Sertoli cells. Also note the large clumps of Leydig cells in the upper part of the image (Hematoxylin and eosin [H&E] stain; *magnification bar*, 100 μm)

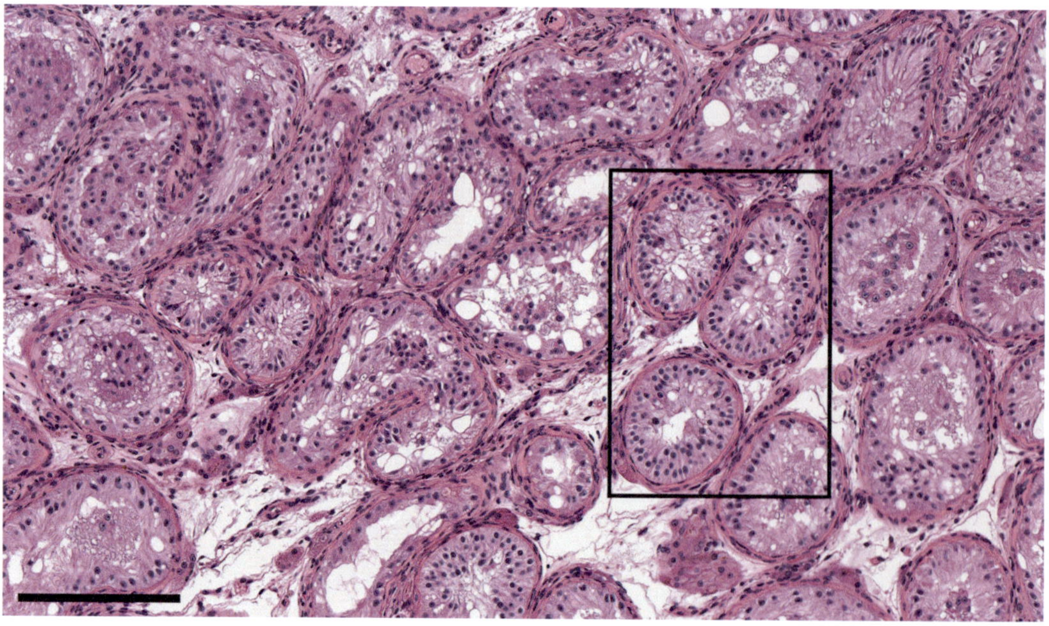

Fig. 12.4 Variably differentiated Sertoli cells in a 29-year-old infertile patient with azoospermia and a bilateral SCO pattern. Note the immature Sertoli cells in the tubules within the *box* (H&E stain; *magnification bar*, 200 μm)

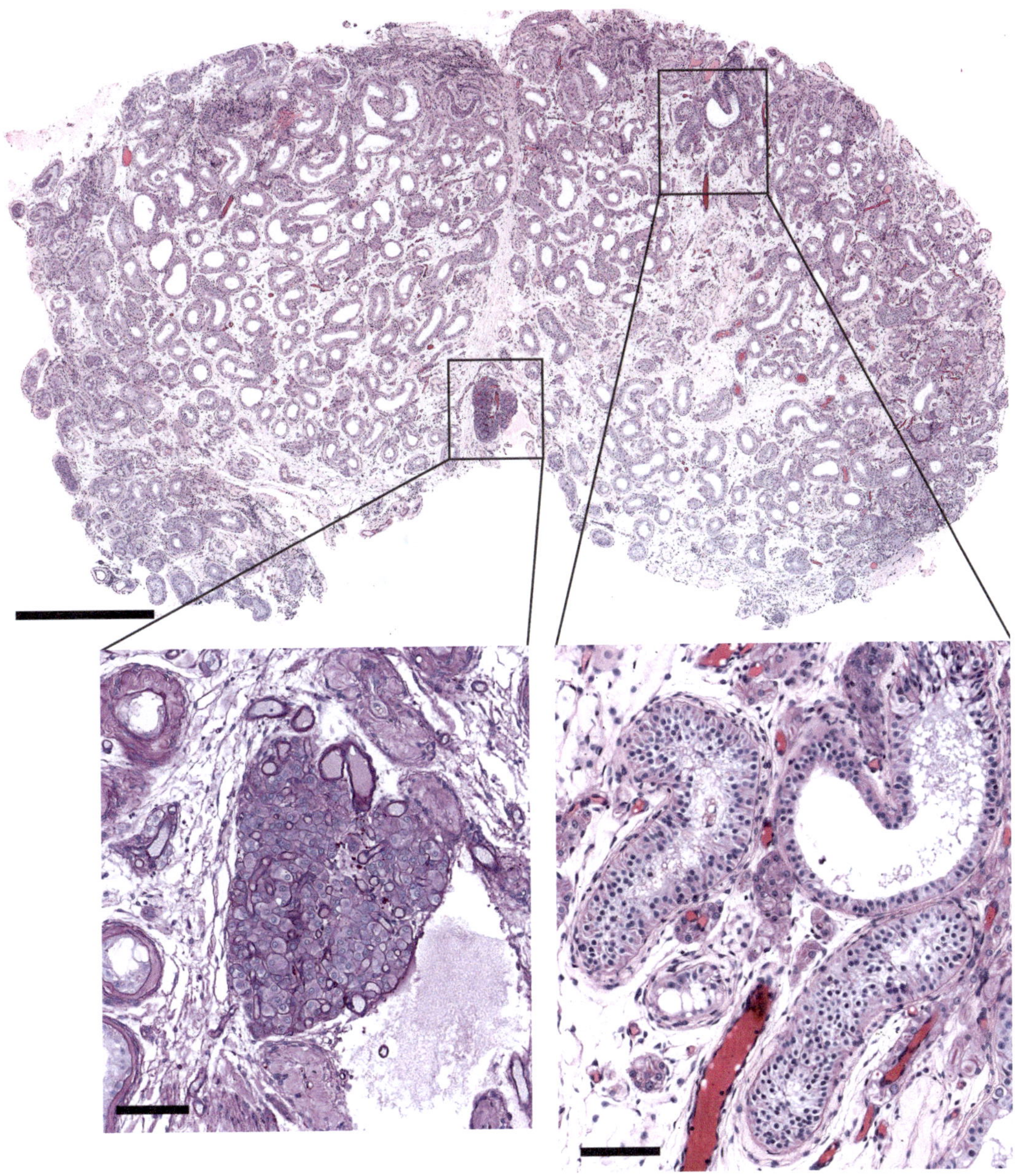

Fig. 12.5 Features of testicular dysgenesis in a biopsy sample from a 35-year-old infertile man. Note the poorly differentiated Sertoli cells (shown in detail in the lower right image) and the large micronodule of Leydig cells with numerous small vessels shown in the *magnified box* on the left (GR fixative, H&E stain. *Magnification bar*: *large image*, 1 mm; *zoomed images*, 100 μm)

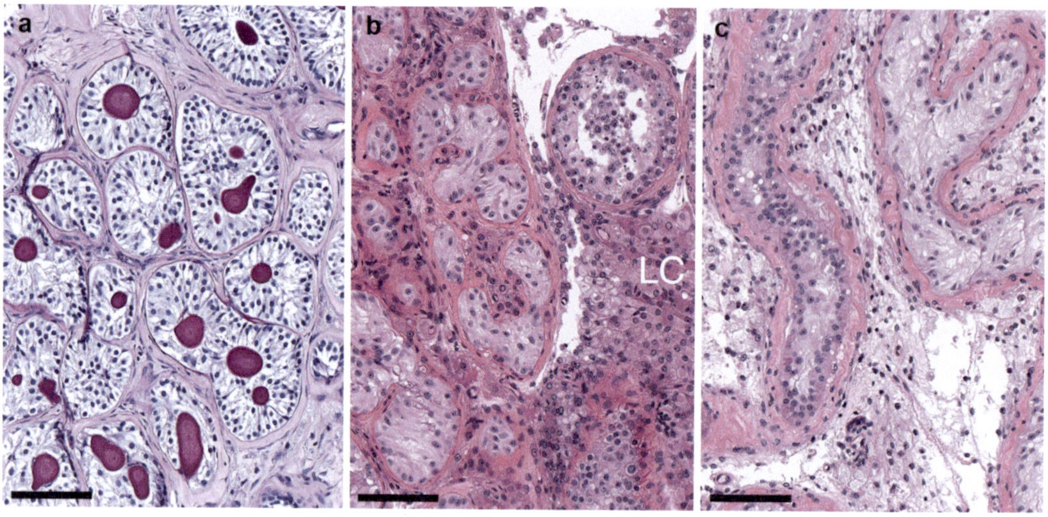

Fig. 12.6 Dysgenetic changes in three different infertile men with cryptorchidism. (**a**) Cluster of poorly developed tubules with an unusually large number of microcalcifications in the right-sided undescended testis of a 43-year-old man. The undifferentiated Sertoli cells are often positioned radially around the calcifications, and there are no Leydig cells within this cluster. The left-sided descended testis (not shown) displayed normal spermatogenesis (Formalin-fixed biopsy, PAS staining). (**b**) Biopsy sample taken from the right-sided undescended testis of a 35-year-old subfertile man. The specimen displays a cluster of small dysgenetic tubules located next to tubules with ongoing spermatogenesis. A large Leydig cell (*LC*) micronodule is visible on the *right*. The left (normally descended) testis displayed a similar pattern but with better maturation of Sertoli cells (not shown) (GR fixative, H&E staining). (**c**) Close-up view of poorly differentiated Sertoli cells in one tubule (on the *left*) side by side with a tubule containing normally matured Sertoli cells. This biopsy sample displays an SCO pattern and was taken from a patient with unilateral cryptorchidism and very poor semen quality. The contralateral descended testis (not shown) had a mixture of SCO and tubules with ongoing complete spermatogenesis (GR fixative, H&E staining. **a–c**, *Magnification bars*, 100 μm)

Features of Dysgenesis in Testes Contralateral to a TGCT

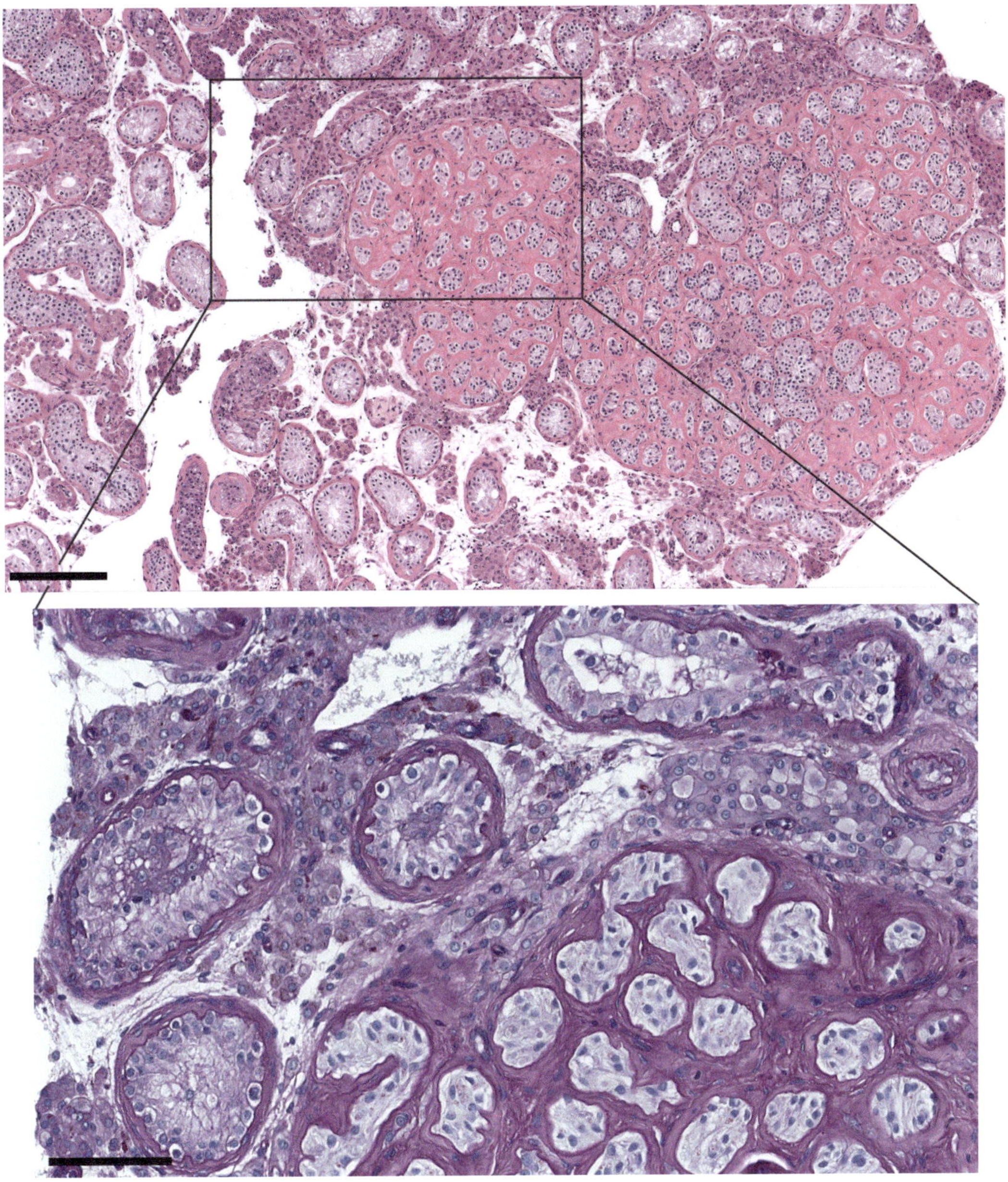

Fig. 12.7 Testis biopsy sample from a 31-year-old infertile man with a nonseminomatous TGCT in the contralateral testicle. The upper H&E-stained image shows a large area containing a cluster of distorted, undifferentiated SCO tubules, which are shown in detail below, side by side with more mature seminiferous tubules containing spermatogonia and Sertoli cells (PAS staining to emphasize the tubule membranes). Also note the presence of Leydig cell hyperplasia (GR fixative. *Magnification bar*: *upper image*, 250 μm; *lower image*, 100 μm)

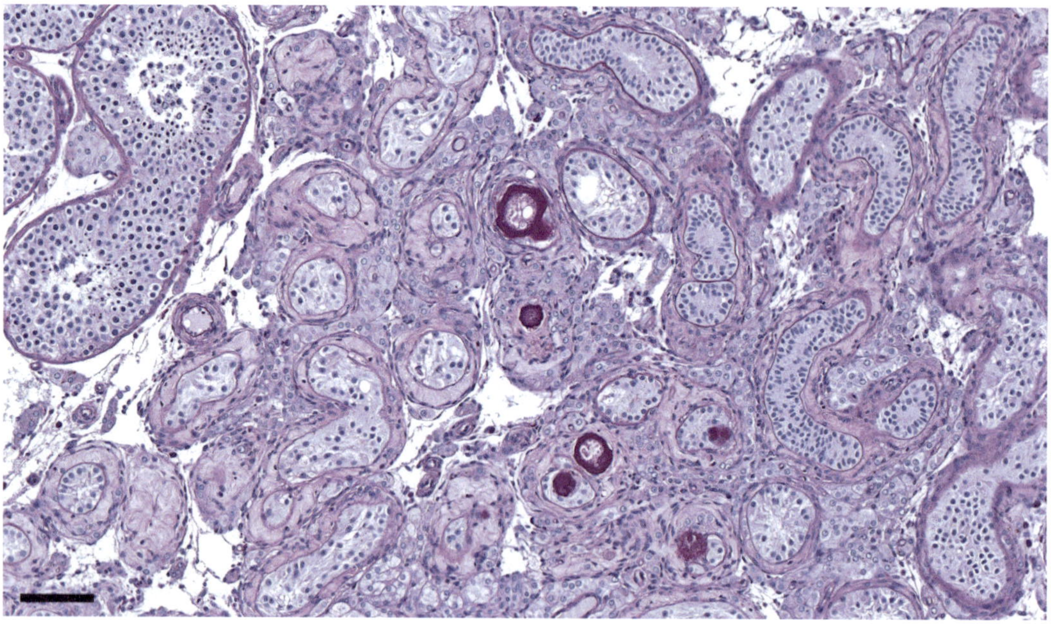

Fig. 12.8 Poorly developed tubules and microliths in a right-sided biopsy sample taken from a 33-year-old patient with a germ cell tumor in the contralateral testis. The sample was PAS stained to emphasize the microcalcifications (GR fixative; *magnification bar*, 100 μm)

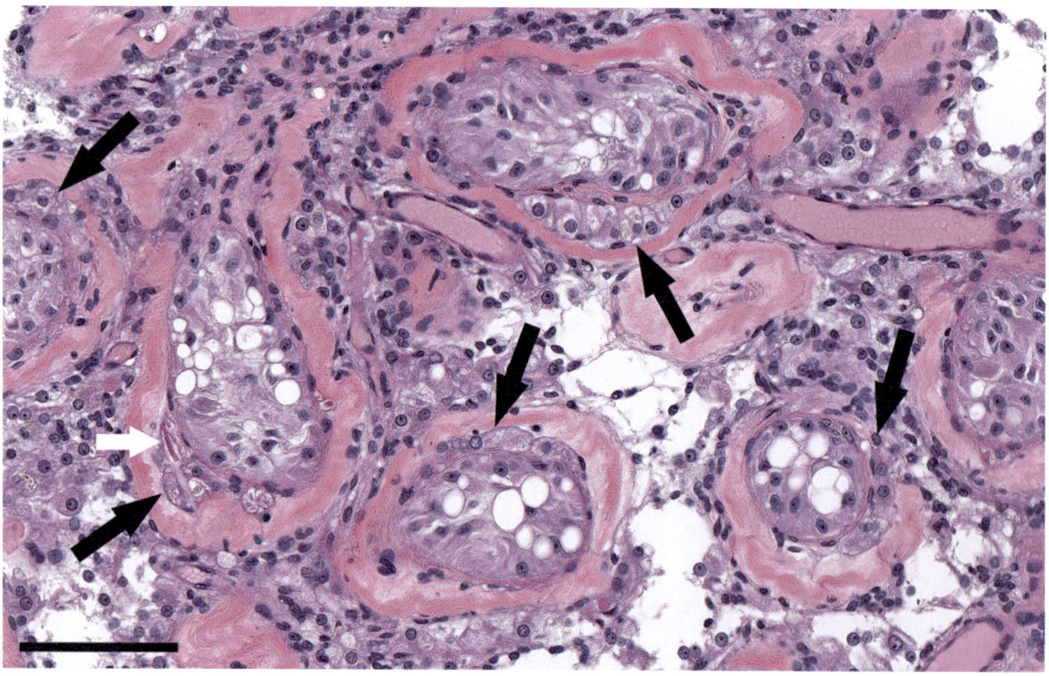

Fig. 12.9 Biopsy specimen from an atrophic testis in an infertile man. The sample shows an SCO pattern, pronounced membrane hyalinization, and hyperplasia of Leydig cells, which appear morphologically heterogeneous. Some Leydig cells, marked by *black arrows*, are trapped inside hyalinized tubular membranes. Note that in the tubule on the left, the Leydig cells contain Reinke crystalloids (indicated by a *white arrow*). A similar pattern with intratubular Leydig cells was described in rats with prenatally induced TDS by exposure to di(*n*-butyl) phthalate [36] (*Magnification bar*, 100 μm)

Testis with CIS in Adolescents with DSD

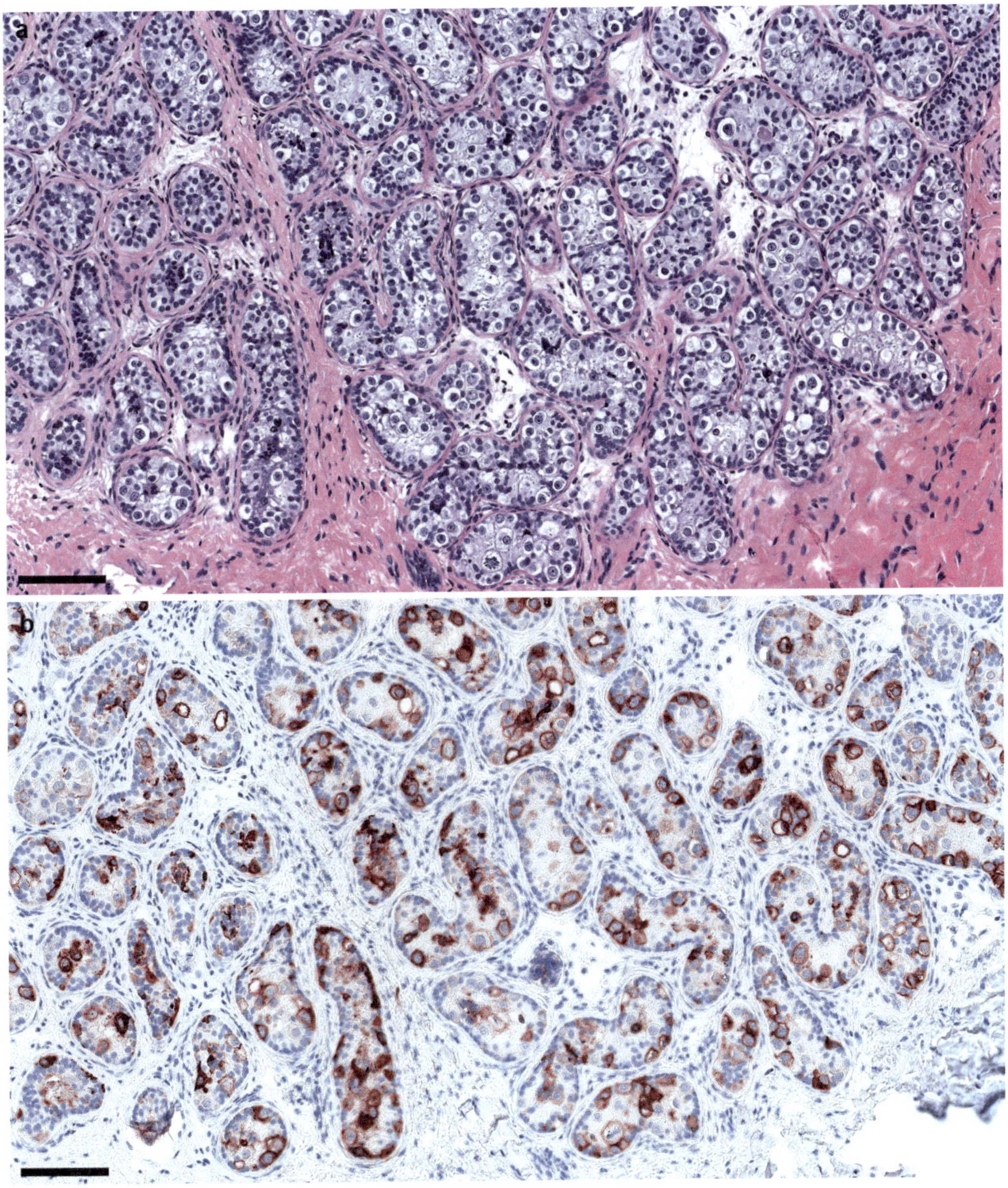

Fig. 12.10 An example of prepubertal CIS in a 13.5-year-old boy with mosaicism for sex chromosome aneuploidy (45,X /46,XY). The diagnosis was established at birth because of genital ambiguity. Gonadal exploration at 1.5 years of age revealed a rudimentary ovary (removed) on one side and a testis with a large number of immature gonocytes on the other. The testis size increased only marginally during childhood; the boy did not enter puberty, and a rebiopsy revealed the typical CIS pattern shown here (*Top*: H&E staining; *bottom*: immunostaining of the same area for placental-like alkaline phosphatase (PLAP), a CIS marker. *Magnification bar*, 100 μm)

Testis with CIS (but No Invasive Tumor) in Adult Men with Subfertility and/or a History of Cryptorchidism or Hypospadias

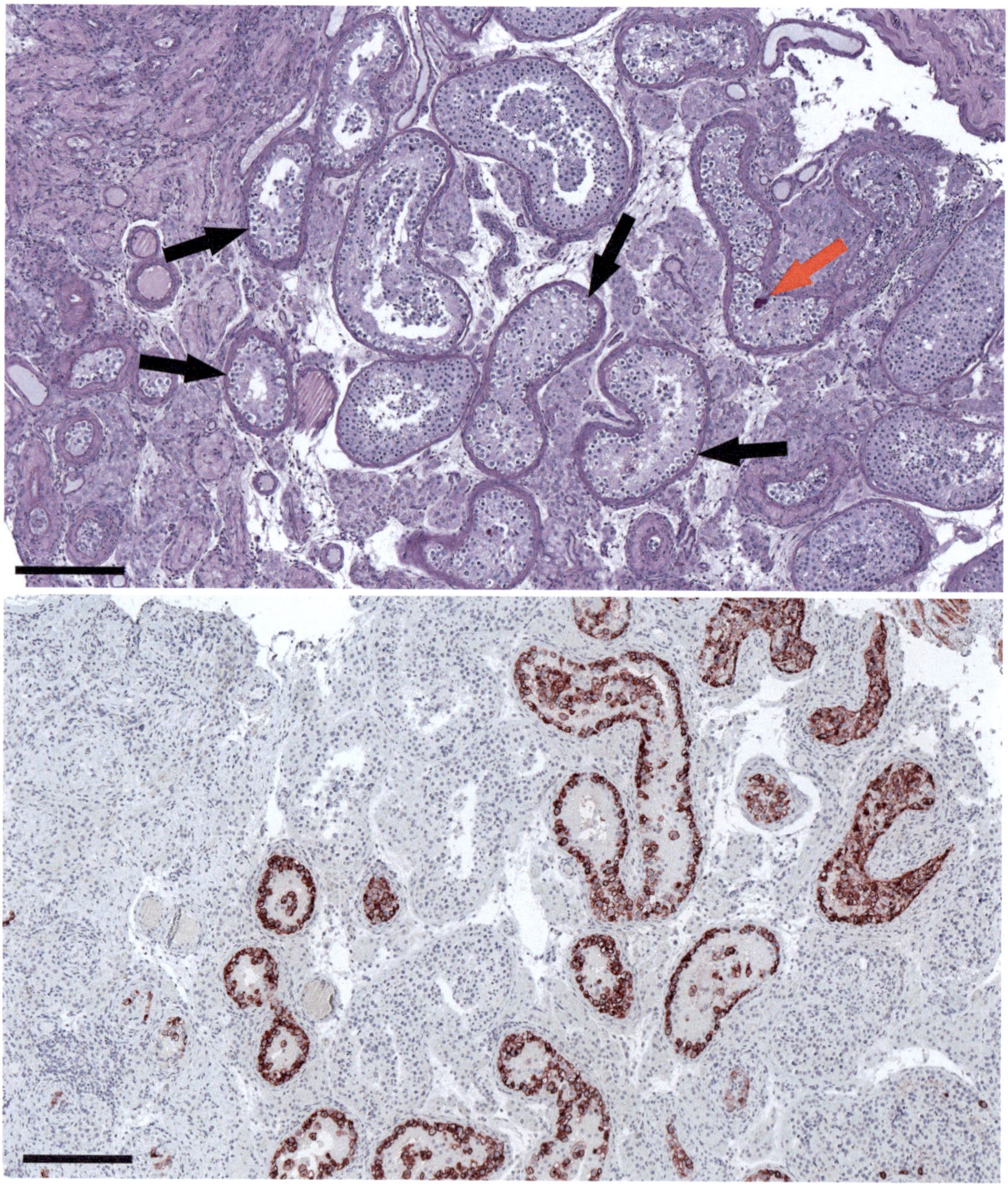

Fig. 12.11 Unilateral CIS in the left testis of a 37-year-old subfertile man with poor semen quality. The *upper panel* illustrates a heterogeneous pattern with a mixture of normal tubules, tubules with CIS (*black arrows*), atrophic hyalinized tubules (on the *left*), and microliths (one marked with a *red arrow*). In some tubules with CIS, a few normal spermatogonia also are present. The *lower panel* shows the same biopsy sample (but a different area), with CIS cells marked by immunostaining for PLAP (GR fixative, PAS stain; *magnification bar*, 250 μm)

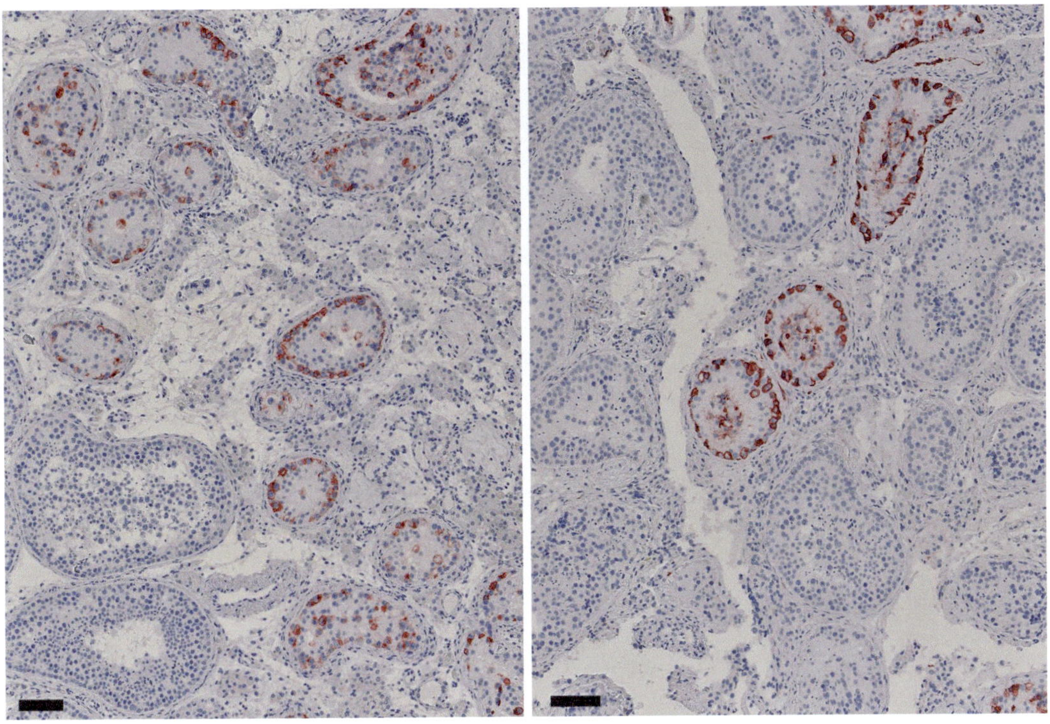

Fig. 12.12 Bilateral CIS in a 29-year-old infertile man with small testes and ultrasonic microlithiasis. A very similar pattern with widespread CIS was found in both biopsy samples, shown here side by side, with CIS cells marked by PLAP immunostaining (GR fixative; *magnification bar*, 100 μm)

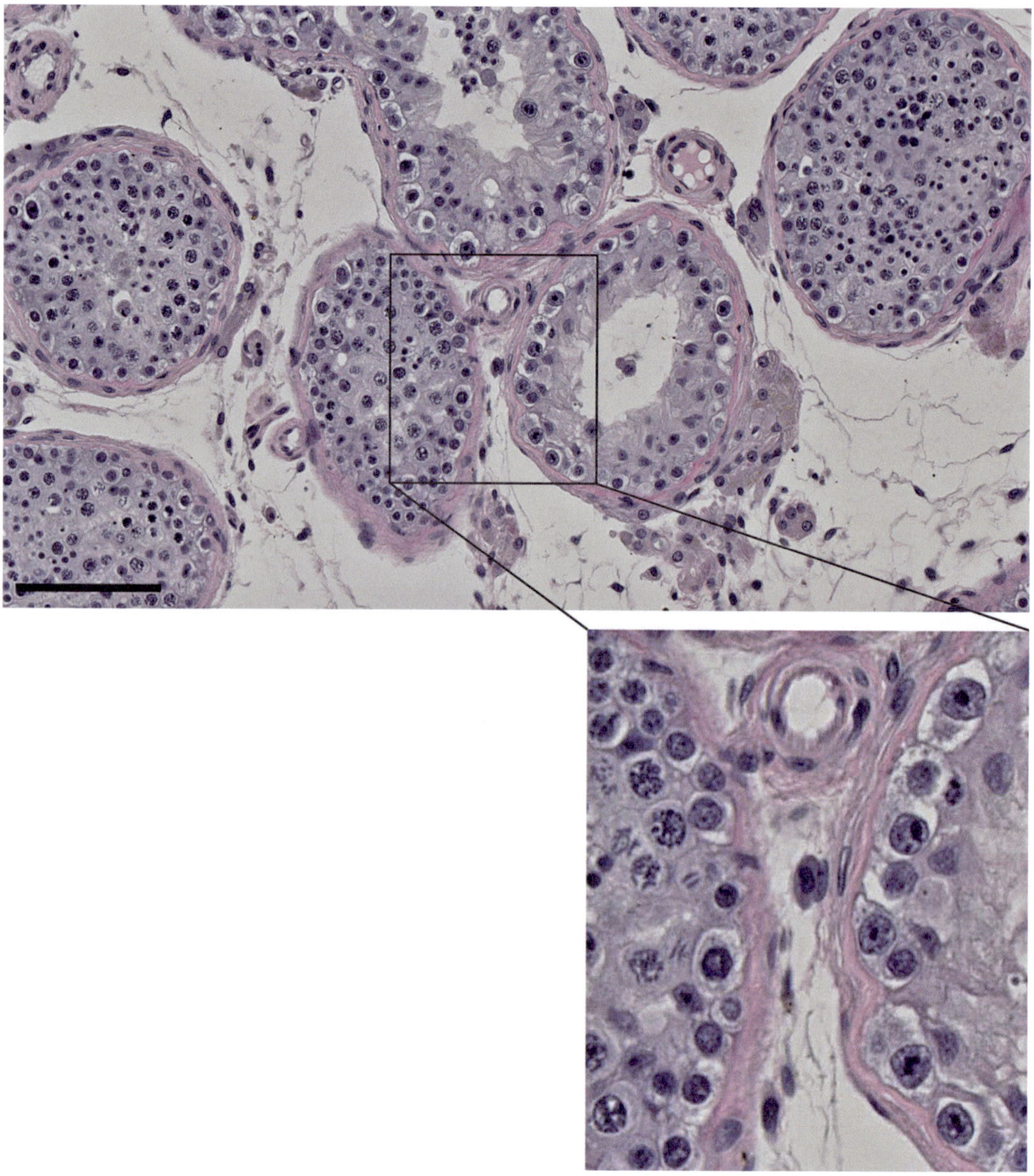

Fig. 12.13 Detailed view (*boxed area*) showing typical features of CIS cells compared with normal germ cells. Note the large nuclei of CIS cells with irregular clumps of chromatin and prominent nucleoli. This biopsy sample was taken from a 31-year-old infertile man with bilateral CIS (GR fixative, H&E stain; *magnification bar*, 100 μm)

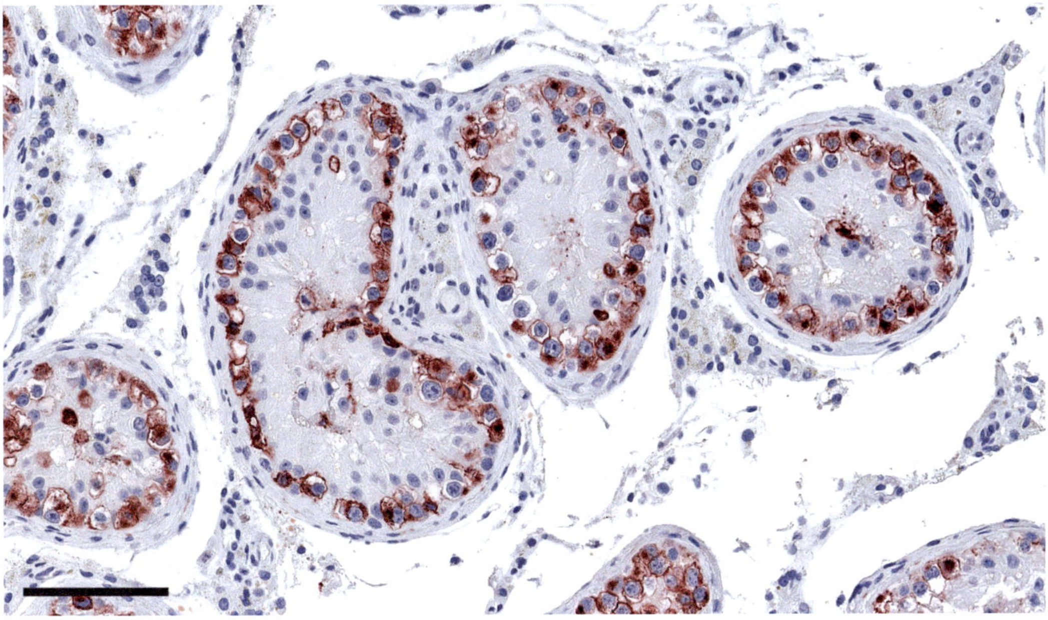

Fig. 12.14 Detailed view showing the typical CIS pattern, with only CIS cells (stained for PLAP) and Sertoli cells present inside the seminiferous tubules. This biopsy sample was taken from a 29-year-old infertile man diagnosed with bilateral widespread CIS testis (GR fixative, H&E stain; *magnification bar*, 100 μm)

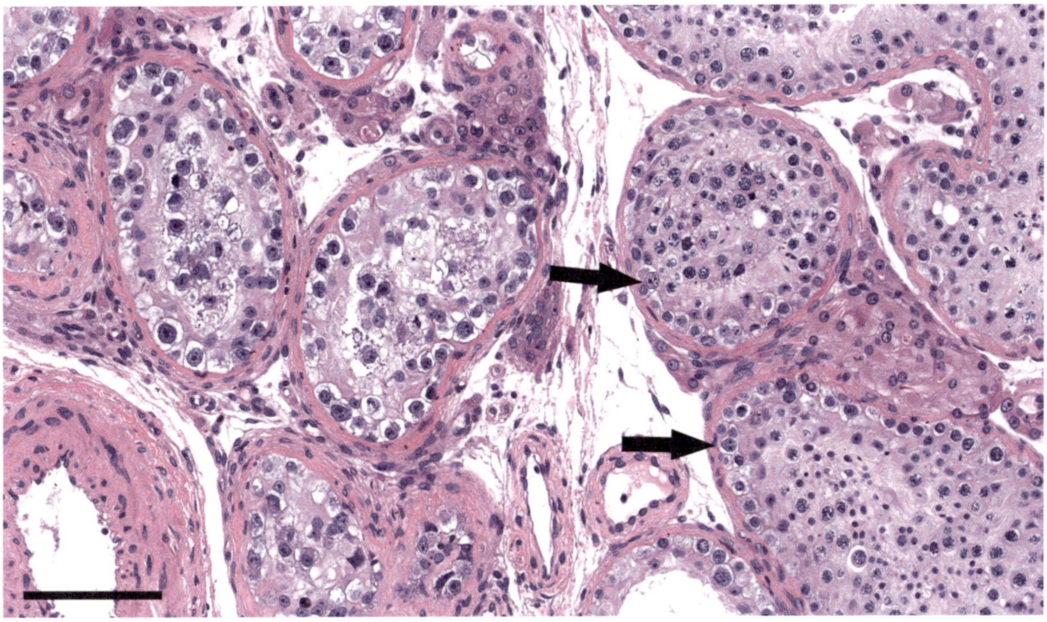

Fig. 12.15 Spread of CIS cells within tubules. The CIS was revealed in a bilateral biopsy specimen taken from a 37-year-old man with ultrasonic microlithiasis and an undescended right testicle. Please note the CIS cells inside the tubules with ongoing spermatogenesis (marked with *arrows*) (GR fixative, H&E stain; *magnification bar*, 100 μm)

Other Features of CIS Testis Contralateral or Adjacent to a Tumor and Technical Aspects of Diagnostic Biopsies

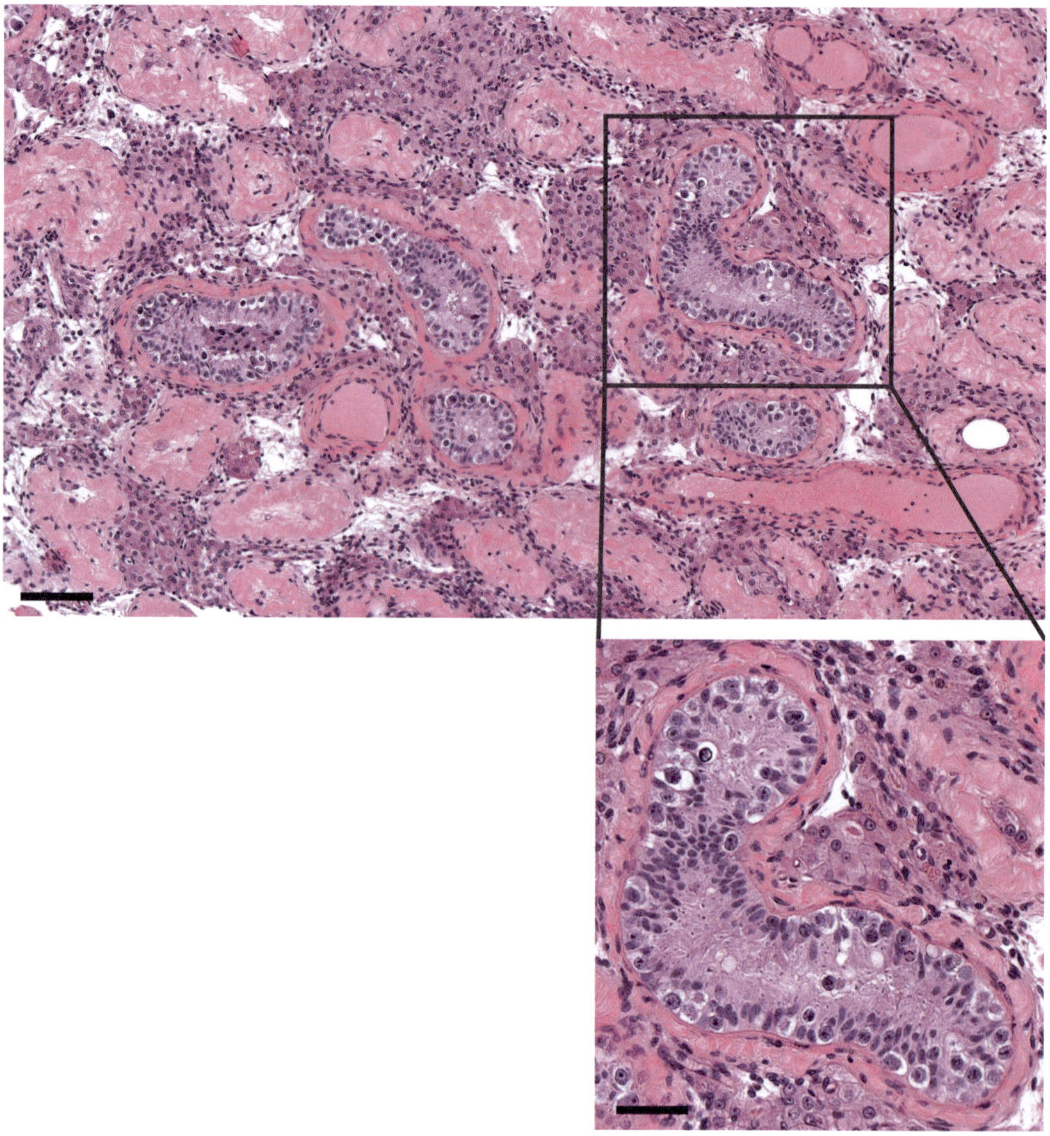

Fig. 12.16 CIS in poorly developed tubules detected in a contralateral biopsy taken from a 34-year-old patient who underwent orchiectomy because of a nonseminoma in the right testicle. The testis also shows signs of atrophy, with many hyalinized tubules. Note the immature, somewhat hyperplastic Sertoli cells on the higher magnification image (GR fixative, H&E stain. *Magnification bar*: *top*, 100 μm; *bottom*, 50 μm)

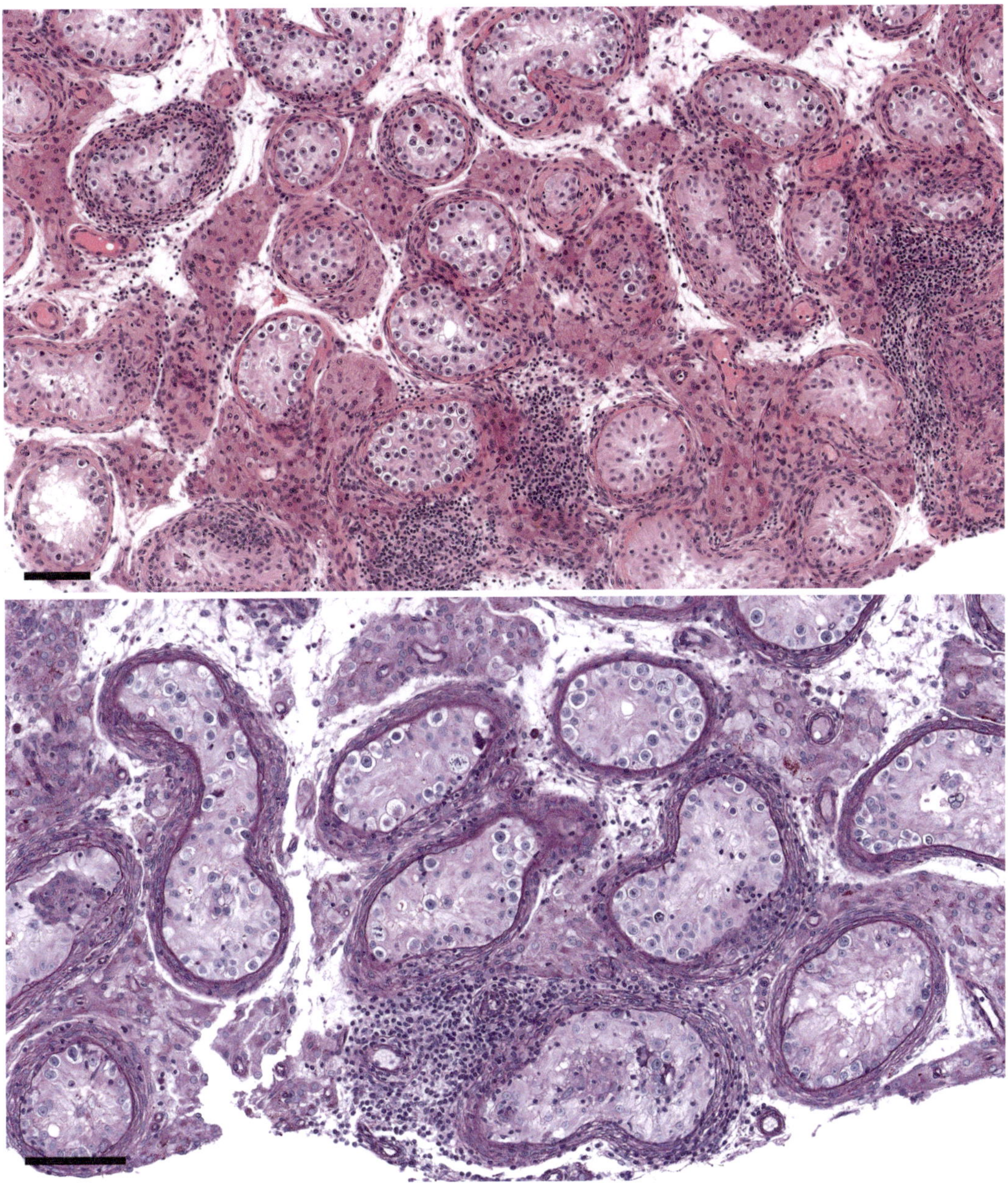

Fig. 12.17 CIS with prominent lymphocytic infiltration. CIS is present in a testis contralateral to a seminoma. An overview is shown in the *upper panel* (H&E stain). The *bottom image* (PAS stain) shows a closer view of inflammatory cells invading the tubules with CIS (*Magnification bar*, 100 μm)

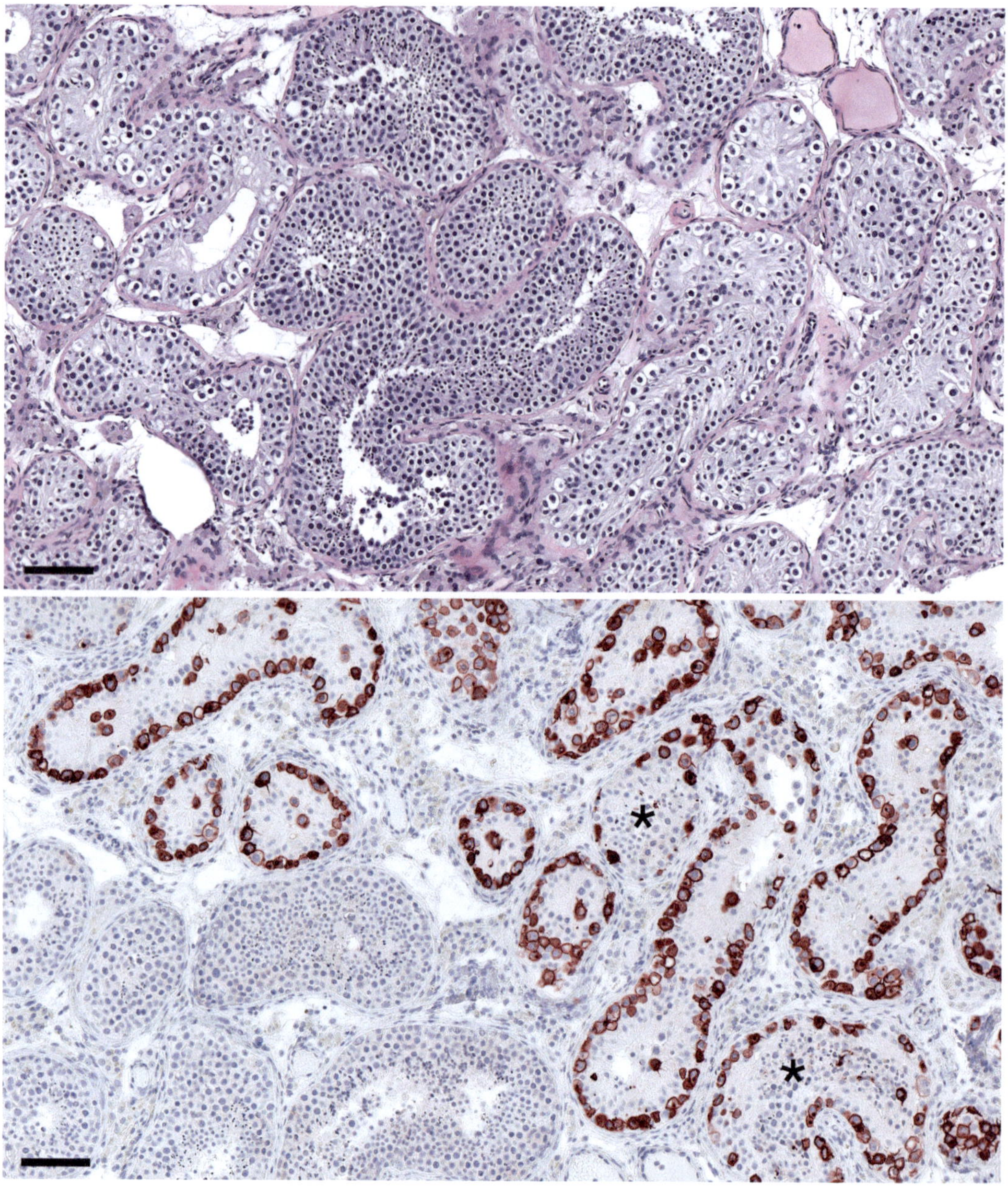

Fig. 12.18 CIS in a very-good-quality formalin-fixed contralateral biopsy sample. The H&E-stained *upper picture* shows an overview of the pattern, whereas the *bottom picture* shows the same biopsy sample immunostained for PLAP. Note the pagetoid spread of CIS cells inside the tubules with ongoing spermatogenesis (*asterisks*) (*Magnification bar*, 100 μm)

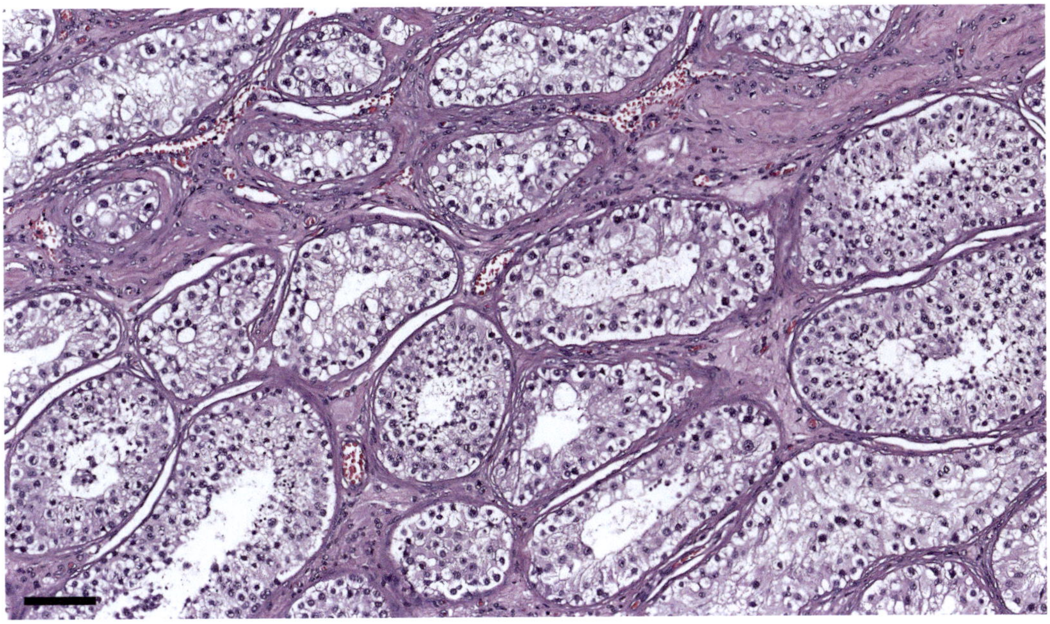

Fig. 12.19 Appearance of CIS in poor-quality formalin-fixed and H&E-stained tissue. This biopsy sample is from a 39-year-old patient diagnosed with CIS and a microinvasive seminoma. In such poorly fixed tissues, it is difficult to identify CIS, even for trained pathologists (*Magnification bar*, 100 μm)

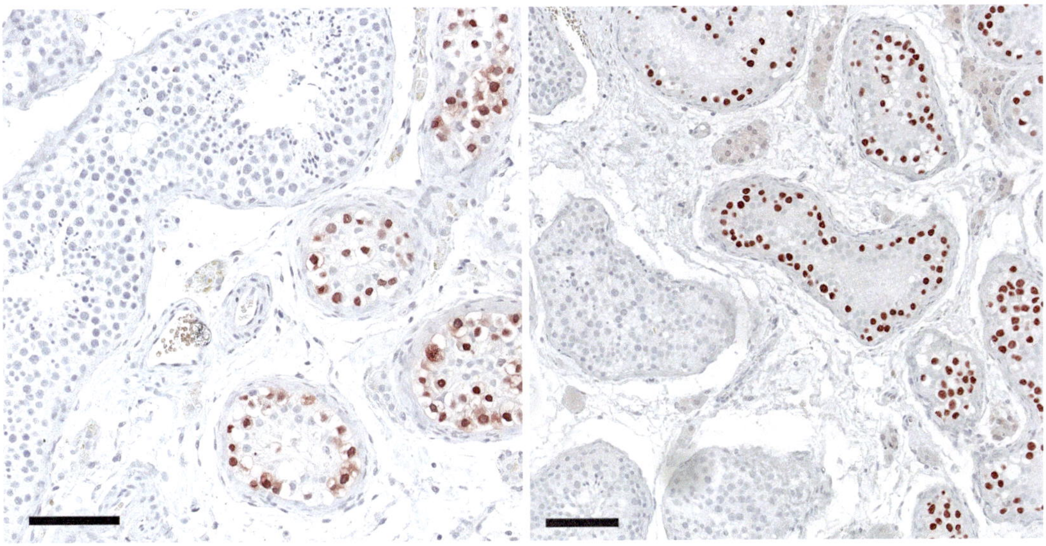

Fig. 12.20 Two transcription factors commonly used as CIS/tumor markers for IHC staining in formalin-fixed tissues. In both cases shown here, the CIS tubules were adjacent to an overt TGCT. *Left*: POU5F1 (also known as OCT4 or OCT3/4) [27]. *Right*: AP2γ (encoded by *TFAP2C*) [28] (*Magnification bar*, 100 μm)

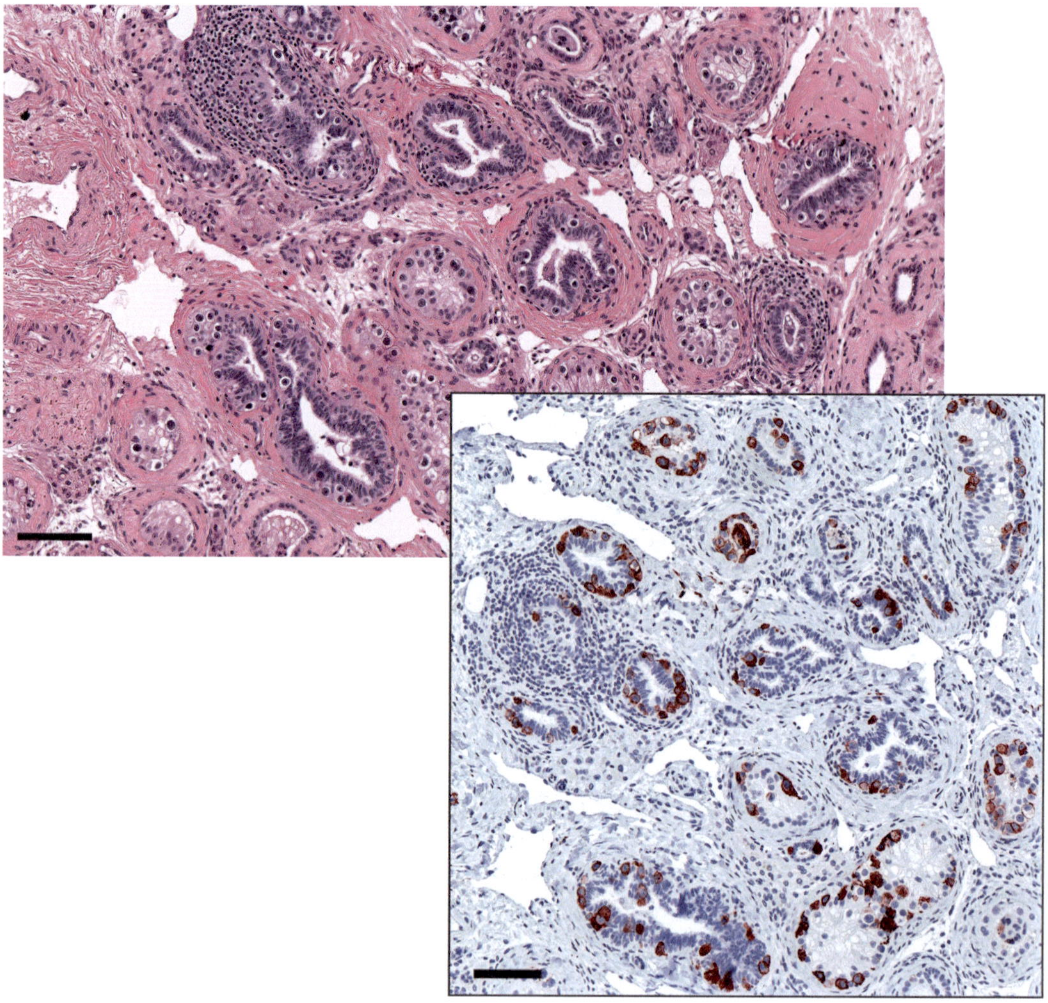

Fig. 12.21 CIS cells invading the rete testis in a tissue sample adjacent to an invasive tumor. A pronounced inflammatory response is seen around the rete tubules. In one of the serial sections of the same tissue sample, shown in the *lower image*, the CIS cells are visualized by IHC staining for PLAP (*Magnification bars*, 100 μm)

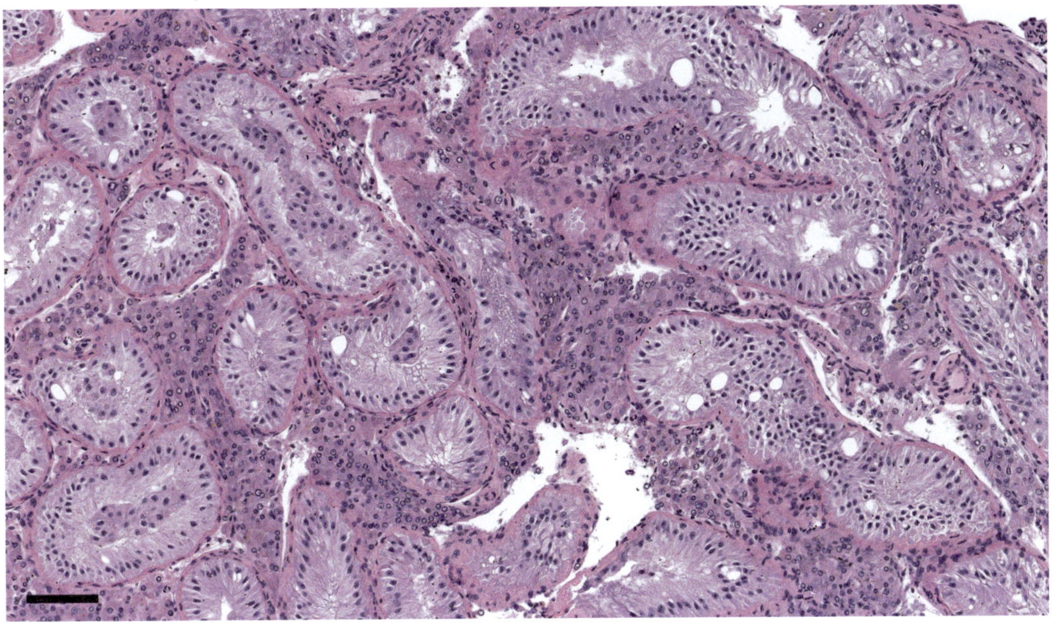

Fig. 12.22 Testis with SCO after treatment of CIS by irradiation. Sertoli cells often have large vacuoles in the cytoplasm. Leydig cell hyperplasia with marked morphologic heterogeneity often is present (GR fixative, H&E stain; *magnification bar*, 100 μm)

References

1. Scully RE. Neoplasia associated with anomalous sexual development and abnormal sex chromosomes. Pediatr Adolesc Endocrinol. 1981;8:203–17.
2. Savage MO, Lowe DG. Gonadal neoplasia and abnormal sexual differentiation. Clin Endocrinol. 1990;32:519–33.
3. Słowikowska-Hilczer J, Romer TE, Kula K. Neoplastic potential of germ cells in relation to disturbances of gonadal organogenesis and changes in karyotype. J Androl. 2003;24:270–8.
4. Cools M, Drop SL, Wolffenbuttel KP, Oosterhuis JW, Looijenga LH. Germ cell tumors in the intersex gonad: old paths, new directions, moving frontiers. Endocr Rev. 2006;27:468–84.
5. Pleskacova J, Hersmus R, Oosterhuis JW, Setyawati BA, Faradz SM, Cools M, et al. Tumor risk in disorders of sex development. Sex Dev. 2010;4:259–69.
6. Scully RE. Gonadoblastoma. A review of 74 cases. Cancer. 1970;25:1340–56.
7. Müller J, Skakkebæk NE. Testicular carcinoma in situ in children with androgen insensitivity (testicular feminisation) syndrome. Br Med J. 1984;288:1419–20.
8. Müller J, Skakkebæk NE, Ritzén EM, Plöen L, Petersen KE. Carcinoma in situ of the testis in children with 45, X/46, XY gonadal dysgenesis. J Pediatr. 1985;106:431–6.
9. Hersmus R, Kalfa N, de Leeuw B, Stoop H, Oosterhuis JW, de Krijger R, et al. FOXL2 and SOX9 as parameters of female and male gonadal differentiation in patients with various forms of disorders of sex development (DSD). J Pathol. 2008;215:31–8.
10. Skakkebæk NE, Rajpert-De Meyts E, Jørgensen N, Carlsen E, Petersen PM, Giwercman A, et al. Germ cell cancer and disorders of spermatogenesis: an environmental connection? APMIS. 1998;106:3–12.
11. Skakkebæk NE, Rajpert-De Meyts E, Main KM. Testicular dysgenesis syndrome: an increasingly common developmental disorder with environmental aspects. Hum Reprod. 2001;16:972–8.
12. Depue RH, Pike MC, Henderson BE. Estrogen exposure during gestation and risk of testicular cancer. J Natl Cancer Inst. 1983;71:1151–5.
13. Weir HK, Marrett LD, Kreiger N, Darlington GA, Sugar L. Pre-natal and peri-natal exposures and risk of testicular germ-cell cancer. Int J Cancer. 2000;87:438–43.
14. Biggs ML, Baer A, Critchlow CW. Maternal, delivery, and perinatal characteristics associated with cryptorchidism: a population-based case–control study among births in Washington State. Epidemiology. 2002;13:197–204.
15. Cook MB, Akre O, Forman D, Madigan MP, Richiardi L, McGlynn KA. A systematic review and meta-analysis of perinatal variables in relation to the risk of testicular cancer – experiences of the son. Int J Epidemiol. 2010;39:1605–18.

16. Sohval AR. Testicular dysgenesis as an etiologic factor in cryptorchidism. J Urol. 1954;72:693–702.
17. Skakkebæk NE. Possible carcinoma-in-situ of the testis. Lancet. 1972;ii:516–7.
18. Kang J-L, Rajpert-De Meyts E, Giwercman A, Skakkebæk NE. The association of testicular carcinoma in situ with intratubular microcalcifications. J Urol Pathol. 1994;2:235–42.
19. Hoei-Hansen CE, Holm M, Rajpert-De Meyts E, Skakkebæk NE. Histological evidence of testicular dysgenesis in contralateral biopsies of 218 patients with testicular germ cell cancer. J Pathol. 2003; 200:370–4.
20. Holm M, Rajpert-De Meyts E, Andersson A-M, Skakkebæk NE. Leydig cell micronodules are a common finding in testicular biopsies from men with impaired spermatogenesis and are associated with decreased testosterone/LH ratio. J Pathol. 2003;199: 378–86.
21. Nistal M, Gonzalez-Peramato P, Regadera J, Serrano A, Tarin V, De Miguel MP. Primary testicular lesions are associated with testicular germ cell tumors of adult men. Am J Surg Pathol. 2006;30:1260–8.
22. Guminska A, Oszukowska E, Kuzanski W, Sosnowski M, Wolski JK, Walczak-Jedrzejowska R, et al. Less advanced testicular dysgenesis is associated by a higher prevalence of germ cell neoplasia. Int J Androl. 2010;33:e153–62.
23. Almstrup K, Hoei-Hansen CE, Wirkner U, Blake J, Schwager C, Ansorge W, et al. Embryonic stem cell-like features of testicular carcinoma in situ revealed by genome-wide gene expression profiling. Cancer Res. 2004;64:4736–43.
24. Rajpert-De Meyts E. Developmental model for the pathogenesis of testicular carcinoma in situ: environmental and genetic aspects. Hum Reprod Update. 2006;12:303–23.
25. Sonne SB, Almstrup K, Dalgaard M, Juncker AS, Edsgard D, Ruban L, et al. Analysis of gene expression profiles of microdissected cell populations indicates that testicular carcinoma in situ is an arrested gonocyte. Cancer Res. 2009;69:5241–50.
26. Burke AP, Mostofi FK. Intratubular malignant germ cells in testicular biopsies: clinical course and identification by staining for placental alkaline phosphatase. Mod Pathol. 1988;1:475–9.
27. de Jong J, Stoop H, Dohle GR, Bangma CH, Kliffen M, van Esser JW, et al. Diagnostic value of OCT3/4 for pre-invasive and invasive testicular germ cell tumours. J Pathol. 2005;206:242–9.
28. Hoei-Hansen CE, Nielsen JE, Almstrup K, Sonne SB, Graem N, Skakkebaek NE, et al. Transcription factor AP-2gamma is a developmentally regulated marker of testicular carcinoma in situ and germ cell tumors. Clin Cancer Res. 2004;10:8521–30.
29. Giwercman A, Marks A, Bailey D, Baumal R, Skakkebaek NE. A monoclonal antibody as a marker for carcinoma in situ germ cells of the human adult testis. APMIS. 1988;96:667–70.
30. Sonne SB, Herlihy AS, Hoei-Hansen CE, Nielsen JE, Almstrup K, Skakkebaek NE, et al. Identity of M2A (D2-40) antigen and gp36 (Aggrus, T1A-2, podoplanin) in human developing testis, testicular carcinoma in situ and germ-cell tumours. Virchows Arch. 2006;449:200–6.
31. West JA, Viswanathan SR, Yabuuchi A, Cunniff K, Takeuchi A, Park IH, et al. A role for LIN28 in primordial germ-cell development and germ-cell malignancy. Nature. 2009;460:909–13.
32. Emerson RE, Ulbright TM. Intratubular germ cell neoplasia of the testis and its associated cancers: the use of novel biomarkers. Pathology. 2010;42:344–55.
33. Oosterhuis JW, Stoop H, Dohle G, Boellaard W, van Casteren N, Wolffenbuttel K, Looijenga LH. A pathologist's view on the testis biopsy. Int J Androl. 2011;34:e14–9; discussion e20.
34. McLachlan R, Rajpert-De Meyts E, Hoei-Hansen C, De Kretser DM, Skakkebaek NE. Histological evaluation of the human testis: approaches to optimizing the clinical value of assessment: mini review. Hum Reprod. 2007;22:2–16.
35. Dieckmann KP, Kulejewski M, Heinemann V, Loy V. Testicular biopsy for early cancer detection – objectives, technique and controversies. Int J Androl. 2011;34:e7–13.
36. Mahood IK, Hallmark N, McKinnell C, Walker M, Fisher JS, Sharpe RM. Abnormal Leydig cell aggregation in the fetal testis of rats exposed to di (n-butyl) phthalate and its possible role in testicular dysgenesis. Endocrinology. 2005;146:613–23.

13 Tumors of the Testis and Epididymis

Ivan Damjanov and Gregor Mikuz

This chapter illustrates the histopathology of the most important tumors involving the testis and epididymis. As guidelines, the most current classification adopted by the experts of the World Health Organization [1] and the recent textbook *Clinical Pathology of Urologic Tumors* [2] is used. To conform to the main goals of this atlas, the text is limited to descriptions of the figures; for additional reading, please refer to Table 13.1 and the references.

I. Damjanov, M.D., Ph.D. (✉)
Department of Pathology, The University of Kansas School of Medicine,
Kansas City, KS, USA
e-mail: idamjano@kumc.edu

G. Mikuz, M.D., Ph.D.
Institute of Pathology, Medical University Innsbruck,
Muellerstrasse 44, 6020, Innsbruck, Austria
e-mail: Gregor.Mikuz@i-med.ac.at

D. Ježek (ed.), *Atlas on the Human Testis*,
DOI 10.1007/978-1-4471-2763-5_13, © Springer-Verlag London 2013

Table 13.1 WHO 2004 classification of testicular and paratesticular tumors*

Germ cell tumors	ICD-O
Intratubular germ cell neoplasia, unclassified	9064/2
Other types	
Tumors of one histologic typic (pure forms)	
Seminoma	9061/3
Seminoma with syncytiotrophoblastic cells	
Spermatocytic seminoma	9063/3
Spermatocytic seminoma with sarcoma	
Embryonal carcinoma	9070/3
Yolk sac tumor	9071/3
Trophoblastic tumors	
Choriocarcinoma	9100/3
Trophoblastic neoplasms other than choriocarcinoma	
Monophasic choriocarcinoma	
Placental site trophoblastic tumor	9104/3
Teratoma	9080/3
Dermoid cyst	9084/0
Monodermal teratoma	
Teratoma with somatic type malignancies	9084/3
Tumors of more than one histologic type (mixed forms)	
Mixed embryonal carcinoma and teratoma	9081/3
Mixed teratoma and seminoma	9085/3
Choriocarcinoma and teratoma/embryonal carcinoma	9101/3
Others	
Sex cord/gonadal stromal tumors	
Pure forms	
Leydig cell tumor	8650/1
Malignant Leydig cell tumor	8650/3
Sertoli cell tumor	8640/1
Sertoli cell tumor lipid rich variant	8641/1
Sclerosing Sertoli cell tumor	
Large cell calcifying Sertoli cell tumor	8642/1
Malignant Sertoli cell tumor	8640/3
Granulosa cell tumor	8620/1
Adult type granulosa cell tumor	8620/1
Juvenile type granulosa cell tumor	8621/1
Tumors of the thecoma/fibroma group	
Thecoma	8600/0
Fibroma	8810/0
Sex cord/gonadal stromal tumor incompletely differentiated	8591/1
Sex cord/gonadal stromal tumor, mixed forms	8592/1
Malignant sex cord/gonadal stromal tumor	8590/3
Tumors containing both germ cell and sex cord/gonadal stromal element	
Gonadoblastoma	9073/1
Germ cell-sex cord/gonadal stromal tumor, unclassified	
Miscellaneous tumors of the testis	
Carcinoid tumor	8240/3
Tumors of ovarian epithelial types	
Serous tumor of borderline malignancy	8442/1
Serous carcinoma	8441/3
Well-differentiated endometrioid carcinoma	8380/3
Mucinous cystadenoma	8470/0
Mucinous cystadenocarcinoma	8470/3
Brenner tumor	8960/3
Nephroblastoma	8680/1
Paraganglioma	
Hematopoietic tumors	9100/3
Tumors of collecting ducts and rete	
Adenoma	8140/1
Carcinoma	8140/3
Tumors of paratesticular structures	
Adenomatoid tumor	9054/0
Malignant mesothelioma	9050/3
Benign mesothelioma	
Well-differentiated papillary mesothelioma	9052/0
Cystic mesothelioma	9055/0
Adenocarcinoma of the epididymis	8140/3
Papillary cystadenoma of the epididymis	8450/0
Melanotic neuroectodermal tumor	9363/0
Desmoplastic small round cell tumor	8806/3
Mesenchymal tumors of the spermatic cord and testicular adnexae	
Secondary tumors of the testis (metastases)	

Reprinted from Eble et al. [1], with permission

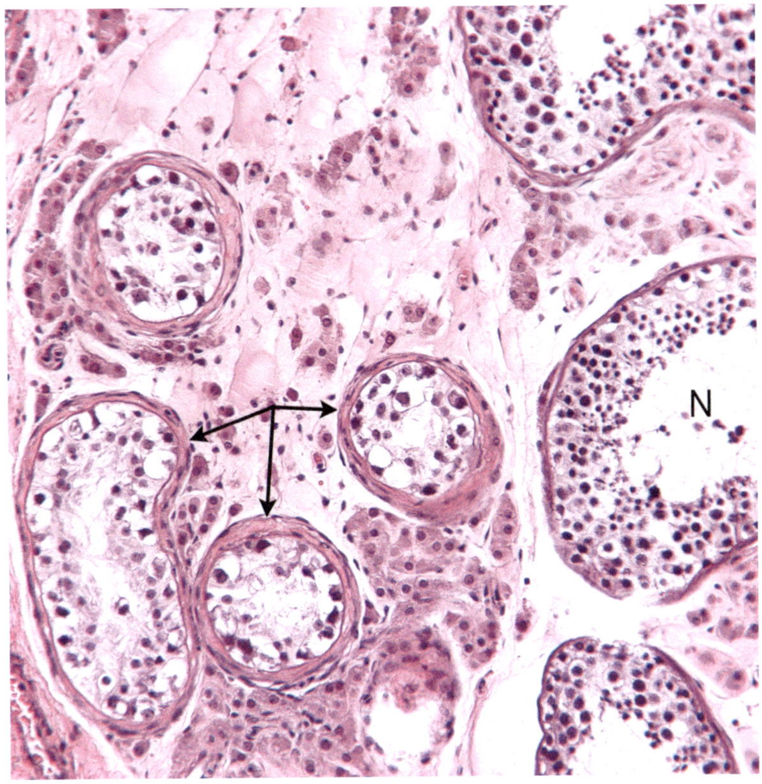

Fig. 13.1 Intratubular germ cell neoplasia, unclassified (IGCNU). The seminiferous tubules containing IGCNU (*arrows*) are adjacent to tubules showing normal spermatogenesis (*N*)

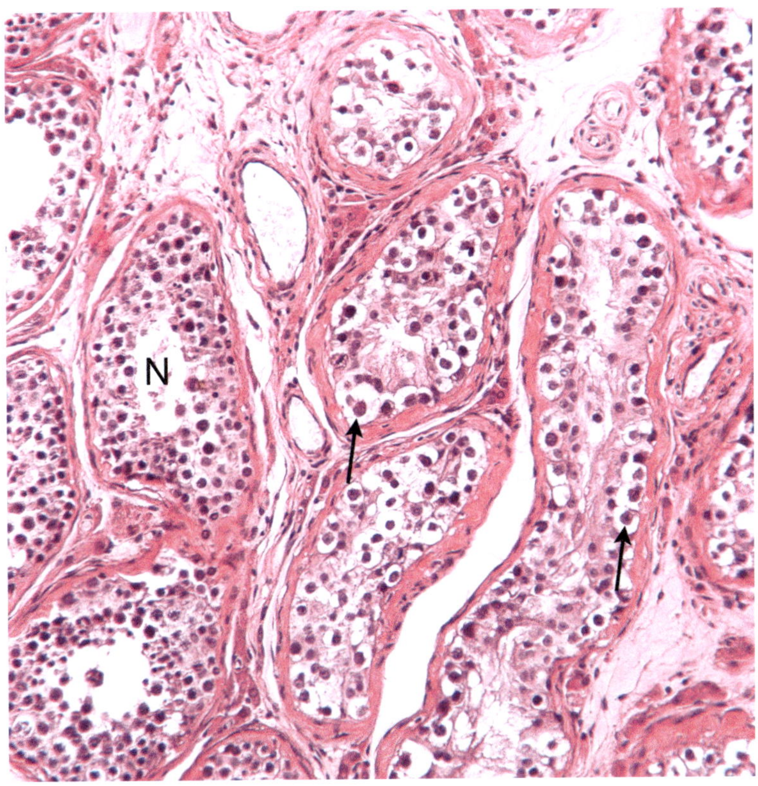

Fig. 13.2 Intratubular germ cell neoplasia, unclassified. Unlike normal tubules (*N*), which have basement membranes of normal thickness and contain normal spermatogenic cells in their lumen, the tubules containing IGCNU (*arrows*) have thickened basement membranes

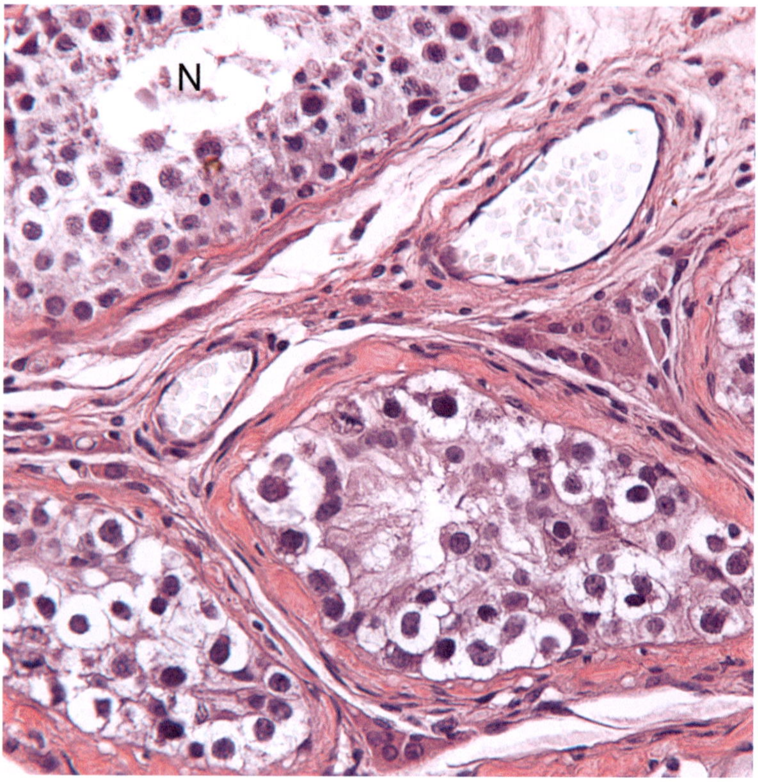

Fig. 13.3 Intratubular germ cell neoplasia, unclassified. The neoplastic germ cells have enlarged, slightly polymorphic, centrally located nuclei surrounded by a clear cytoplasm, giving them the appearance of fried eggs. Unlike normal tubules (*N*), which contain normal spermatogenic cells, tubules containing IGCNU do not show spermatogenesis

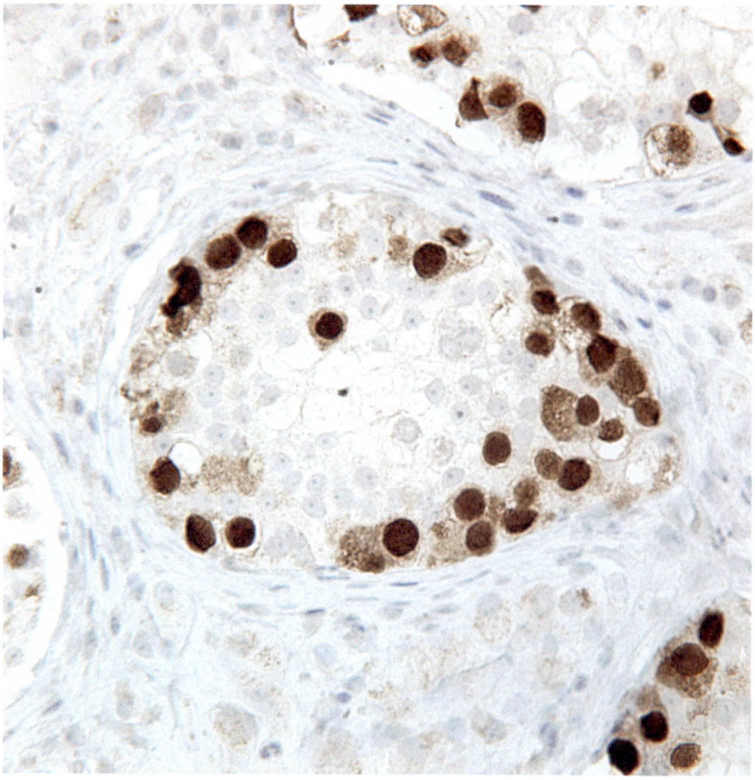

Fig. 13.4 Intratubular germ cell neoplasia, unclassified. Immunohistochemistry with antibody to OCT3 shows strong reactivity of the nuclei of the neoplastic cells, which appear brown in this figure

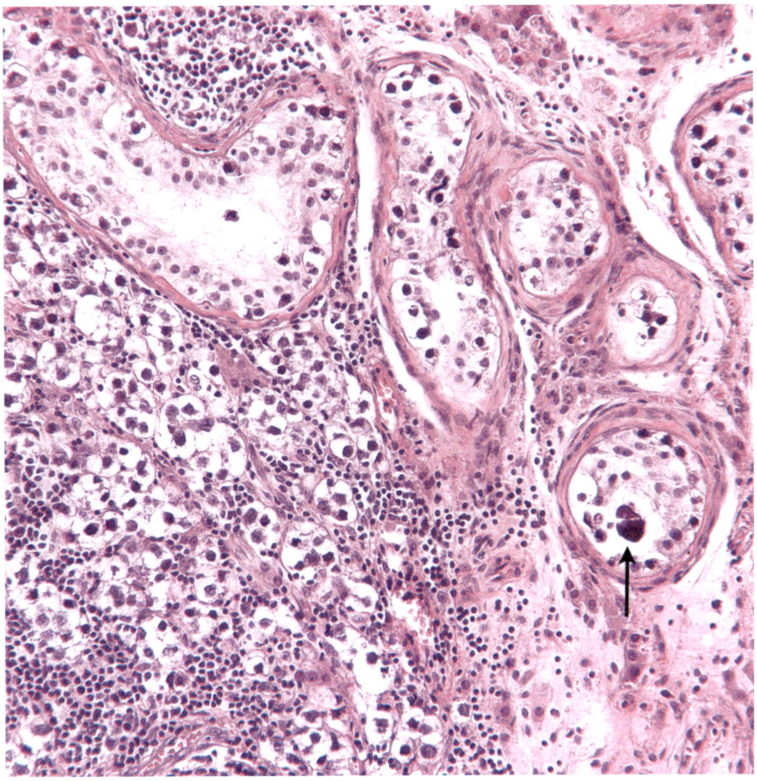

Fig. 13.5 Seminoma. A seminoma, illustrated in the *left lower part* of the figure, is adjacent to tubules with thickened basement membranes containing IGCNU and showing no signs of spermatogenesis. One of the tubules contains a calcification (*arrow*)

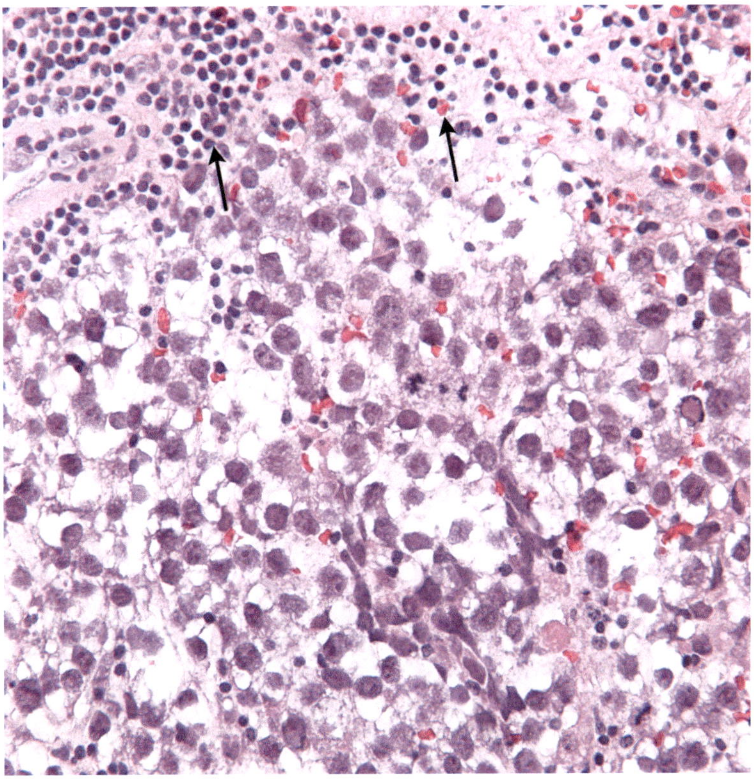

Fig. 13.6 Seminoma. The tumor is composed of polygonal cells that have round nuclei surrounded by clear cytoplasm. The fibrous septa contain lymphocytes (*arrows*)

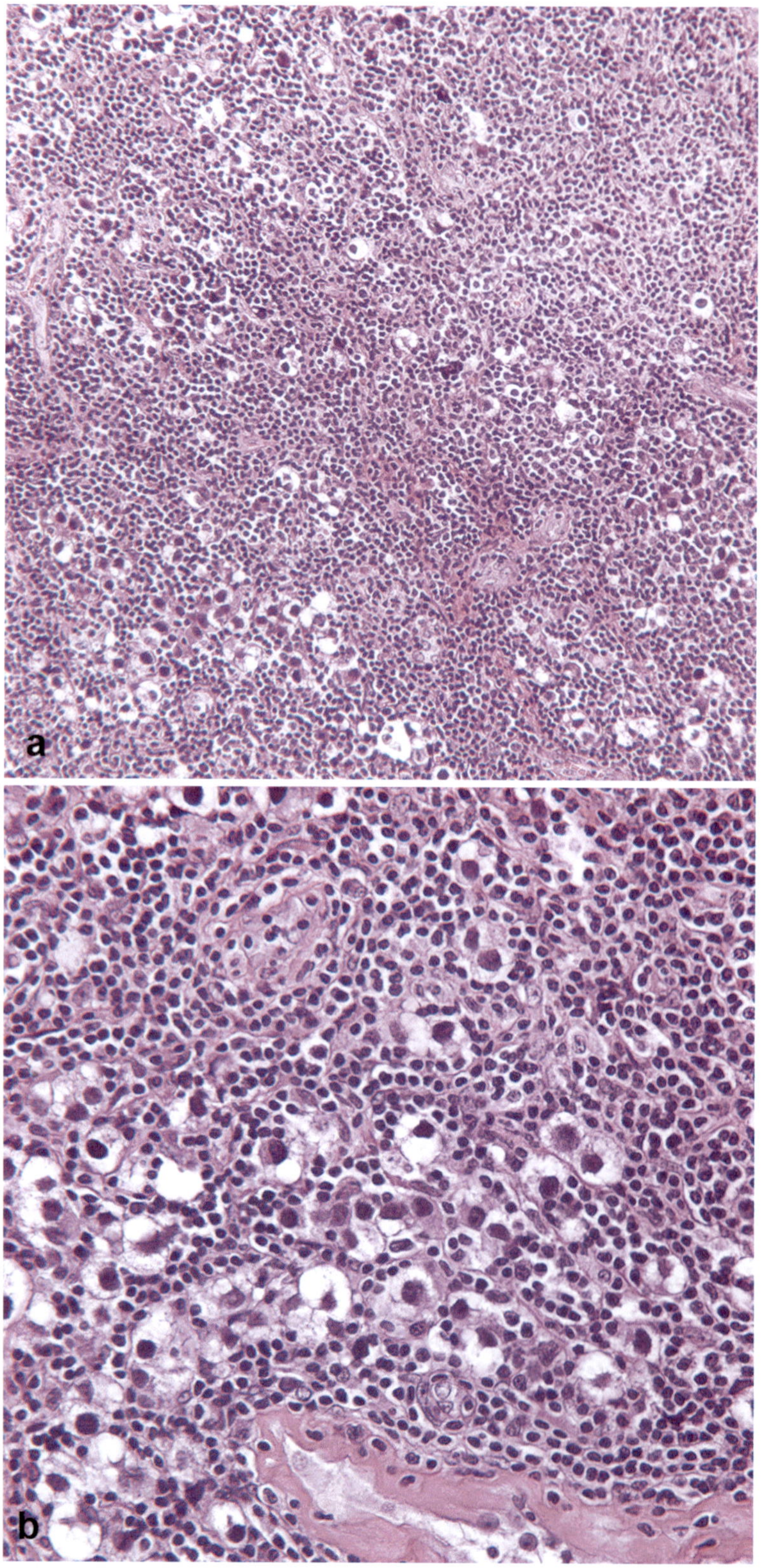

Fig. 13.7 Seminoma. (**a**) Scattered seminoma cells, recognizable even at this low magnification, are seen in a sea of lymphocytes. This figure demonstrates that the lymphoid reaction to seminoma cells may be very prominent. (**b**) Seminoma cells are surrounded by numerous lymphocytes. Despite a vigorous lymphocytic response, the tumor cells do not seem to be affected and have retained their usual appearance

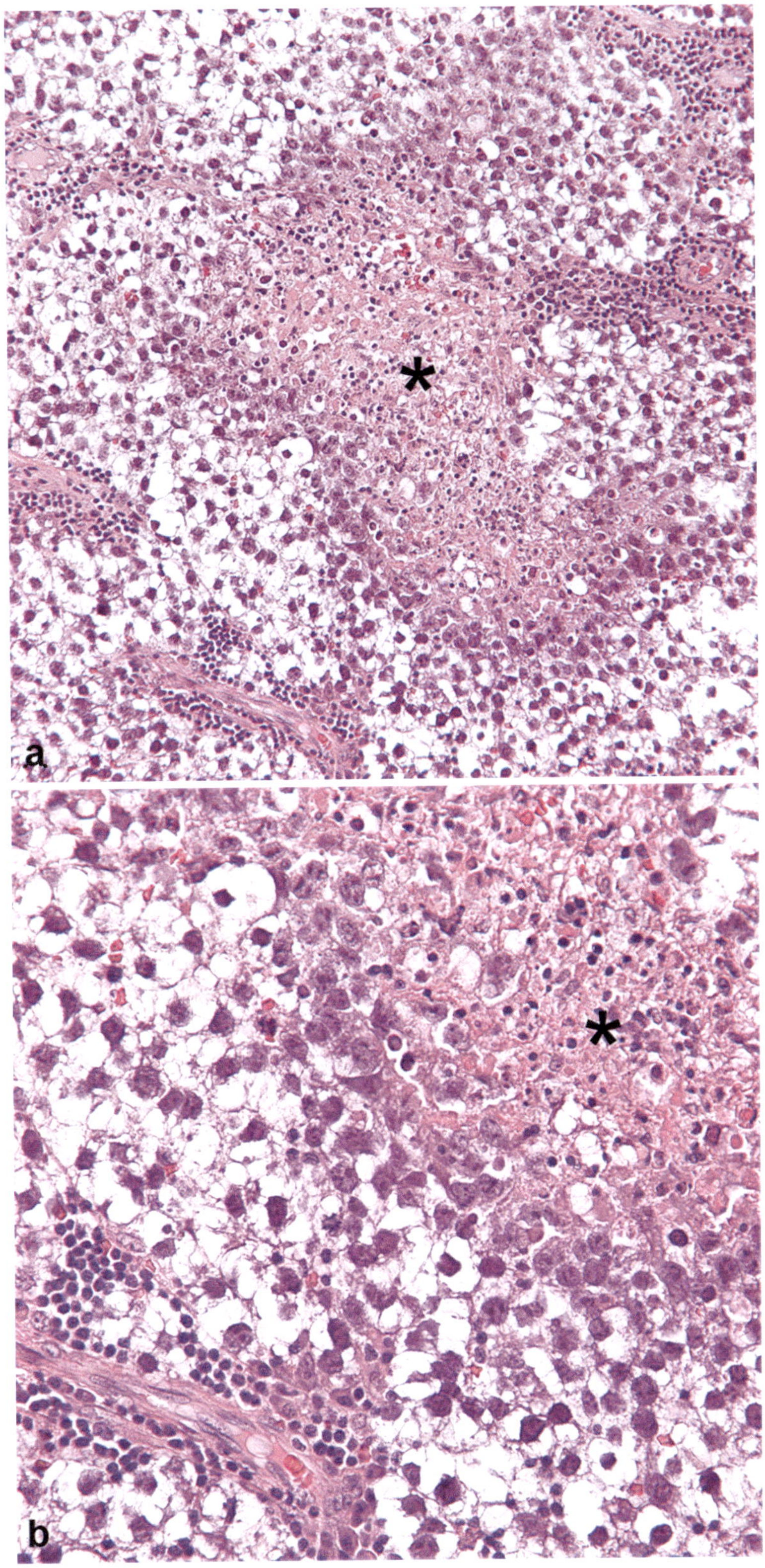

Fig. 13.8 Seminoma. (**a**) This tumor shows a prominent area of necrosis in the middle of the figure (*asterisk*). (**b**) Higher magnification view of a seminoma, showing an area of necrosis (*asterisk*). The remaining tumor cells show no signs of injury. In the *lower left corner*, there is a blood vessel in a fibrous septum containing lymphocytes

Fig. 13.9 Seminoma. (**a**) This tumor shows marked edema and scattered syncytiotrophoblastic giant cells (*arrows*). (**b**) This tumor shows edema and hemorrhage separating the tumor cells from one another and a multinucleated syncytiotrophoblastic giant cell (*arrow*)

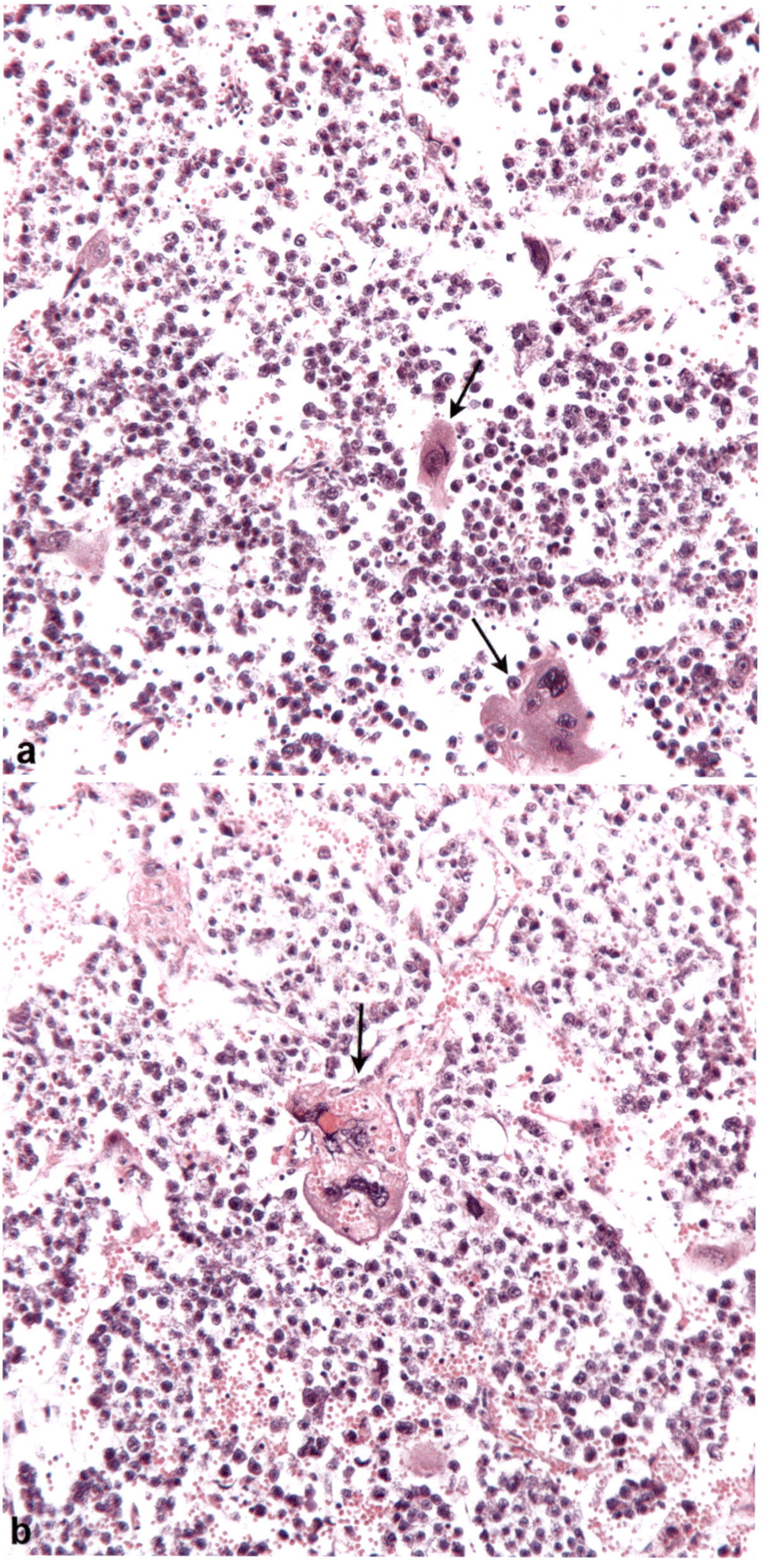

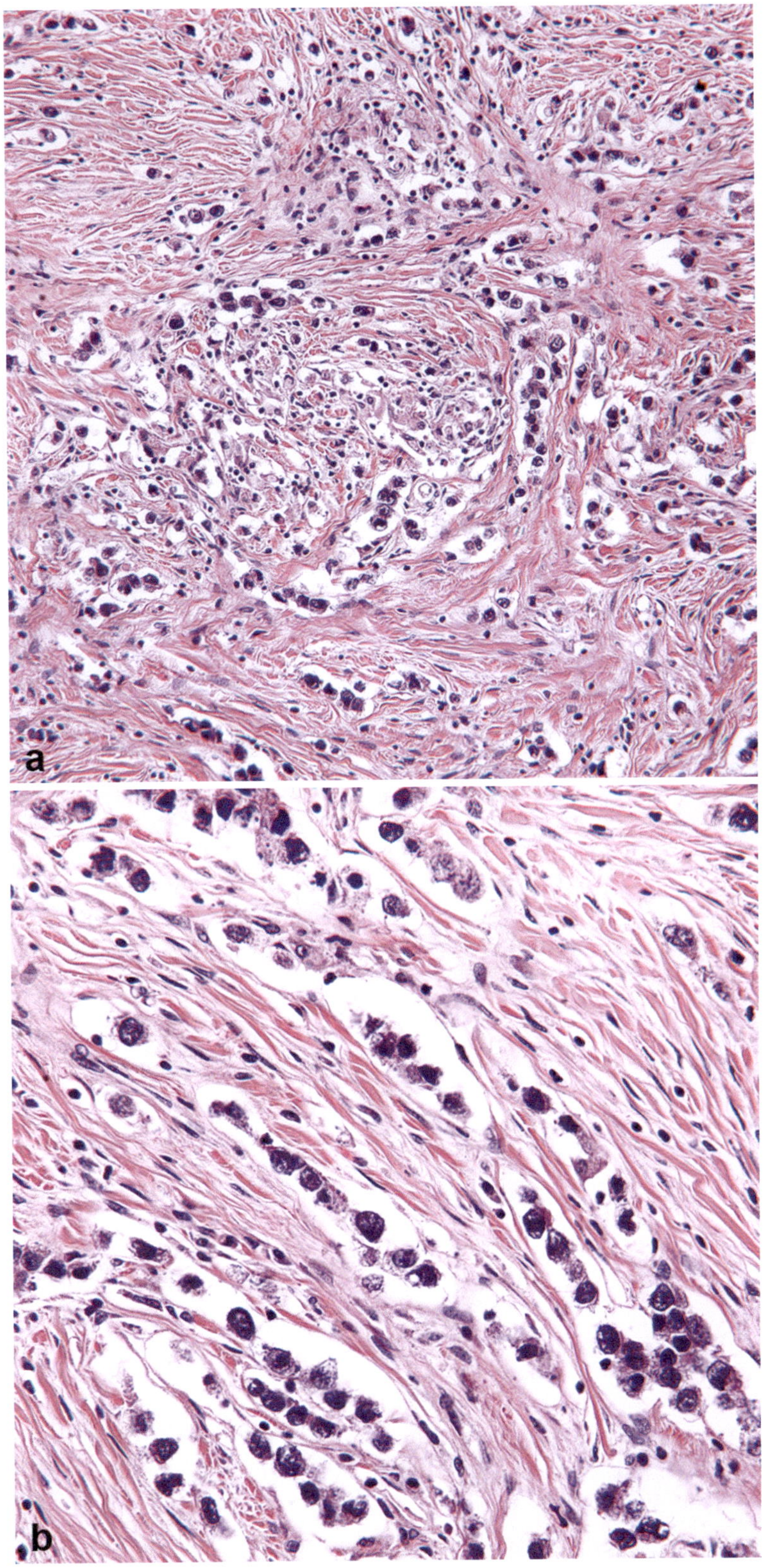

Fig. 13.10 Seminoma. (**a**) This tumor shows prominent fibrosis surrounding a strand of tumor cells and scant lymphocytes. (**b**) This tumor shows prominent fibrosis encasing seminoma cells arranged in single files

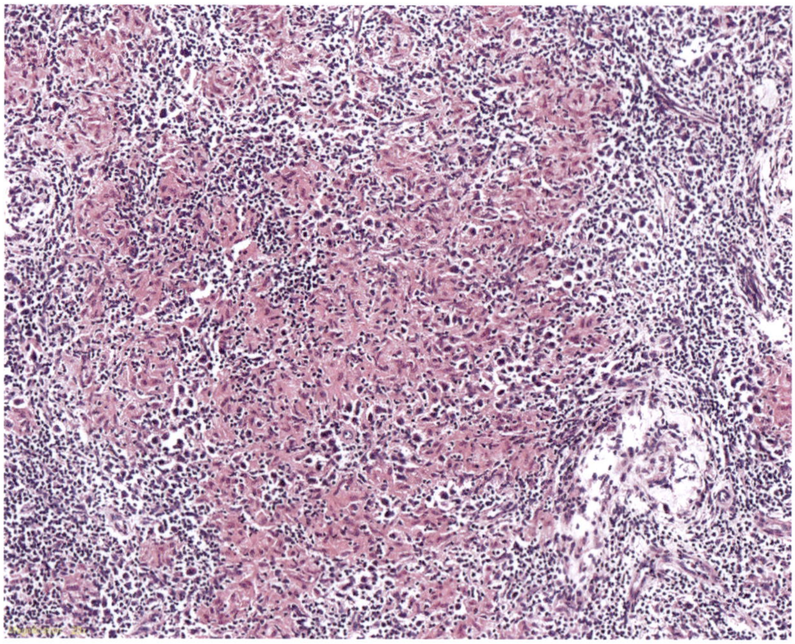

Fig. 13.11 Seminoma. Confluent granulomas obliterate the central part of this tumor. Around the granulomas, there are lymphocytes and well-preserved scattered seminoma cells

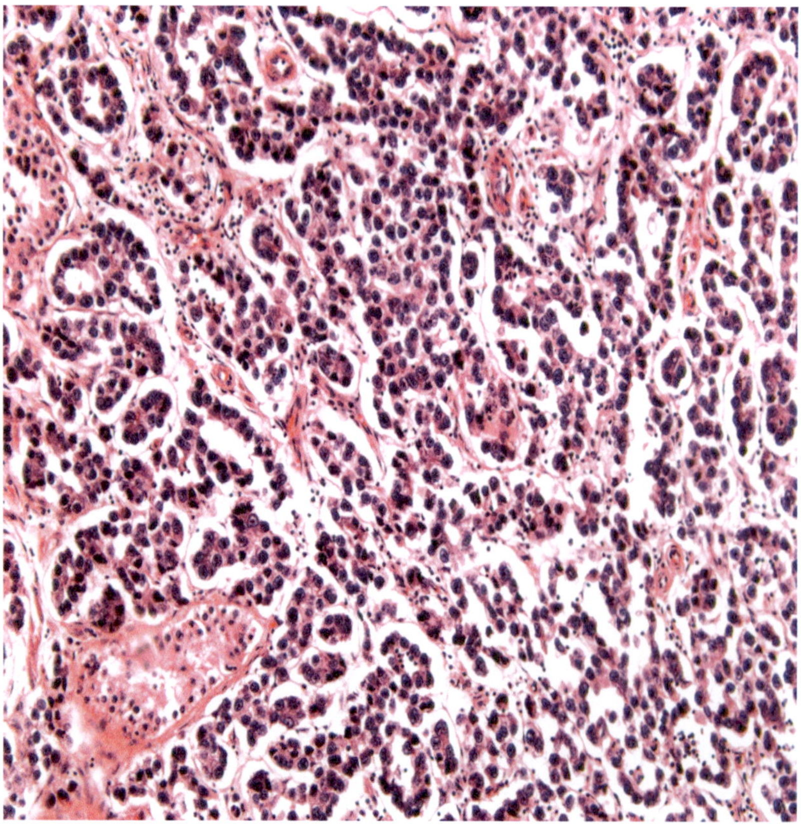

Fig. 13.12 Seminoma. This "tubular" seminoma is composed of tumor cells forming tubular or gland-like structures

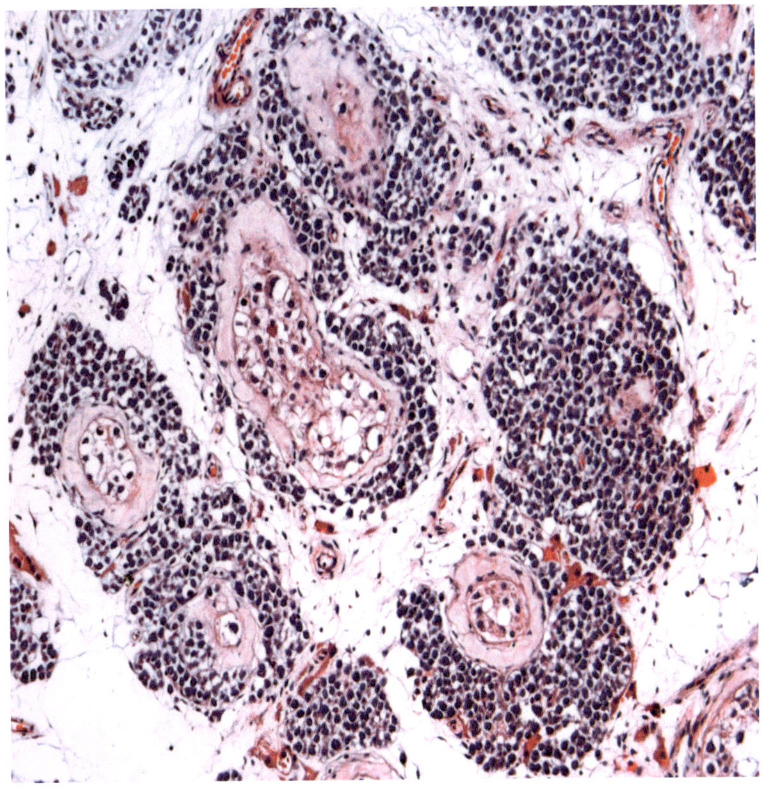

Fig. 13.13 Seminoma. Tumor cells are seen in the lumen of the seminiferous tubules as well as around the tubules, where they form collarets

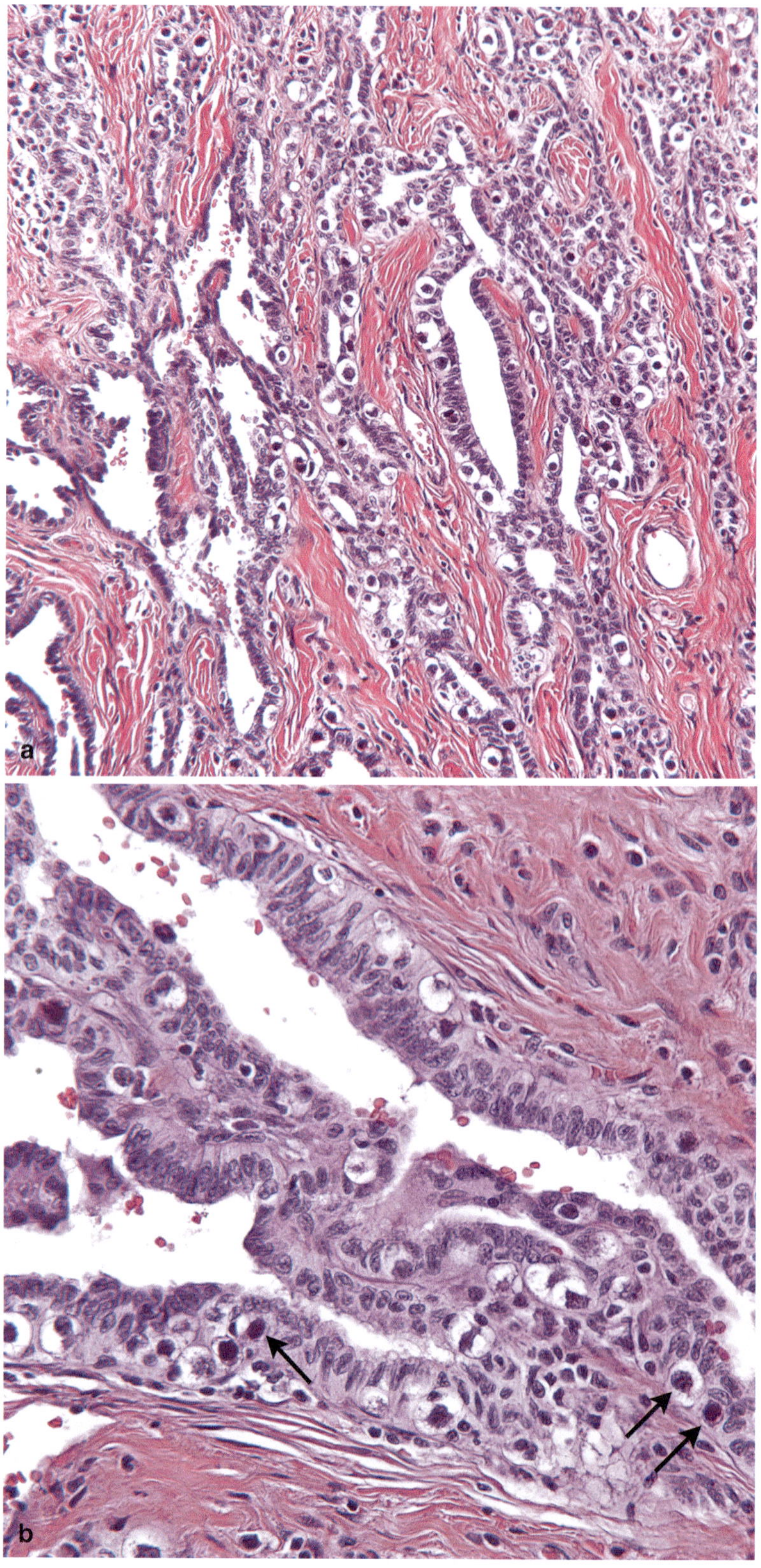

Fig. 13.14 Seminoma. (**a**) This tumor invades the epithelium of the rete testis in a "pagetoid" manner. (**b**) Single seminoma cells (*arrows*) stand out because of their clear cytoplasm and round nuclei

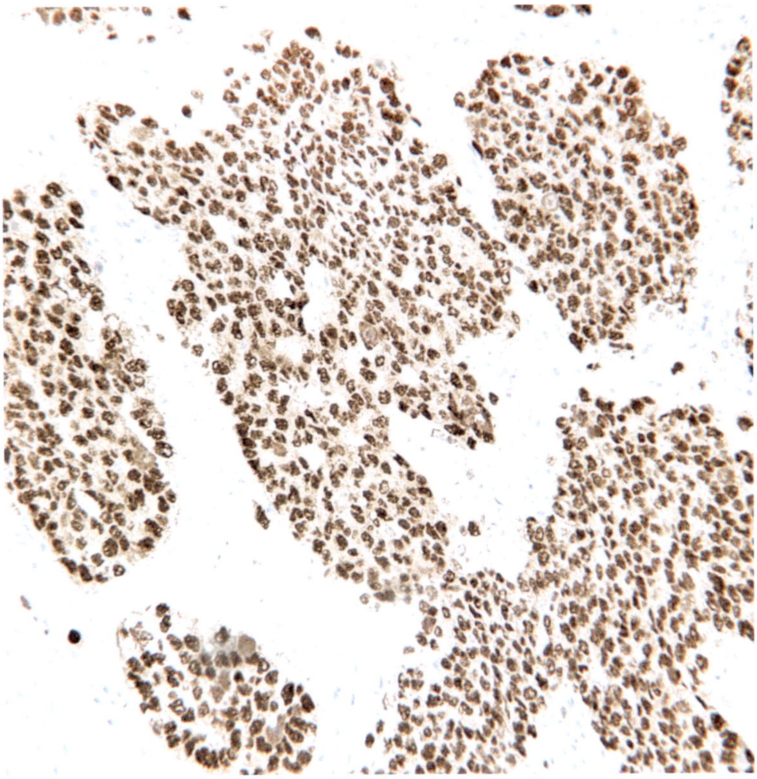

Fig. 13.15 Seminoma. Immunohistochemistry with the antibody to antigen OCT3 shows strong nuclear reactivity in all tumor cells, which are arranged in nests. The unreactive areas between the groups of tumor cells correspond to stroma infiltrated by lymphocytes

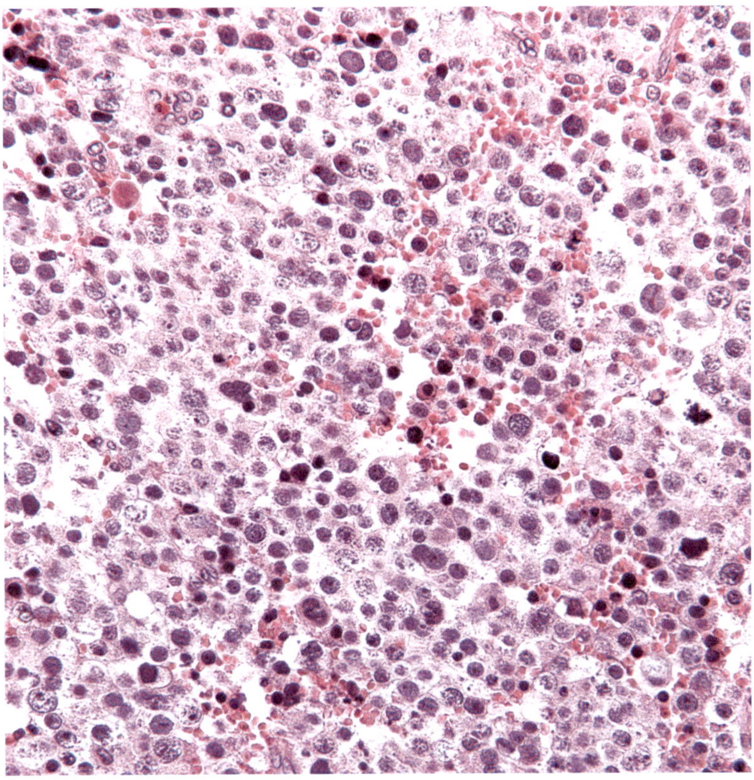

Fig. 13.16 Spermatocytic seminoma. The tumor consists of small, large, and giant cells

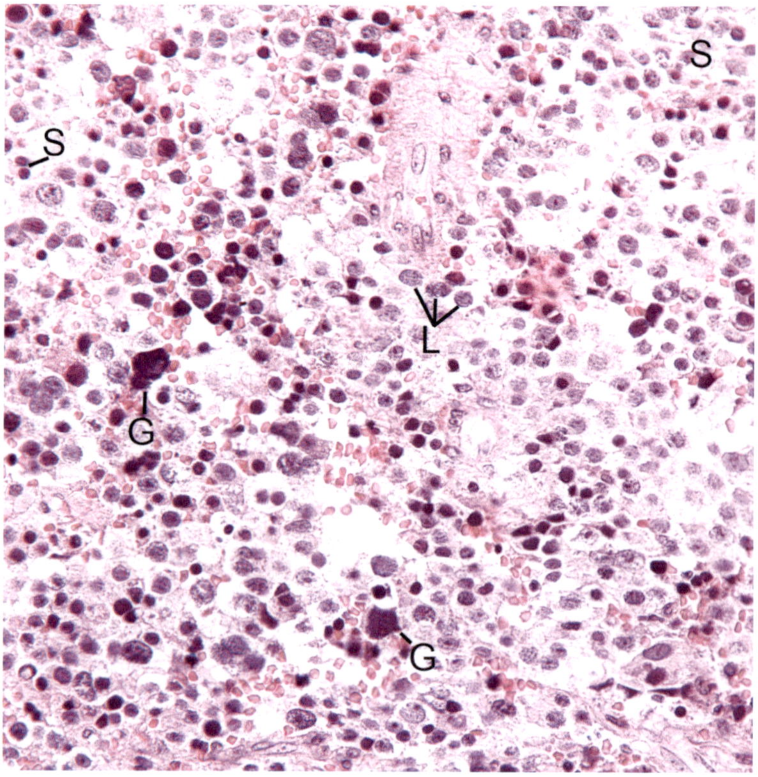

Fig. 13.17 Spermatocytic seminoma. Small cells (*S*) have nuclei resembling lymphocytes, large cells (*L*) have vesicular nuclei, and the giant cells (*G*) stand out because of their nuclei and hyperchromasia

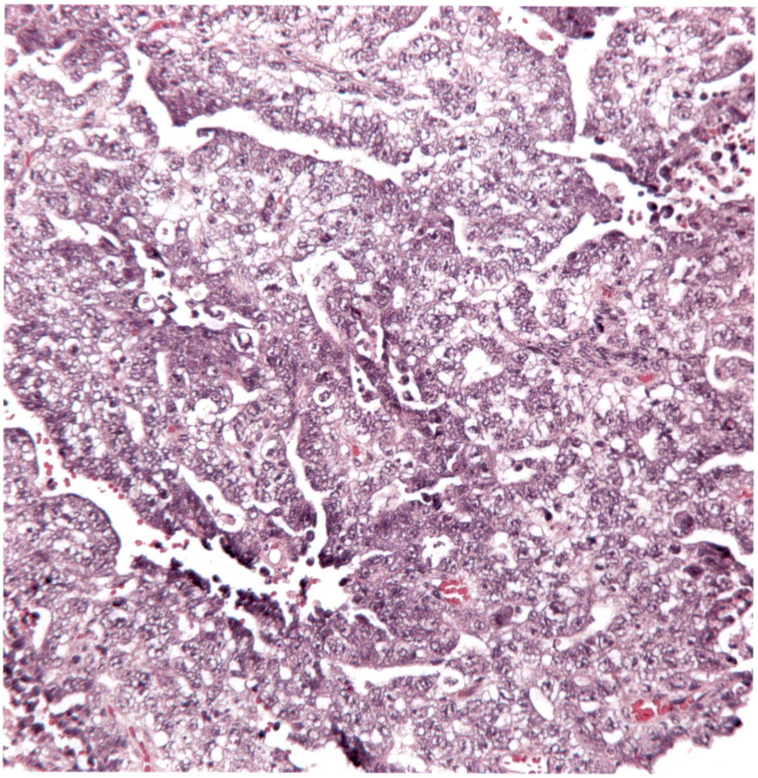

Fig. 13.18 Embryonal carcinoma. The entire tumor is composed of undifferentiated cells that have overlapping nuclei and scant cytoplasm. Tumor cells form solid nests with some slit spaces, allowing for the focal formation of short papillae

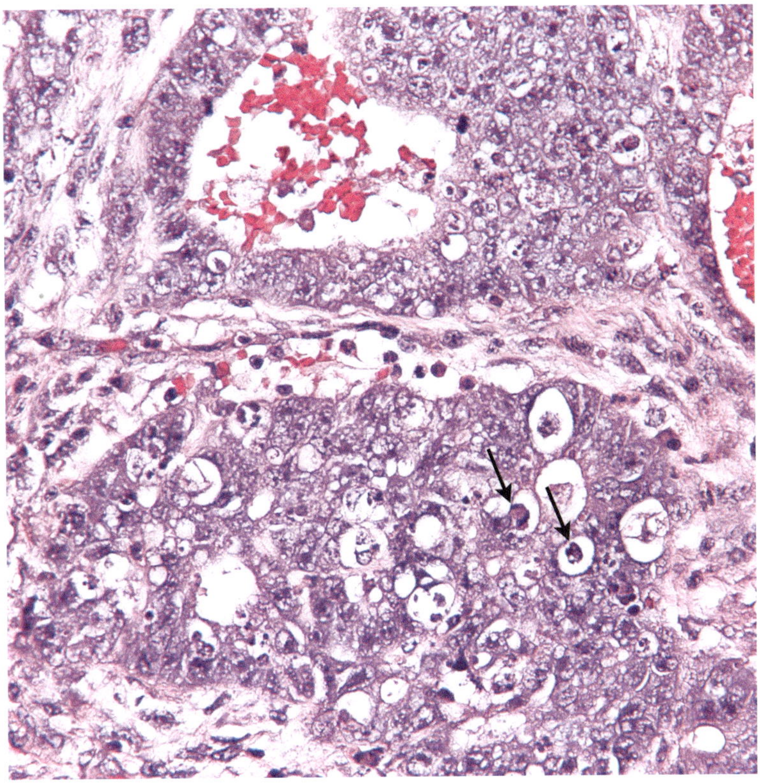

Fig. 13.19 Embryonal carcinoma. Tumor cells form nests that may be solid (lower nest) or cavitated (upper nest). In the lower nest, there are single cells undergoing apoptosis and vacuolization (*arrows*). The central cavity of the upper nest contains blood

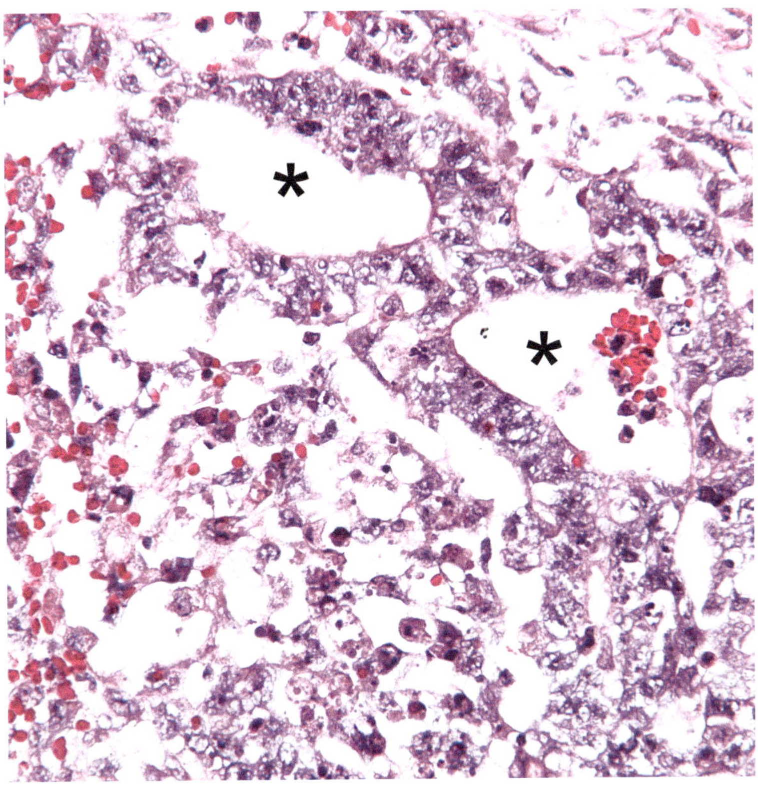

Fig. 13.20 Embryonal carcinoma. Tumor cells surround the central cavities (*asterisks*), forming gland-like structures. In the *lower part* of the picture, the cells are arranged loosely and are undergoing apoptosis. At the periphery on the left side, there is hemorrhage, which also is visible in the lumen of one of the two gland-like structures

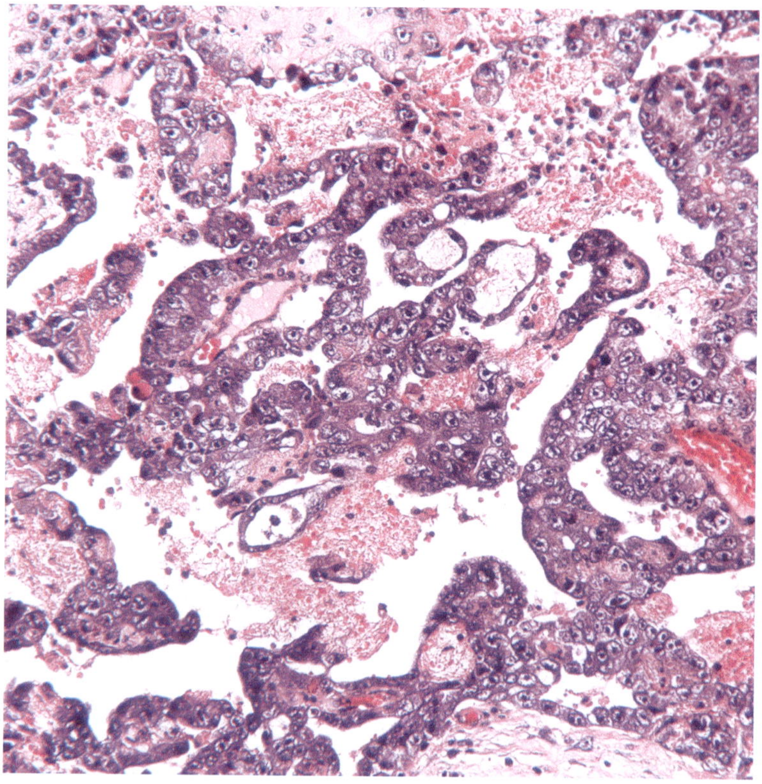

Fig. 13.21 Embryonal carcinoma. Tumor cells form interanastomosing strands and short papillae

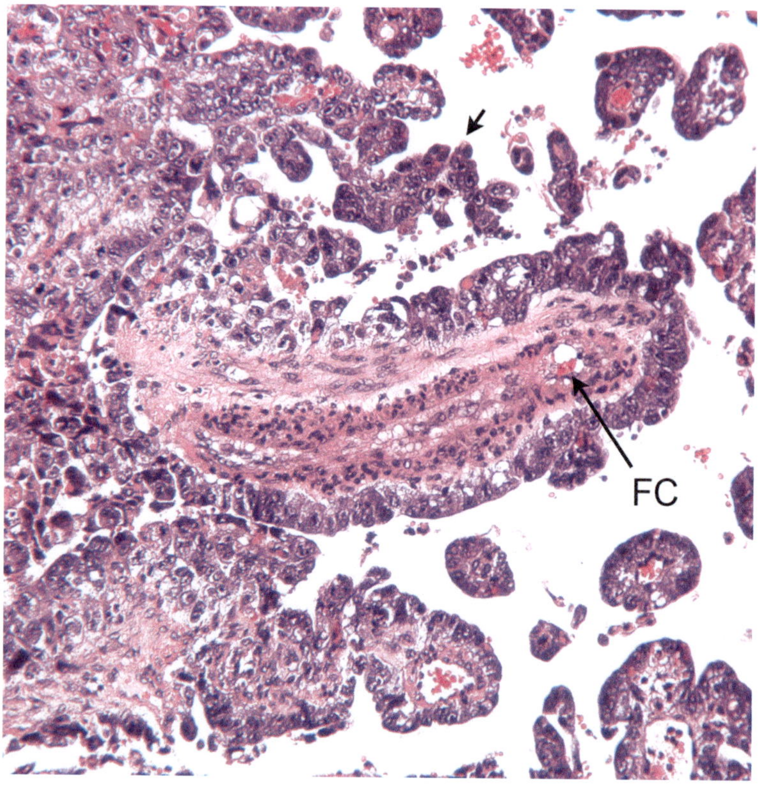

Fig. 13.22 Embryonal carcinoma. Tumor cells line well-formed papillae containing vascular fibrous cores (*FC*). Shorter papillary projections without such fibrovascular cores (*short arrow*) also are seen projecting into the cavitary spaces of the tumor

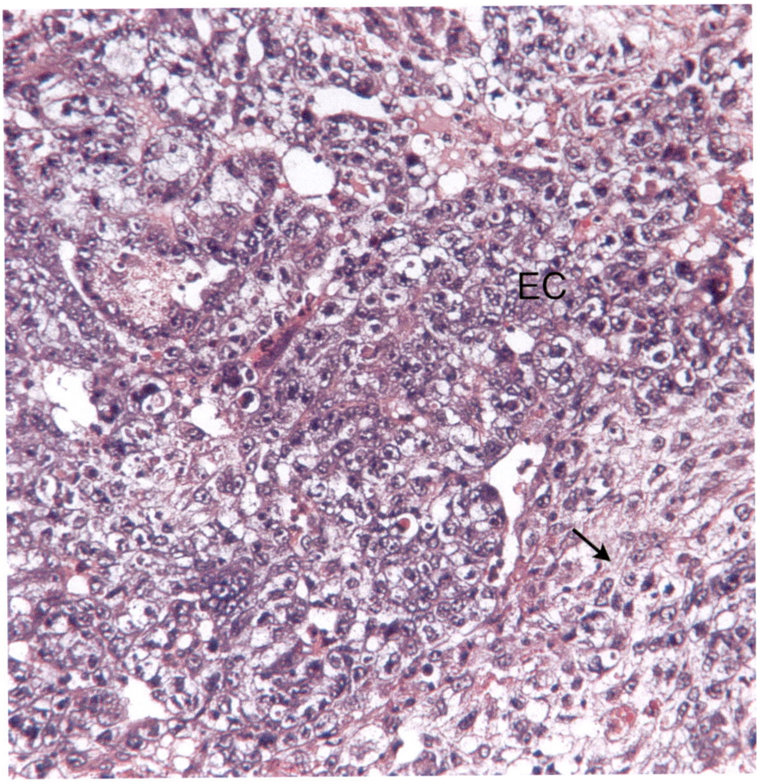

Fig. 13.23 Embryonal carcinoma. The central portion of this figure is composed of undifferentiated embryonal carcinoma cells (*EC*). Surrounding these nests are cells that might represent embryonal carcinoma cells or some of their more differentiated derivatives. For example, the cells in the upper left corner are forming papillae while retaining the nuclear features of embryonal carcinoma cells. The cells in the lower right corner (*arrow*) appear different from embryonal carcinoma cells and most likely are yolk sac cells. The true nature of all these cells cannot be determined with certainty without additional immunohistochemical studies

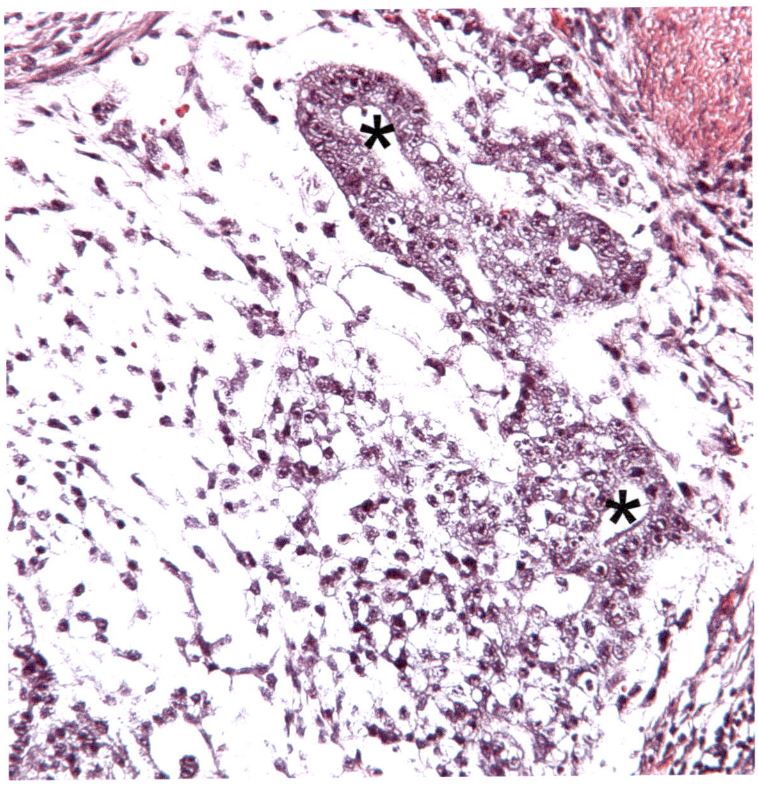

Fig. 13.24 Embryonal carcinoma. This photograph was taken from a malignant nonseminomatous germ cell tumor (NSGCT), or teratocarcinoma. The tumor cells form gland-like structures composed of embryonal carcinoma cells (*asterisks*) surrounded by loose (magma reticulare-like) areas probably corresponding to yolk sac cells

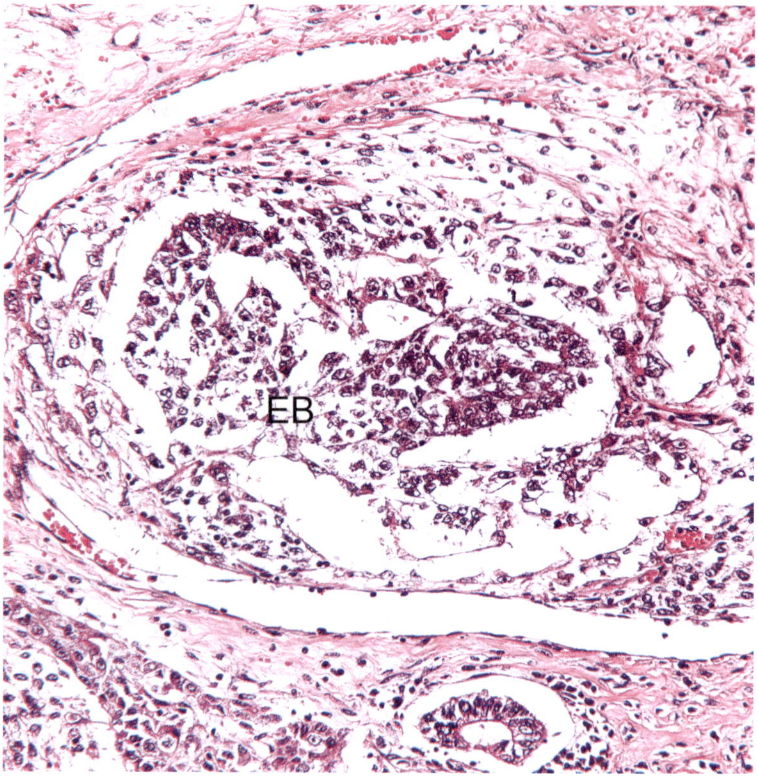

Fig. 13.25 Nonseminomatous germ cell tumor. Groups of embryonal carcinoma in this NSGCT are surrounded by loose yolk sac areas corresponding to ill-formed embryoid bodies (*EB*)

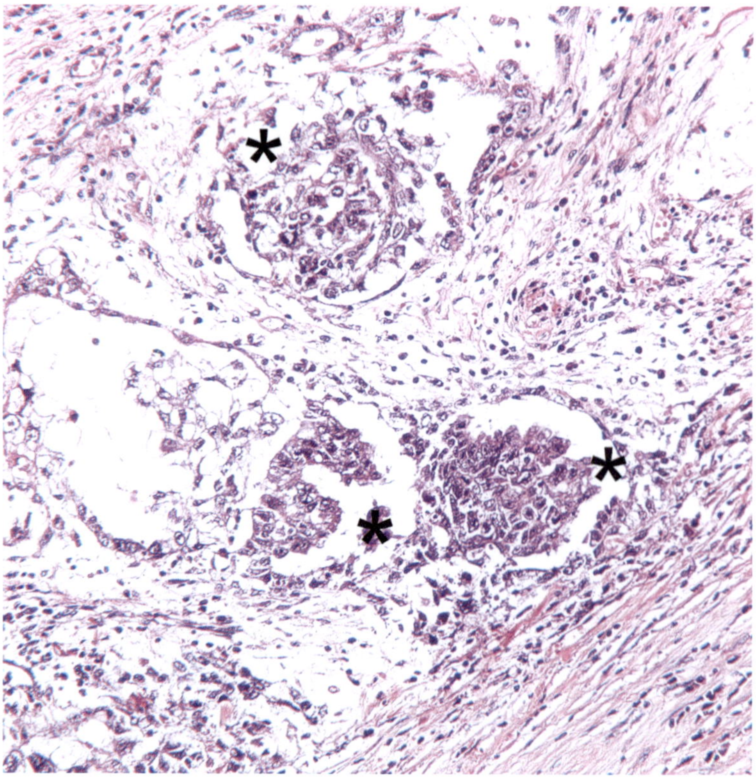

Fig. 13.26 Nonseminomatous germ cell tumor. This tumor contains embryonic carcinoma cells forming embryoid bodies (*asterisks*) dispersed in a loose connective tissue stroma

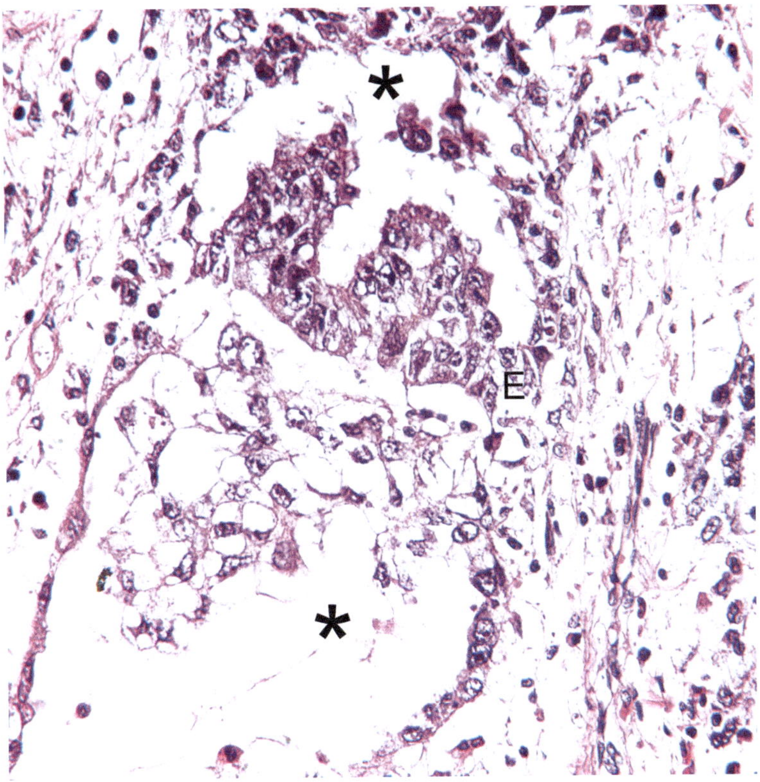

Fig. 13.27
Nonseminomatous germ cell tumor. This higher power magnification detail from Fig. 13.26 shows the embryoid body with an upper and lower cavity (*asterisks*) separated by a layer of cells corresponding to the embryonic disc (*E*)

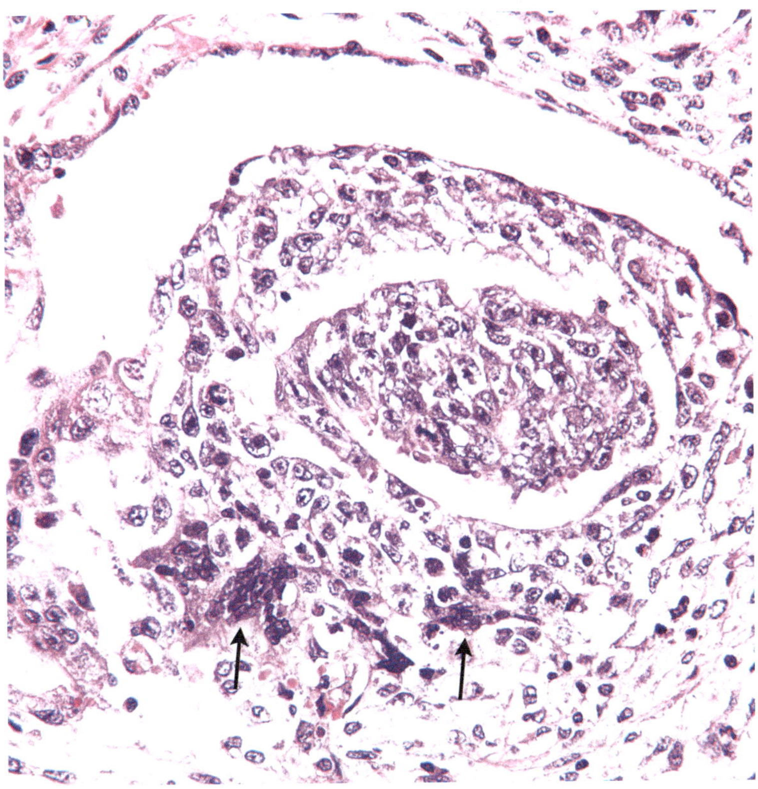

Fig. 13.28
Nonseminomatous germ cell tumor. The grouping of embryonal carcinoma cells surrounded by cavities and loosely arranged cells is reminiscent of early embryogenesis, but it is too disorderly to be properly characterized or compared with normal postimplantation human embryos. Syncytiotrophoblastic cells (*arrows*) are seen at the periphery of this nest

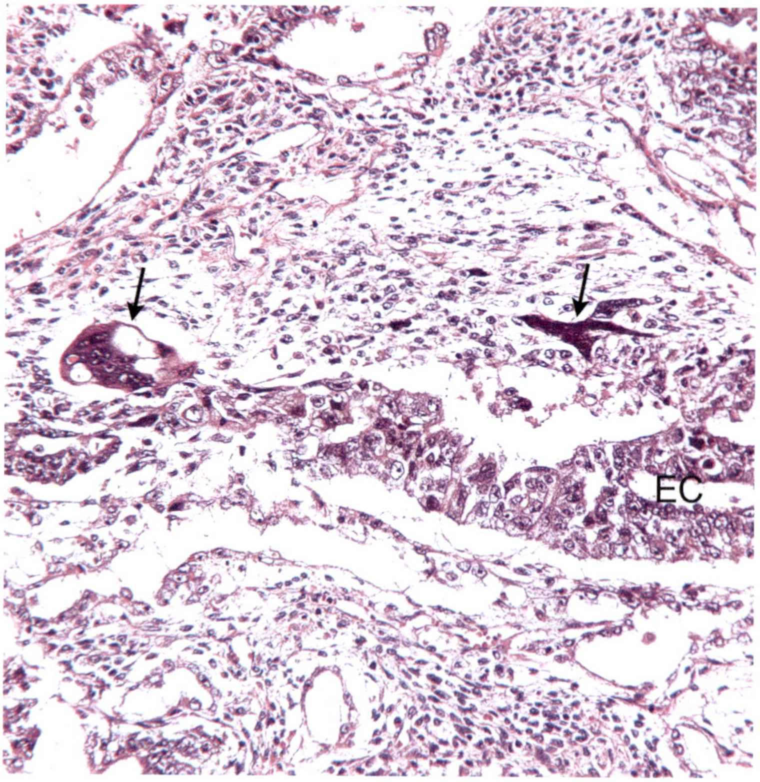

Fig. 13.29
Nonseminomatous germ cell tumor. A solid nest of embryonal carcinoma cells (*EC*) seems to be floating in a cavity formed by the loose stroma. Among the latter cells are syncytiotrophoblastic giant cells (*arrows*). Although the whole structure bears some resemblance to an early-implanting human embryo, it is too disorderly to be compared with a normal embryo

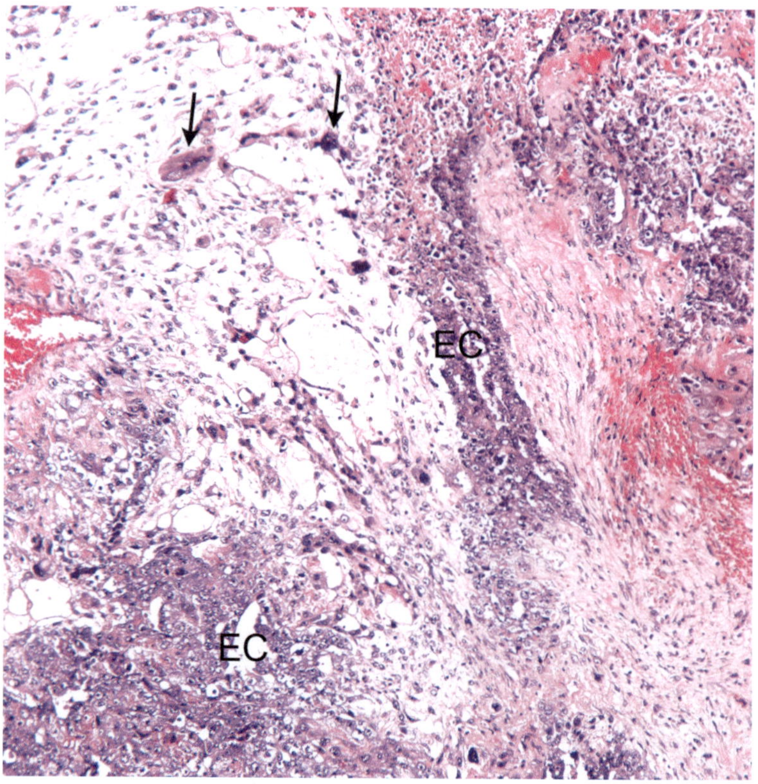

Fig. 13.30
Nonseminomatous germ cell tumor. This tumor consists of embryonal carcinoma cells (*EC*), trophoblastic cells, and possibly yolk sac cells. The centrally located embryonal carcinoma cells form a readily identifiable elongated nest, with syncytiotrophoblastic giant cells (*arrows*) at its periphery. Embryonal carcinoma cells are separated by a space from the loosely arranged cells above and below them, probably corresponding to yolk sac cells

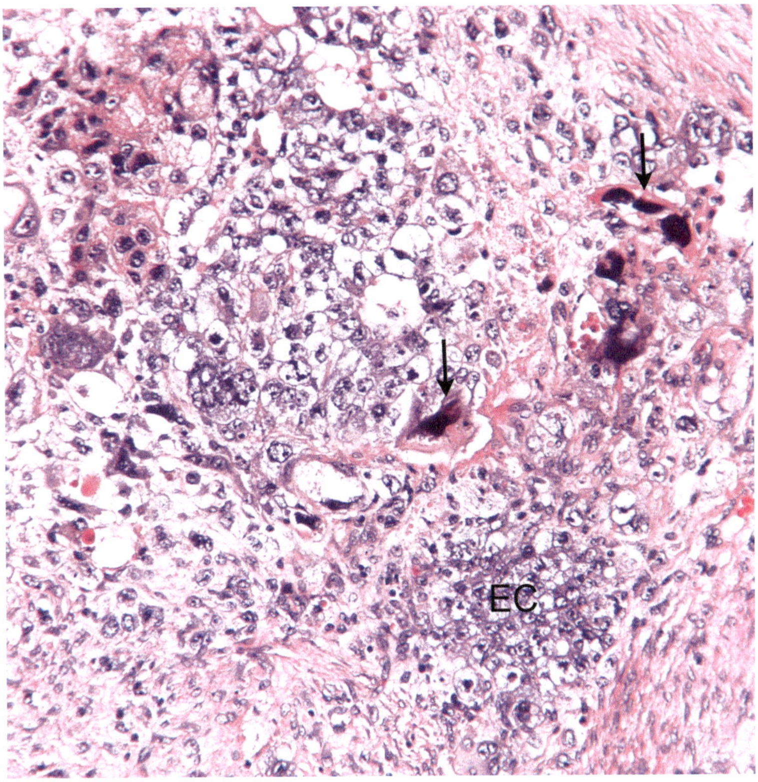

Fig. 13.31 Nonseminomatous germ cell tumor. Embryonal carcinoma, trophoblastic, and possibly yolk sac cells are present in a tumor associated with elevated serum levels of human chorionic gonadotropin and α-fetoprotein. Embryonal carcinoma cells (*EC*) are arranged in compact nests or nests with slit spaces, whereas the yolk sac areas are composed of loosely arranged cells on the left side of the figure. Also present are scattered syncytiotrophoblastic giant cells (*arrows*)

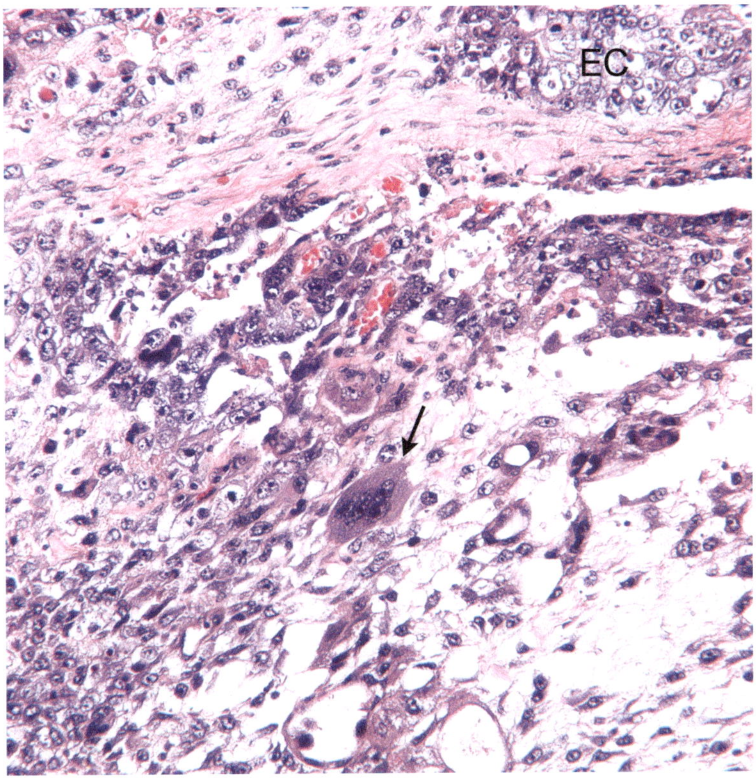

Fig. 13.32 Nonseminomatous germ cell tumor. Embryonal carcinoma cells (*EC*) are intermixed with nondescript stromal cells and hyperchromatic multinucleated syncytiotrophoblastic giant cells (*arrow*)

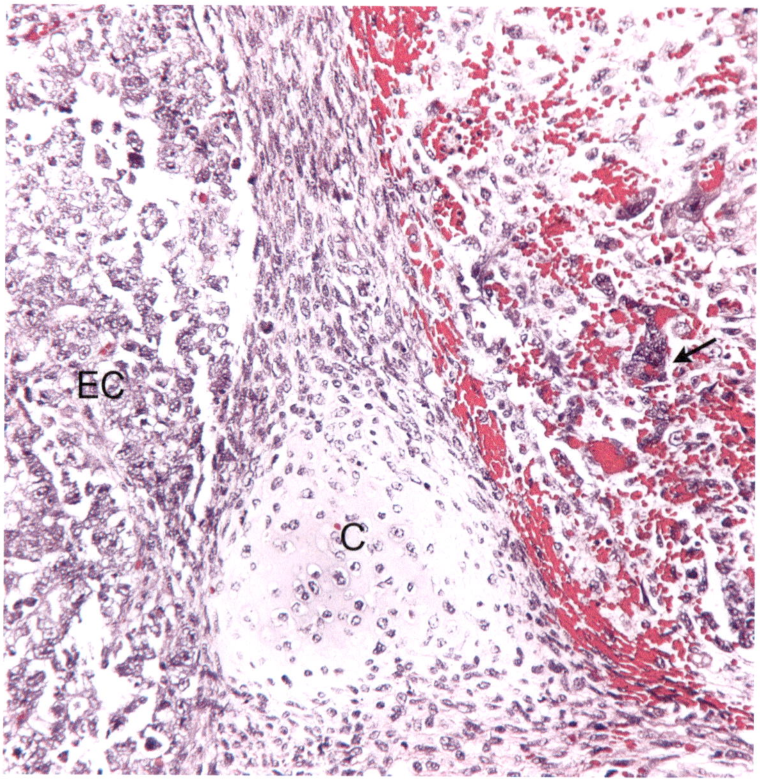

Fig. 13.33 Nonseminomatous germ cell tumor. Embryonal carcinoma cells (*EC*) are separated from the multinucleated syncytiotrophoblastic cells (*arrow*) by condensed stromal cells, forming a nest of cartilage (*C*)

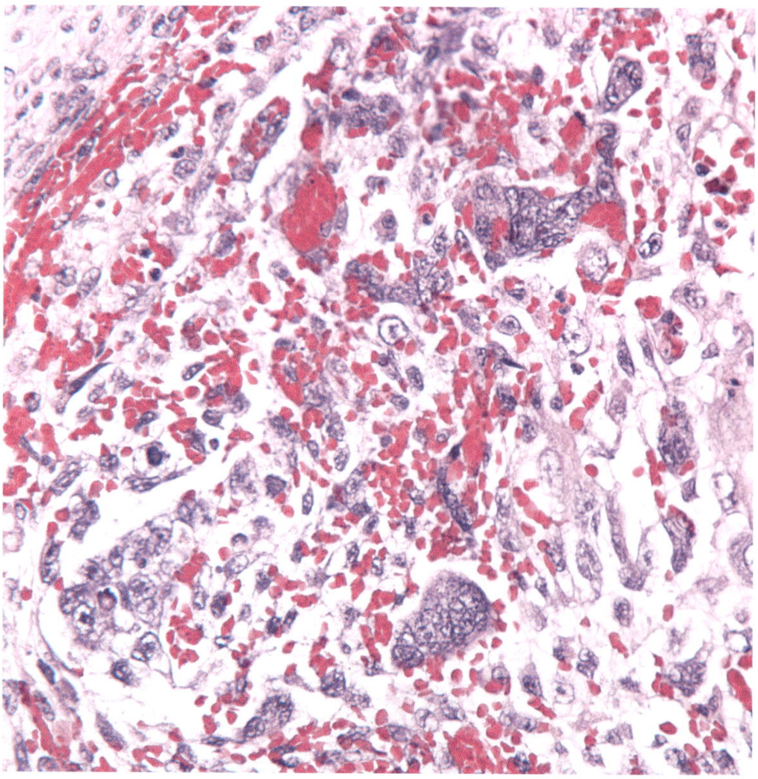

Fig. 13.34 Nonseminomatous germ cell tumor. In the tumor, there is an area composed of cytotrophoblastic and syncytiotrophoblastic cells. The entire field is hemorrhagic, which is typical of trophoblastic areas

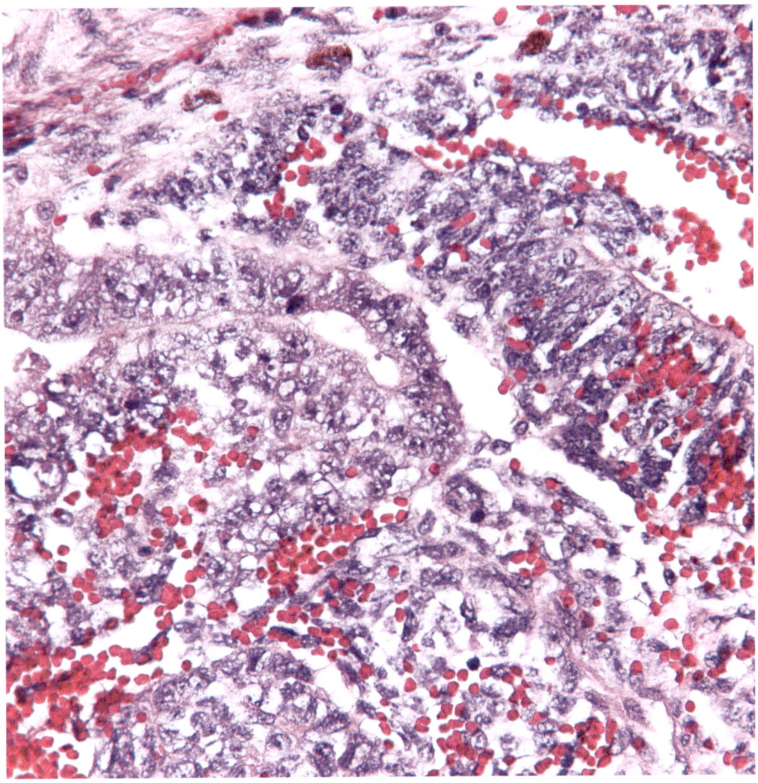

Fig. 13.35 Nonseminomatous germ cell tumor. Embryonal carcinoma cells also are admixed with extravasated red blood cells, illustrating the fact that not all hemorrhagic areas of NSGCT contain trophoblastic cells

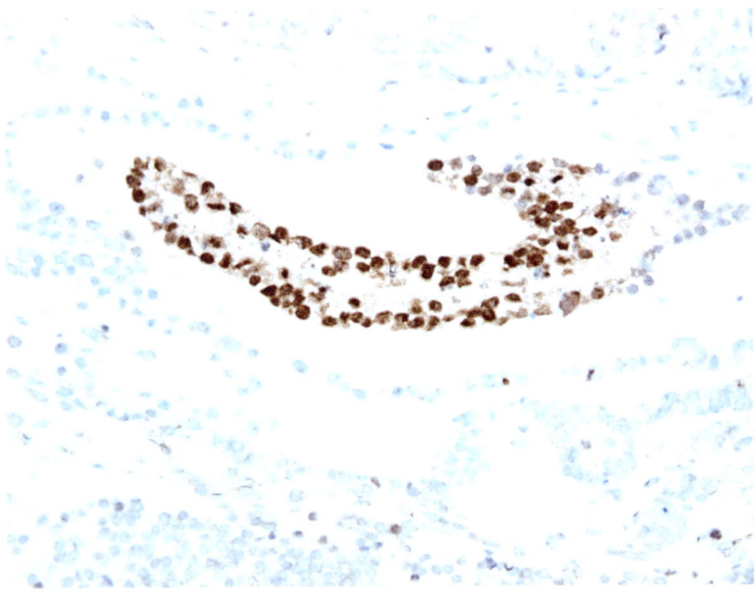

Fig. 13.36 Nonseminomatous germ cell tumor. Immunohistochemistry with antibody to OCT3 reacts selectively with a group of embryonal carcinoma cells. This figure shows that OCT3 is an excellent marker for embryonal carcinoma cells

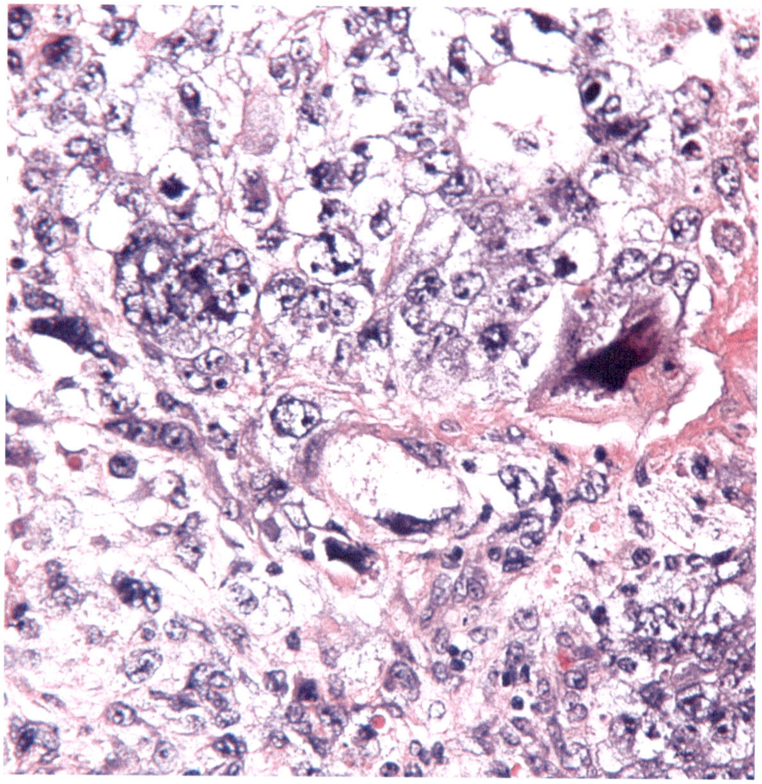

Fig. 13.37 Choriocarcinoma. Pure choriocarcinomas are rare and account for 1 in 1,000 testicular germ cell tumors. This tumor is composed of mononuclear cytotrophoblastic cells and multinucleate syncytiotrophoblastic cells. No embryonal carcinoma cells, yolk sac components, or other tissues were identified

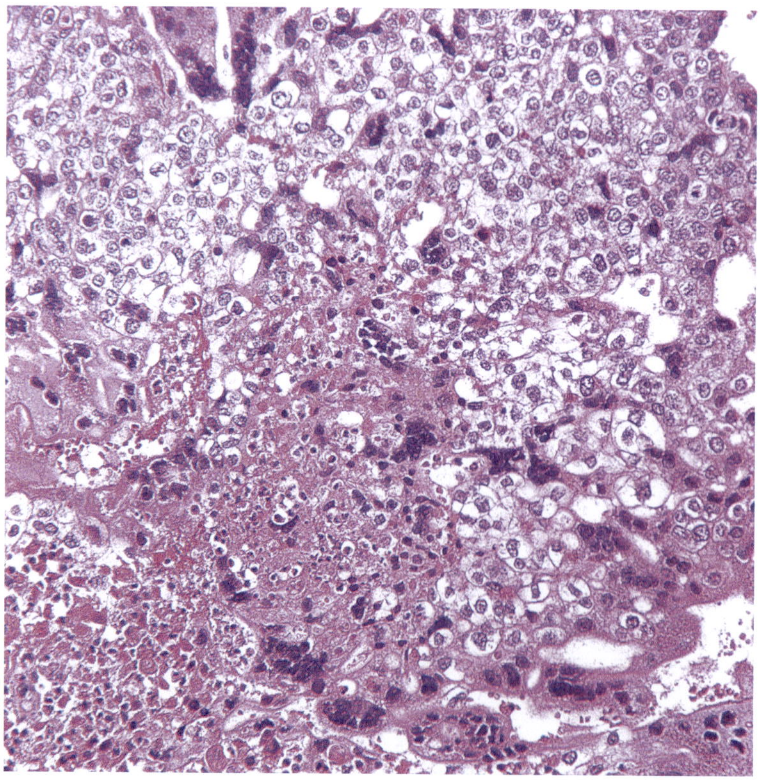

Fig. 13.38 Choriocarcinoma. This tumor is composed predominantly of mononuclear cytotrophoblastic cells and scattered syncytiotrophoblastic cells. No embryonal carcinoma cells, yolk sac components, or other tissues were identified

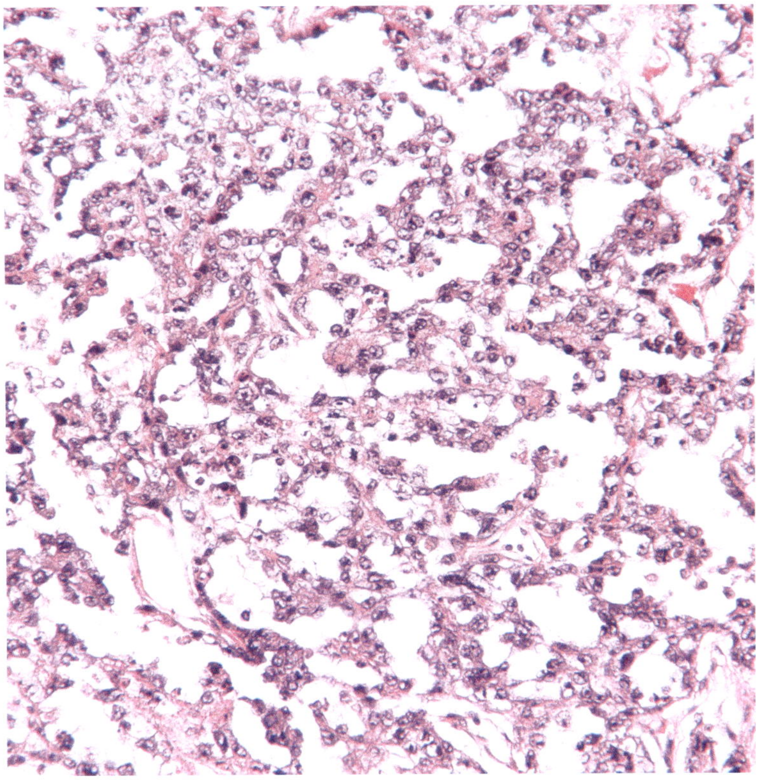

Fig. 13.39 Yolk sac carcinoma. Yolk sac carcinoma components of NSGCT may present several histologic patterns. This figure illustrates the labyrinthine pattern formed by anastomosing strands of flat or cuboidal cells

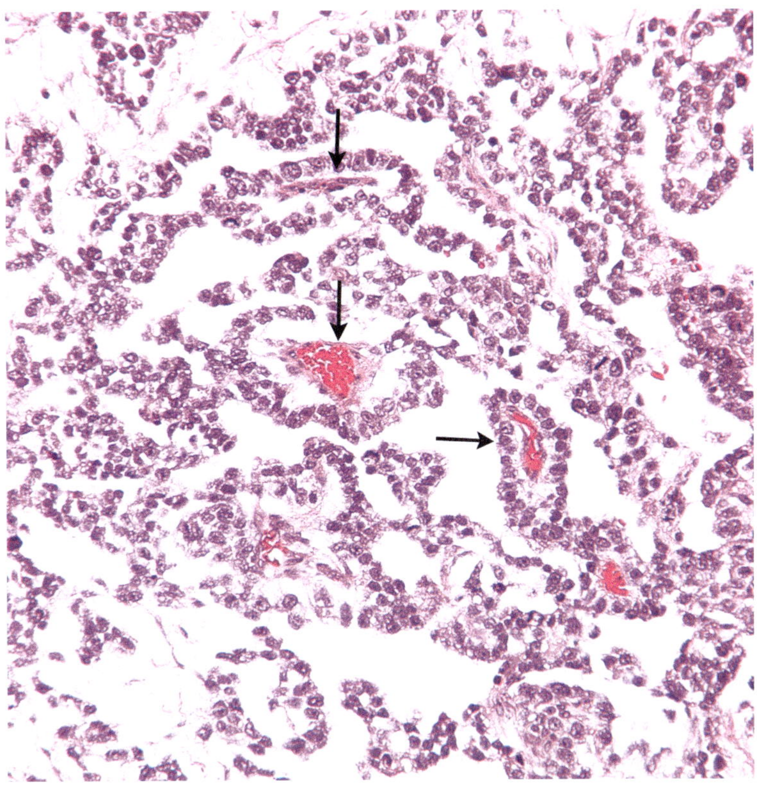

Fig. 13.40 Yolk sac carcinoma. This figure illustrates a labyrinthine pattern with micropapillary features. The micropapillae (*arrows*) have fibrovascular cores

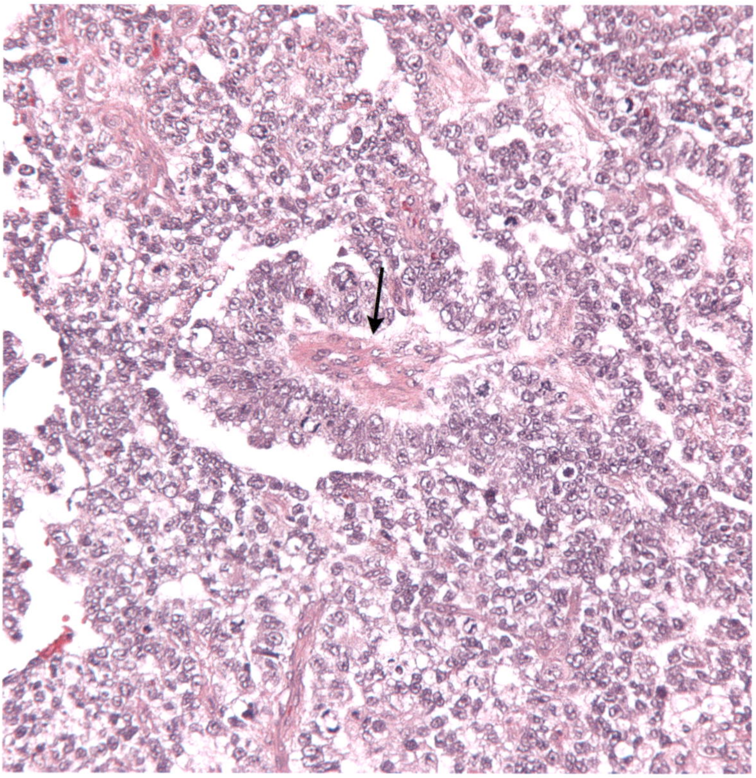

Fig. 13.41 Yolk sac carcinoma. This tumor consists of solid nests, but focally, it also forms papillae (*arrow*) and papillary components

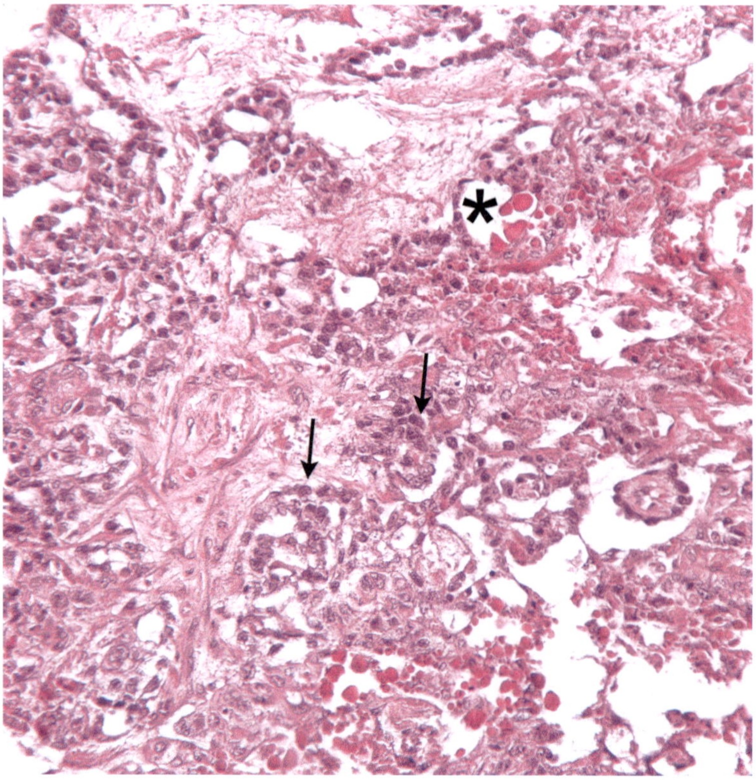

Fig. 13.42 Yolk sac carcinoma. An endodermal sinus pattern is composed of complex cords and solid groups of cuboidal cells focally forming glomeruloid Schiller–Duval bodies (*arrows*). Also present are a fibrous stroma and eosinophilic globules (*asterisk*)

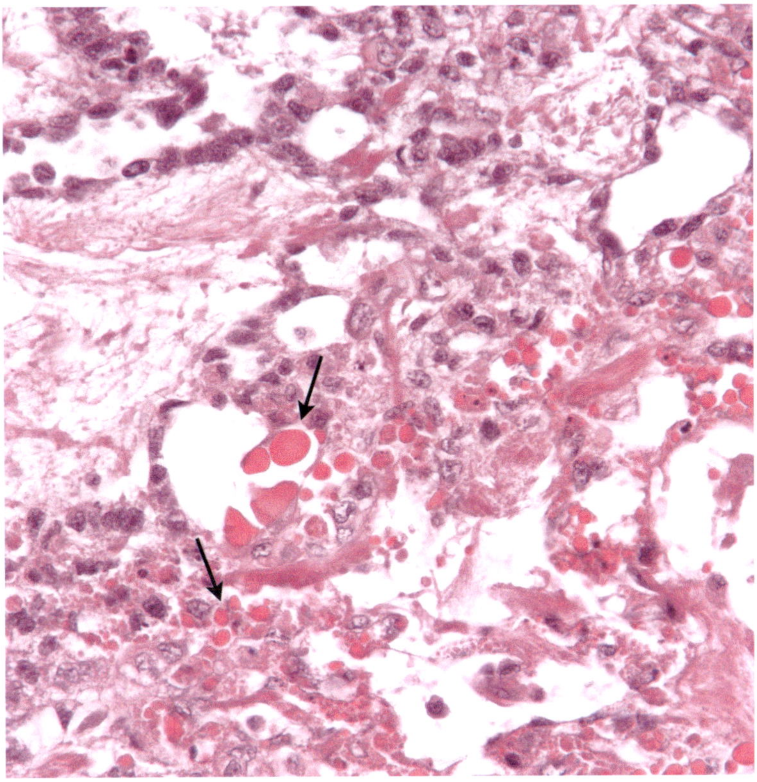

Fig. 13.43 Yolk sac carcinoma. An endodermal sinus pattern with eosinophilic globules of hyalinized basement membrane material (*arrows*) is present

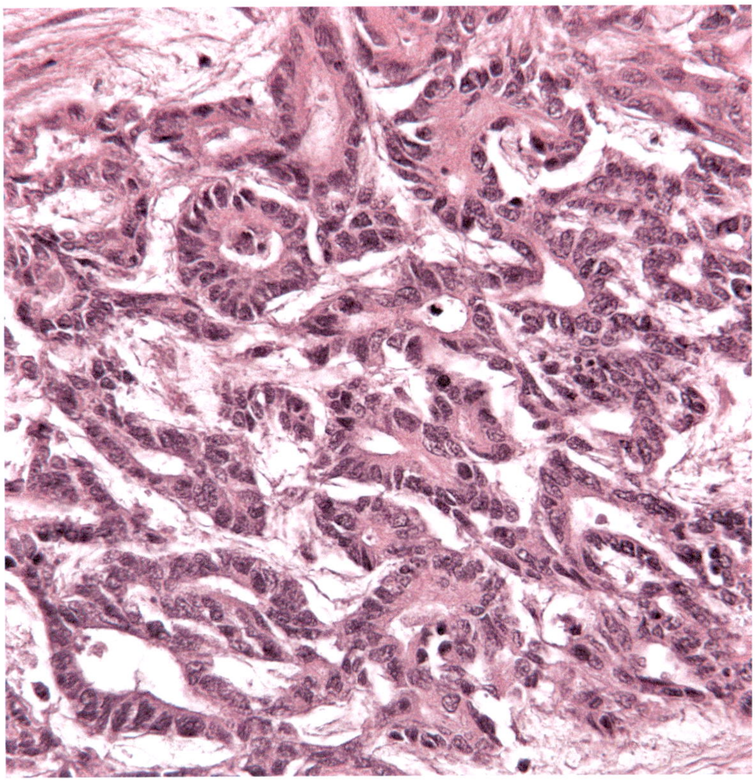

Fig. 13.44 Yolk sac carcinoma. This tumor shows a glandular pattern

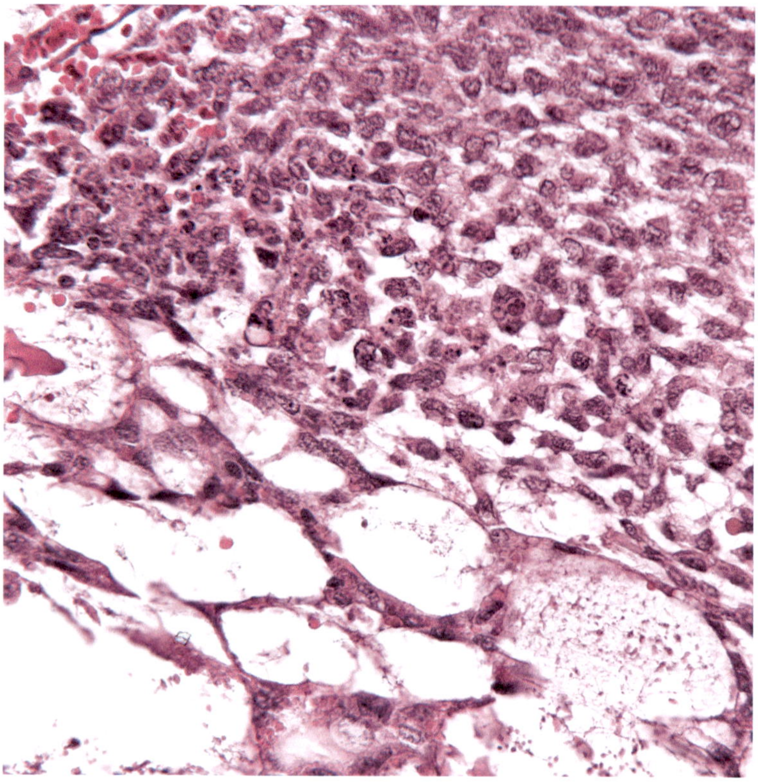

Fig. 13.45 Yolk sac carcinoma. A solid hepatoid pattern is characterized by solid nests or cords of clear cells and prominent hyaline globules

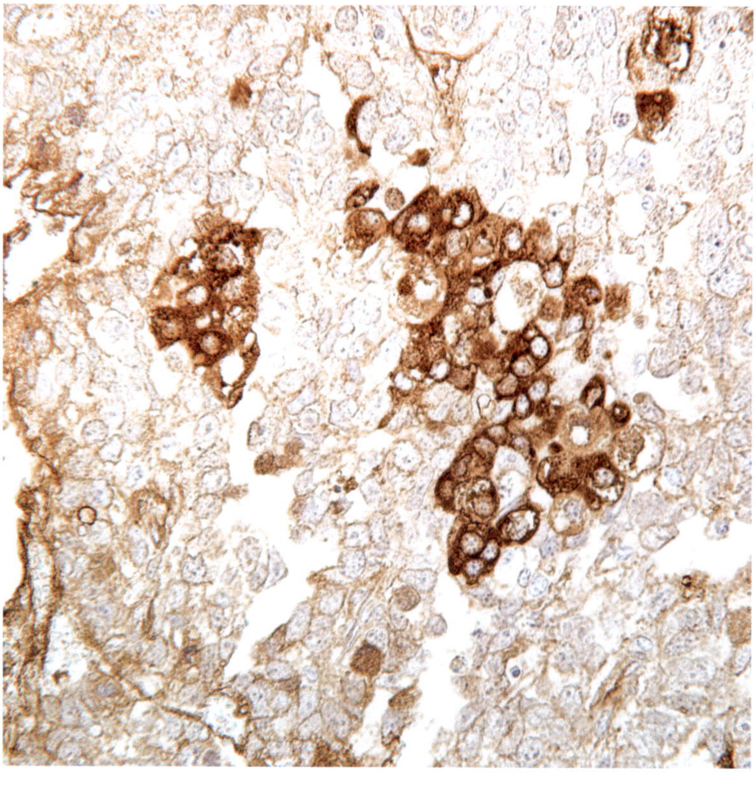

Fig. 13.46 Yolk sac carcinoma. Solid nests of tumor cells stain focally with antibodies to α-fetoprotein

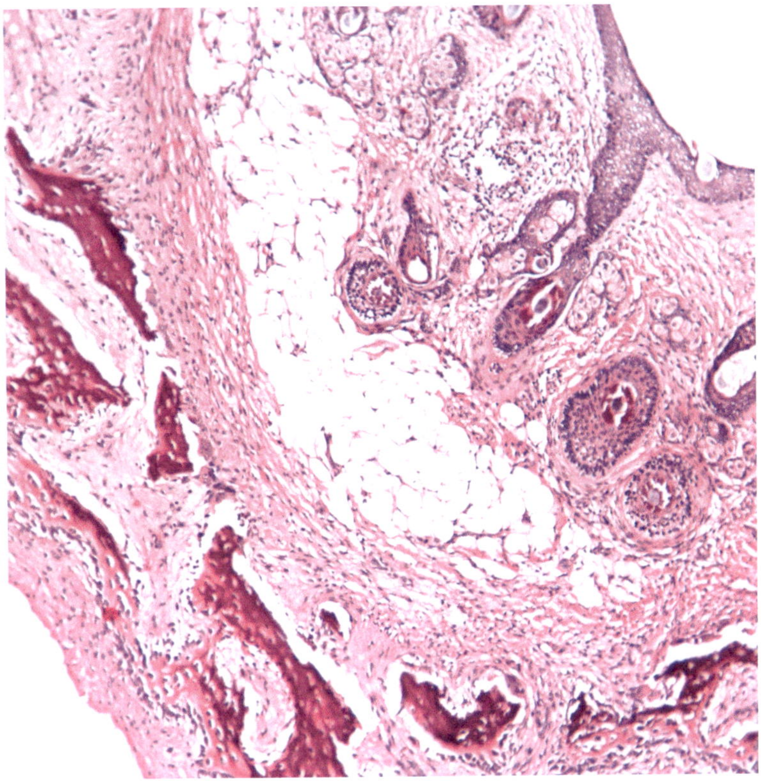

Fig. 13.47 Teratoma. This tumor is composed of mature somatic tissue, including squamous epithelium, fat, and bone

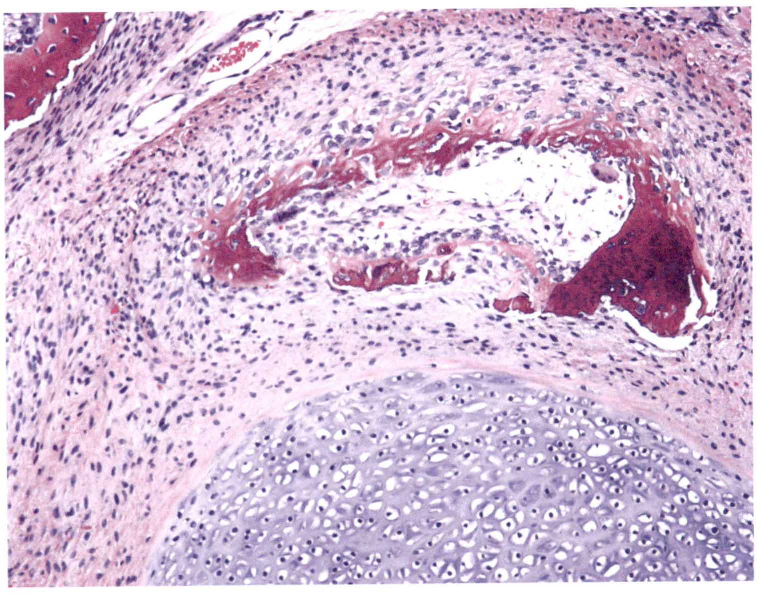

Fig. 13.48 Teratoma. This tumor is composed of newly formed bone and cartilage. The cartilage appears hypercellular and most likely is still growing, although it appears benign

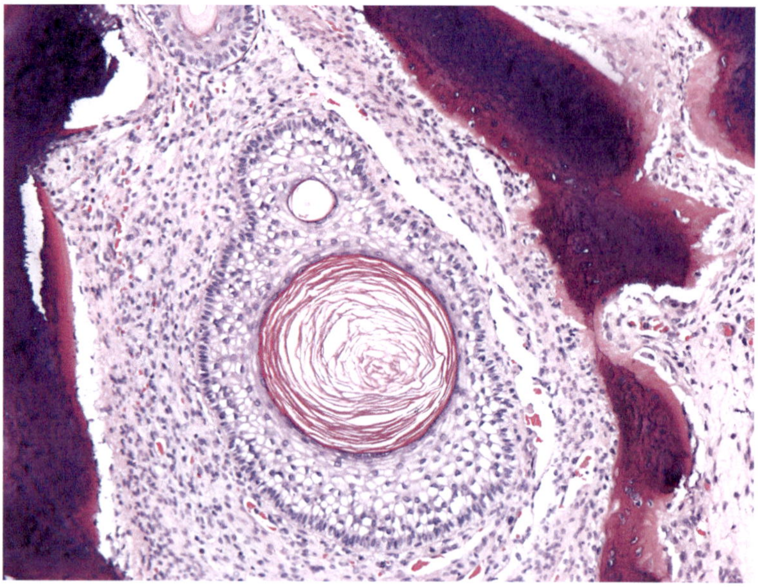

Fig. 13.49 Teratoma. This tumor is composed of calcified bone surrounding a nest of keratinizing squamous epithelium

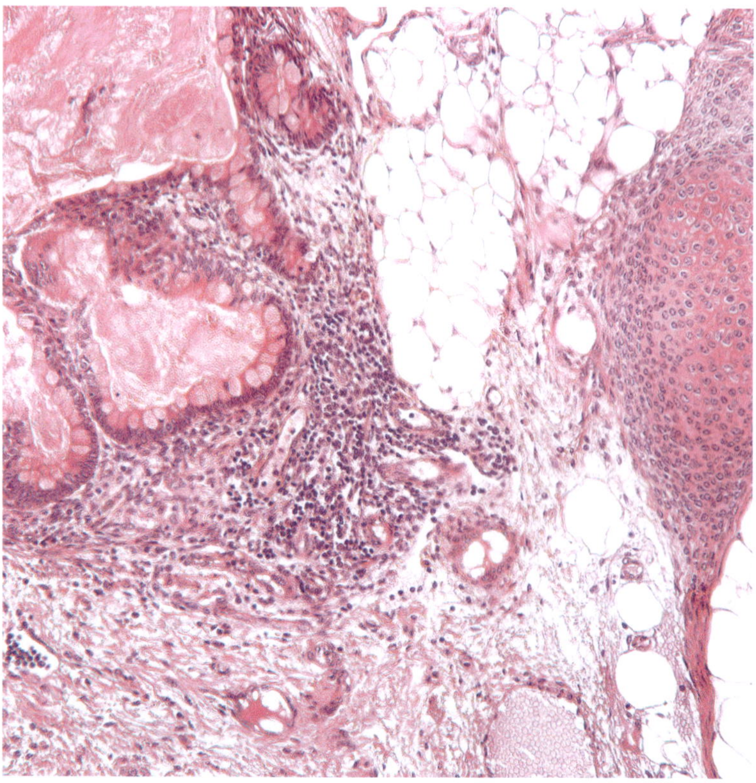

Fig. 13.50 Teratoma. This tumor is composed of mucus-secreting glands surrounded by fibrous tissue containing lymphocytes. Fat tissue and immature cartilage (*right side*) also are present

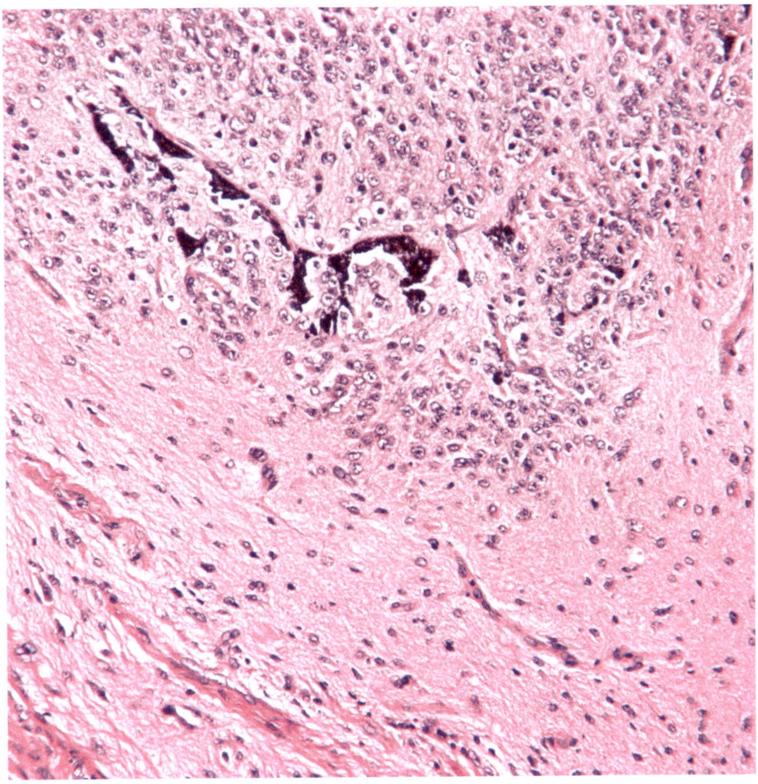

Fig. 13.51 Teratoma. This tumor is composed of a neuropil that appears locally hypercellular and shows foci of calcification, which appear as bluish strands

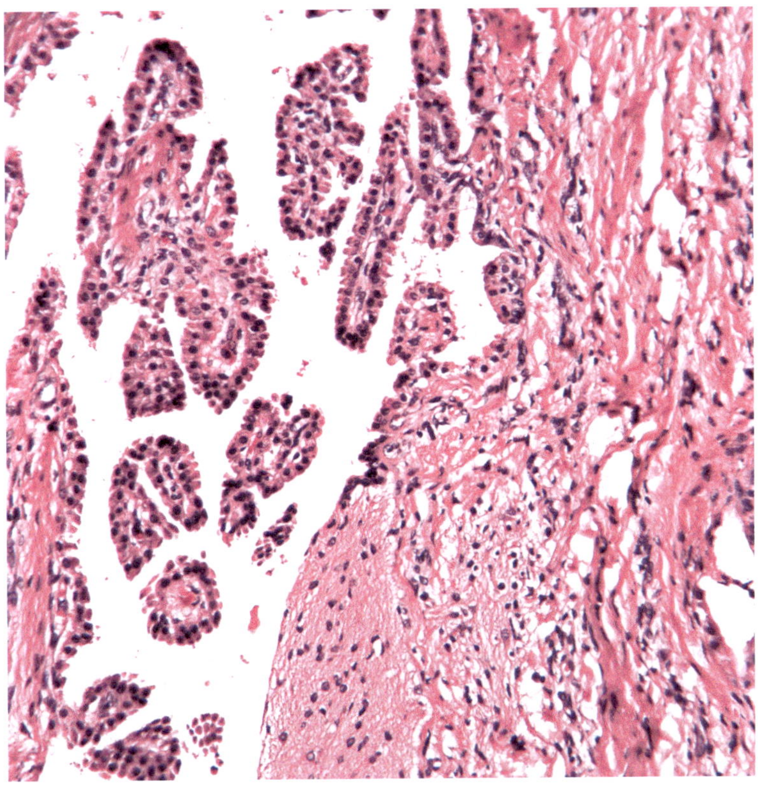

Fig. 13.52 Teratoma. This tumor is composed predominantly of choroid plexus forming papillae lined by cuboidal epithelium and focally attached to the neuropil, best seen in the lower part of the figure

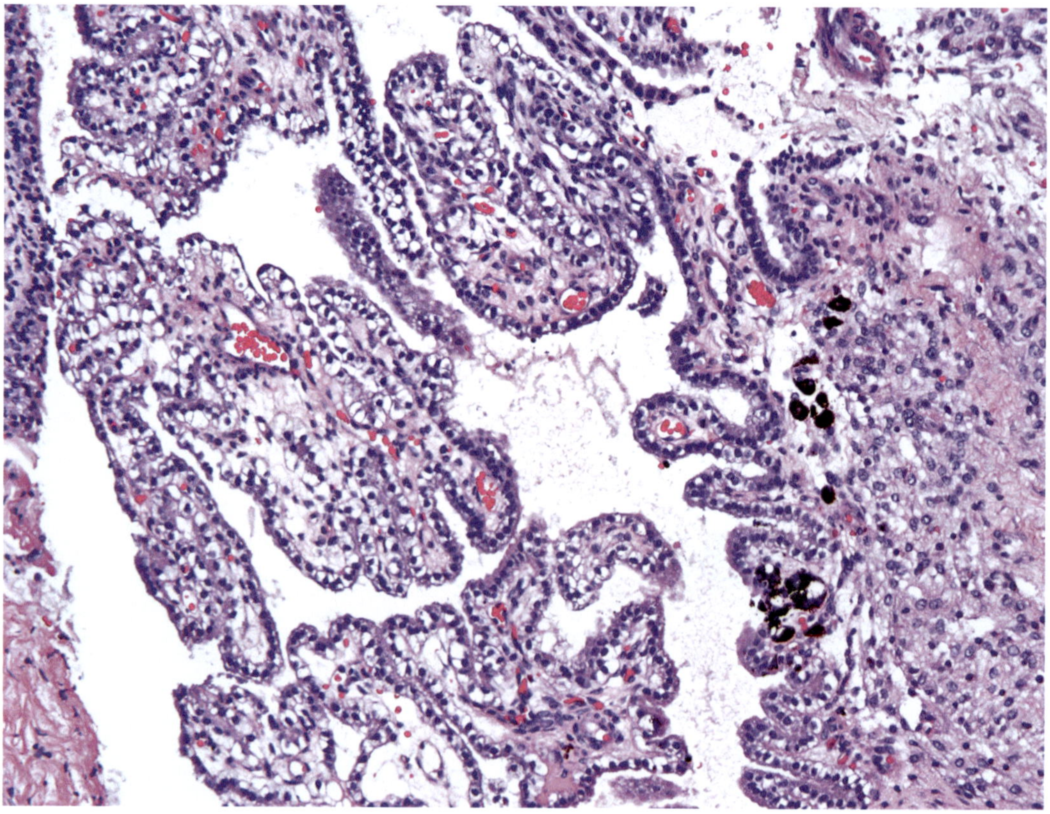

Fig. 13.53 Teratoma. This tumor is composed of papillary structures that focally show melanotic pigmentation, typical of eye pigment

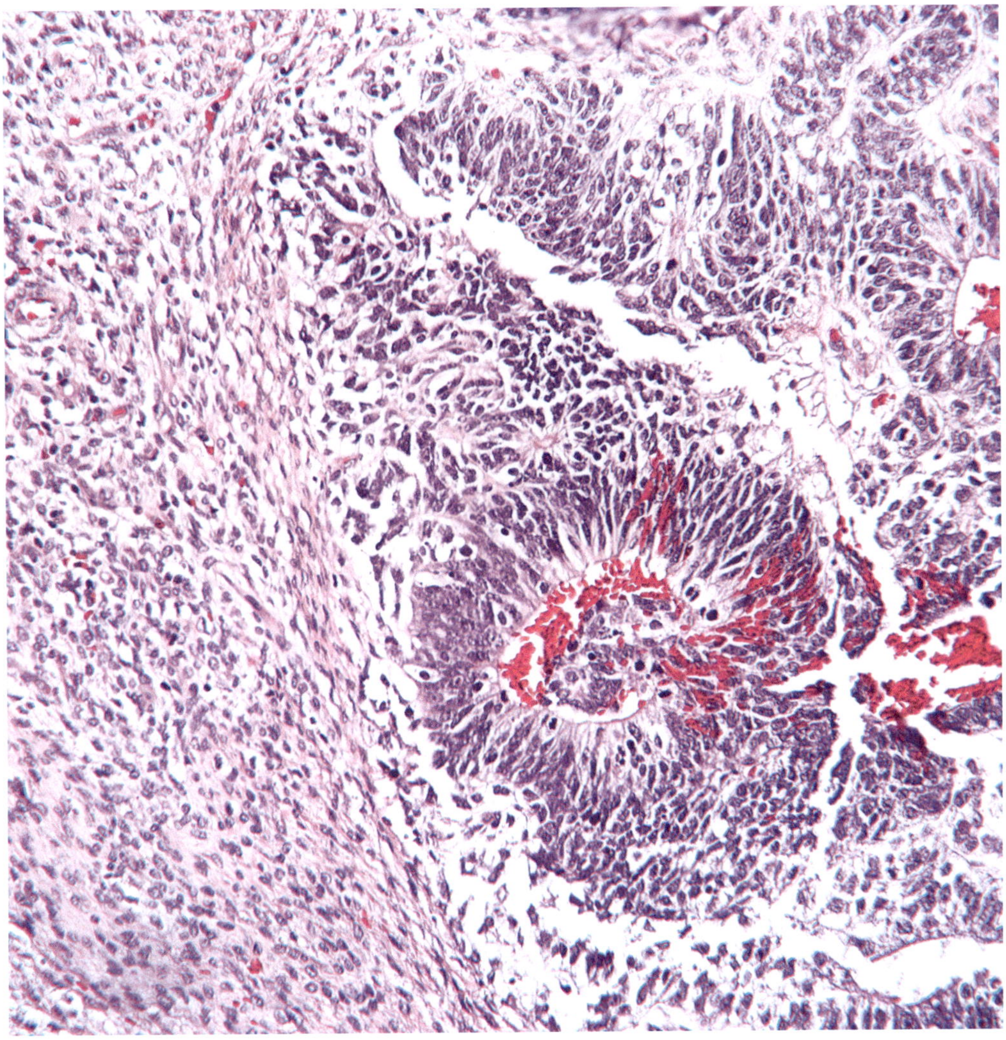

Fig. 13.54 Immature teratoma. This tumor contains numerous immature nerve cell precursors arranged in neuroblastic tubelike rosettes and nests

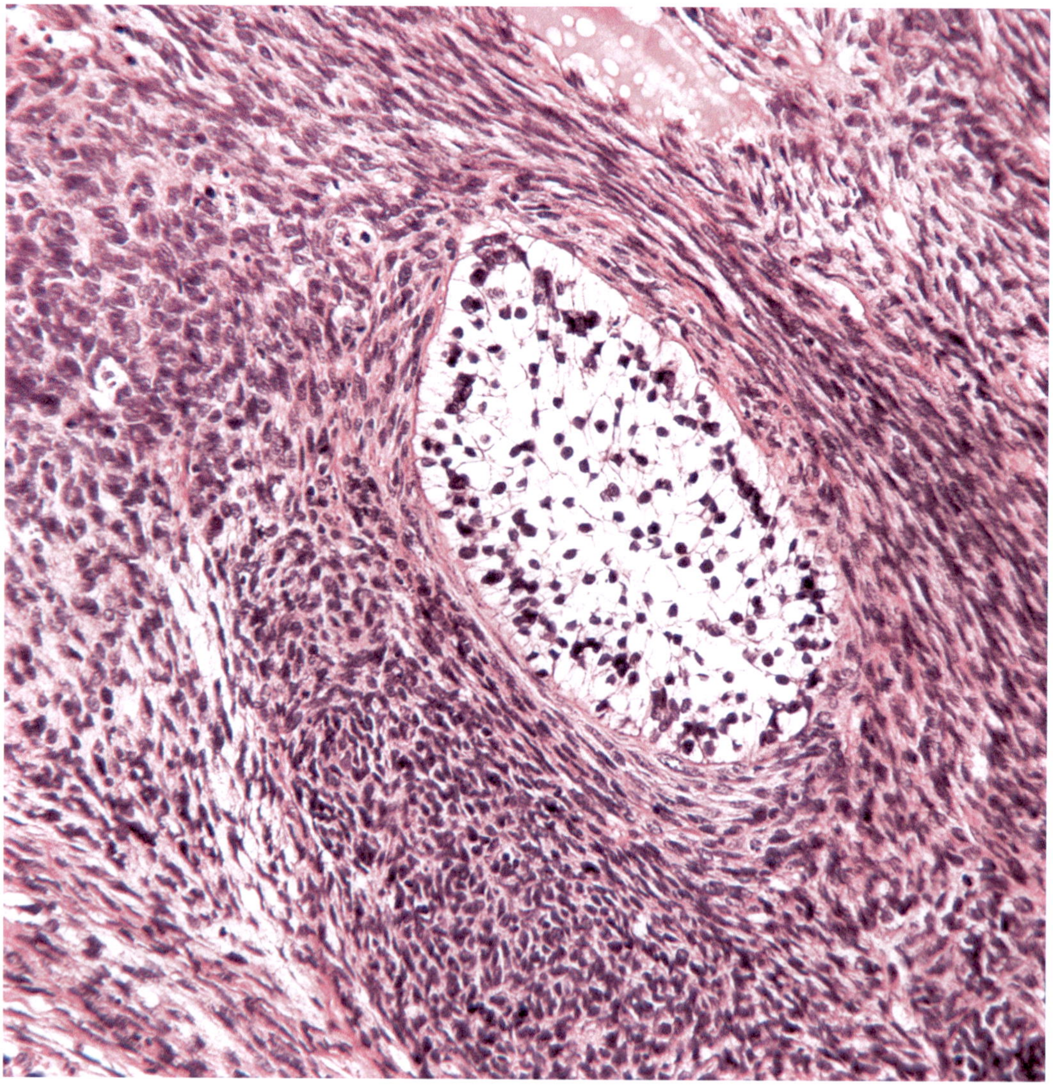

Fig. 13.55 Immature teratoma. This component of the tumor shown in Fig. 13.54 contains a nest of glycogen-rich fetal-like squamous epithelium surrounded by nondescript hypercellular ("fetal-like") stroma

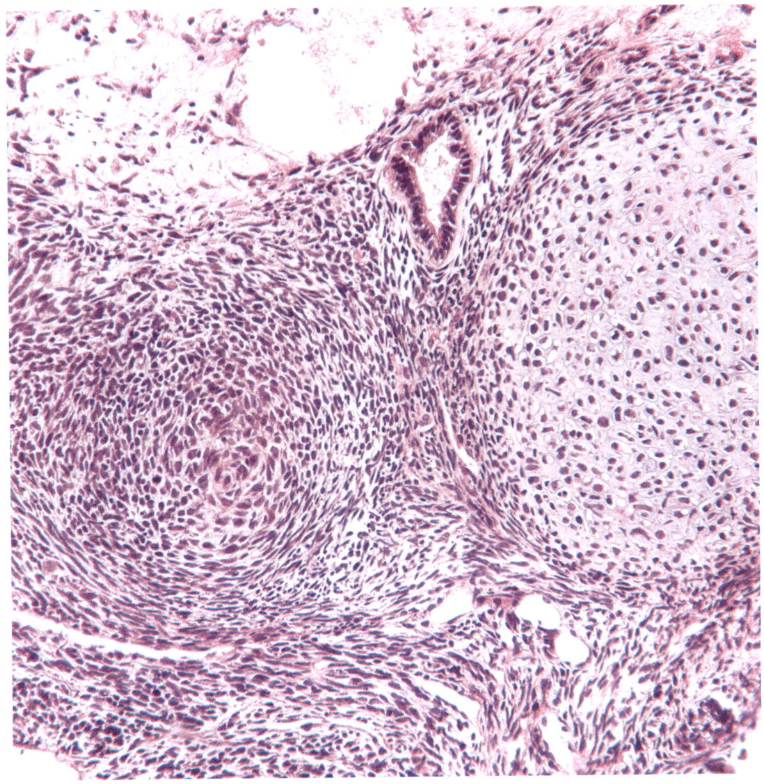

Fig. 13.56 Immature teratoma. This component of the tumor shown in Fig. 13.54 contains two nests of cartilage formation, one of which (to the *left*) is barely recognizable as fetal cartilage

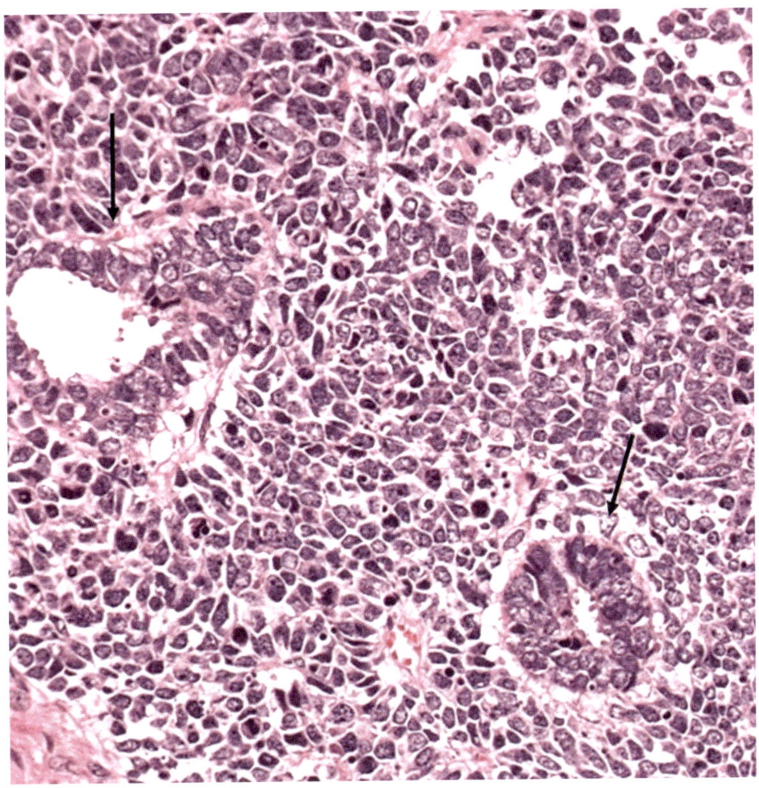

Fig. 13.57 Teratoma with somatic-type malignancy. These areas correspond to a nephroblastoma consisting of immature tubules (*arrows*) and blastematous stroma

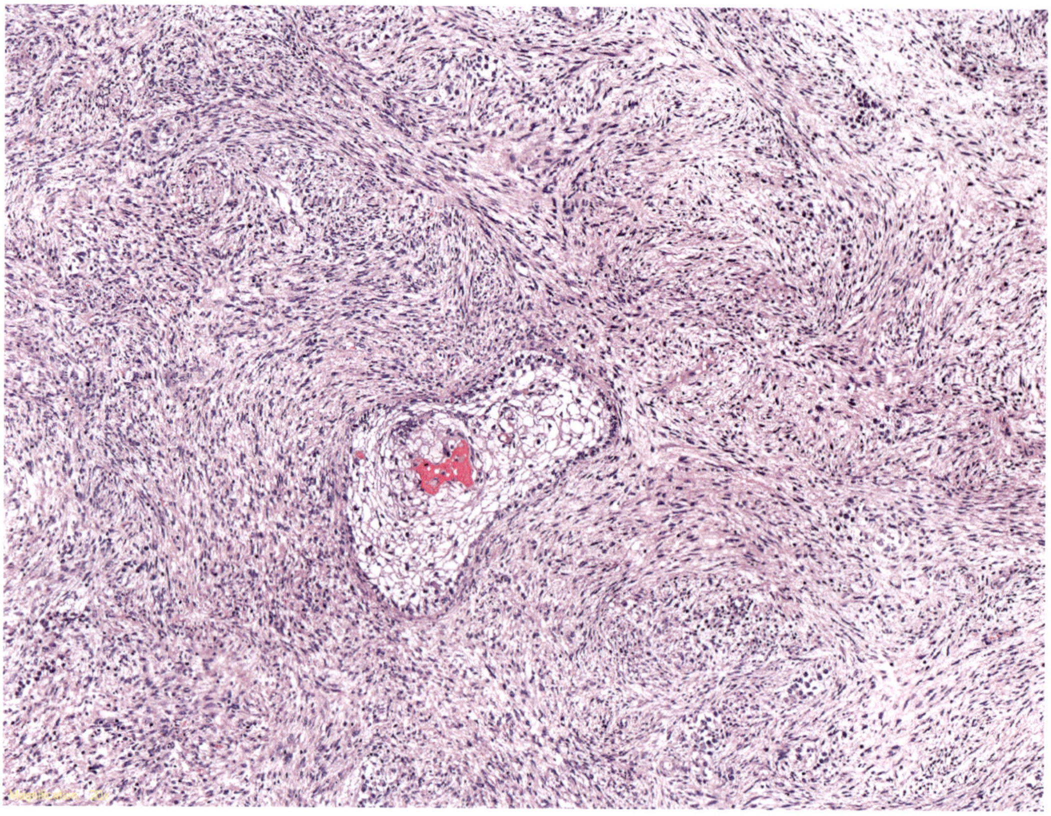

Fig. 13.58 Teratoma with somatic-type malignancy. This tumor contains a nest of squamous epithelium surrounded by leiomyosarcoma-like stromal tissue

Fig. 13.59 Yolk sac tumor of infancy and childhood. This tumor consists of loosely arranged interanastomosing strands of flattened epithelial cells arranged in a reticular pattern

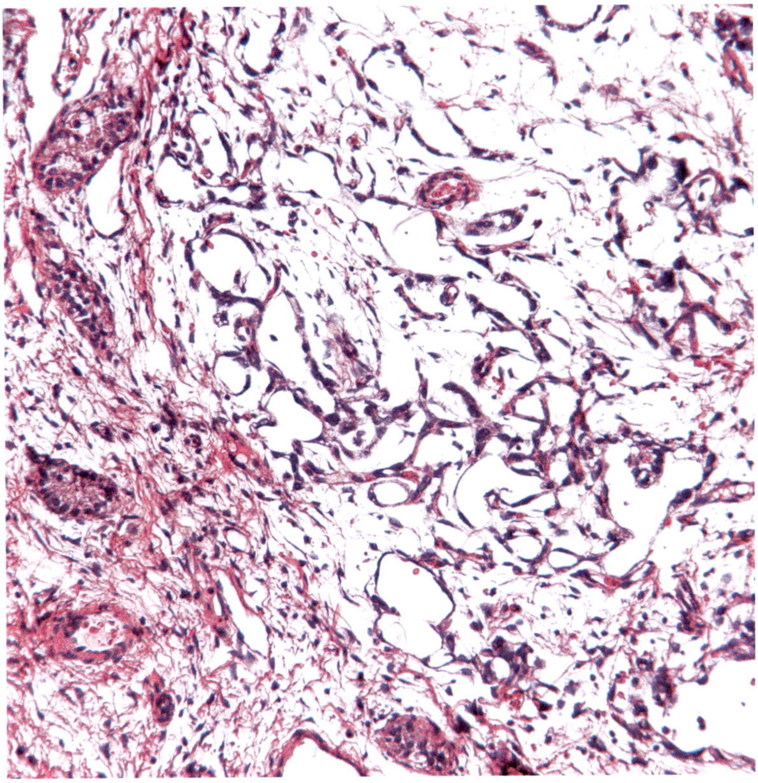

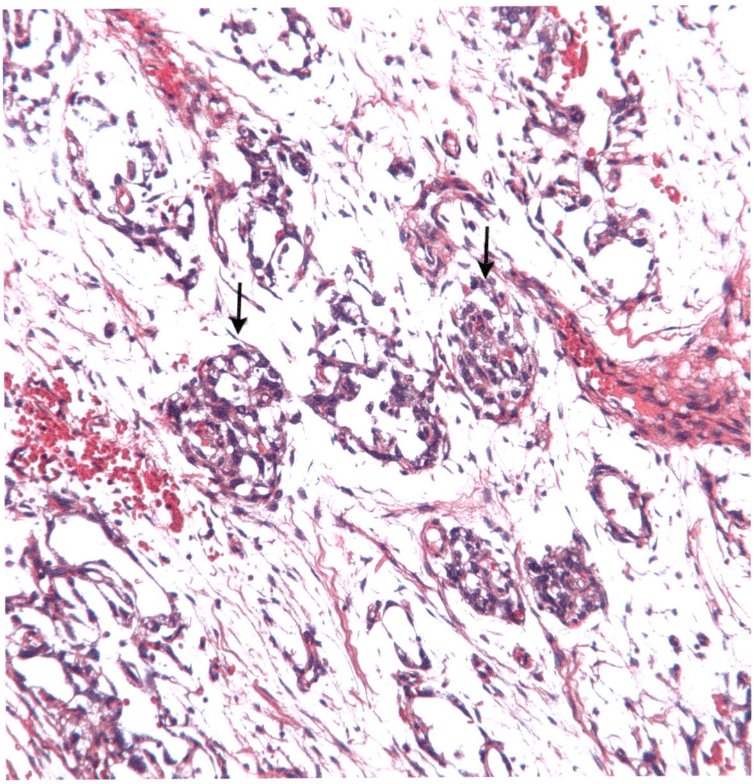

Fig. 13.60 Yolk sac tumor of infancy. This tumor grows in a reticular pattern but also contains numerous glomeruloid structures (Schiller–Duval bodies, *arrows*)

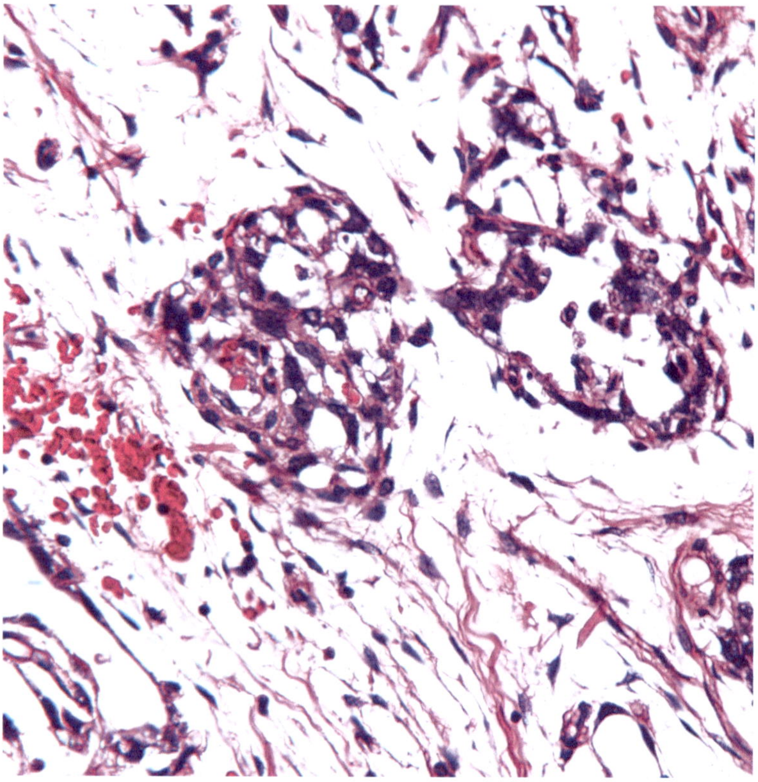

Fig. 13.61 Yolk sac tumor of infancy and childhood. Higher magnification of Fig. 13.60 shows the glomeruloid structures (Schiller–Duval bodies)

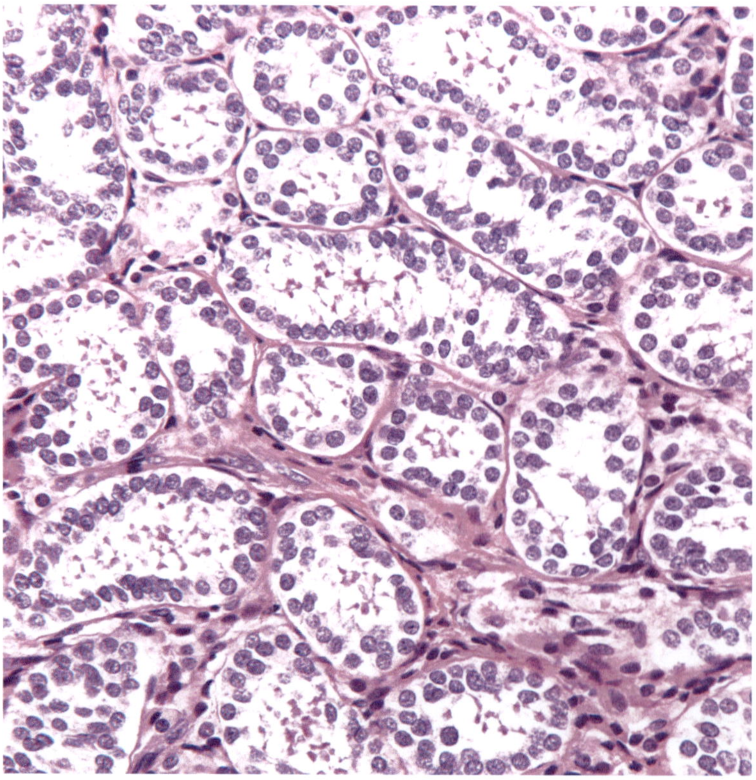

Fig. 13.62 Sertoli cell tumor. This tumor forms tubules lined by fetal-like Sertoli cells

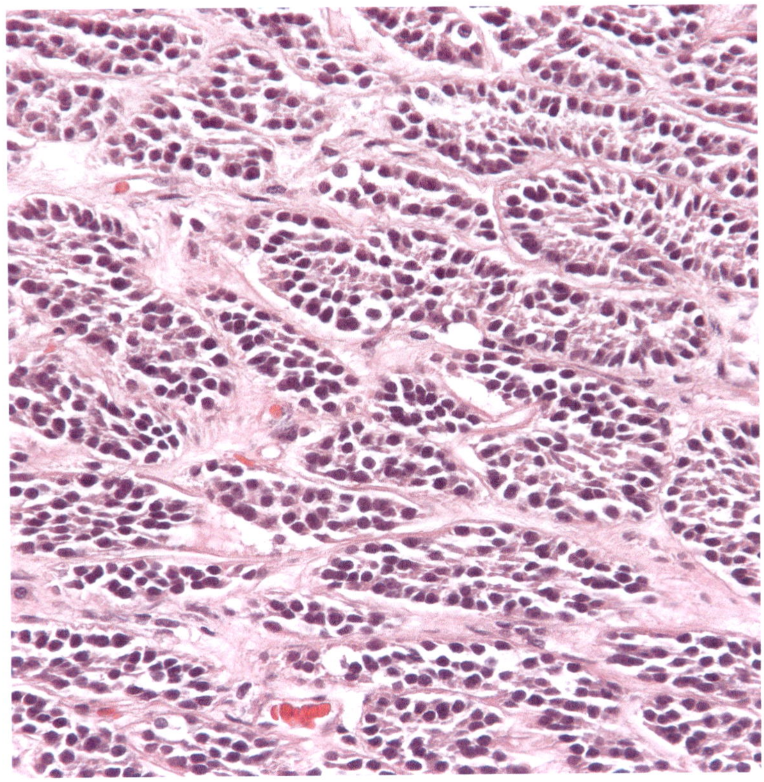

Fig. 13.63 Sertoli cell tumor. This tumor forms cords of cells and nests surrounded by basement membrane-like hyalinized fibrous tissue

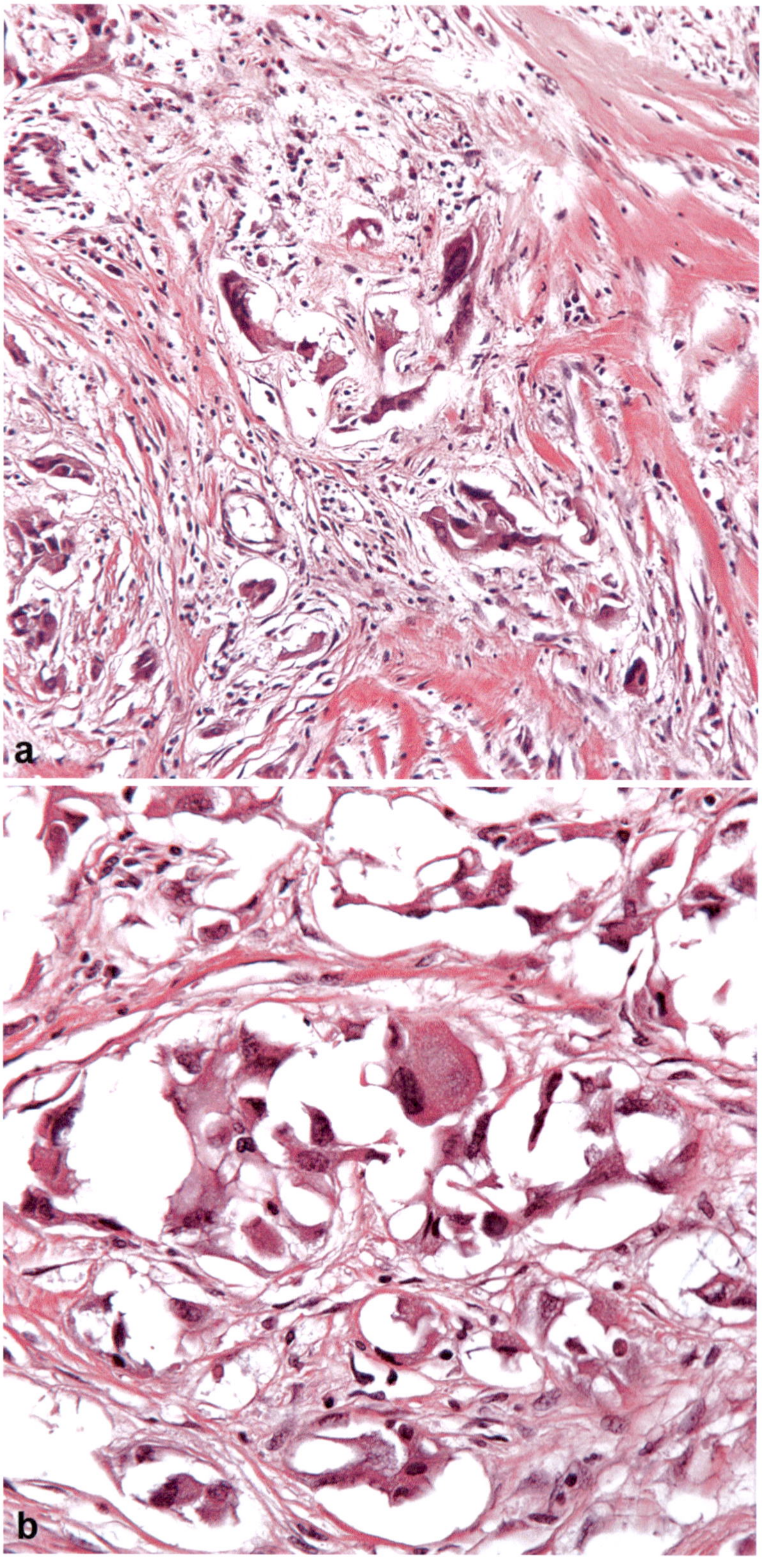

Fig. 13.64 Sertoli cell tumor. (**a**) This large-cell calcifying variant of Sertoli cell tumor is typically fibrotic, focally calcified, and composed of large cells irregularly arranged in groups within loose connective stroma. (**b**) This higher magnification view of a large-cell calcifying Sertoli cell tumor shows large cells with abundant cytoplasm and elongated nuclei forming abortive nests

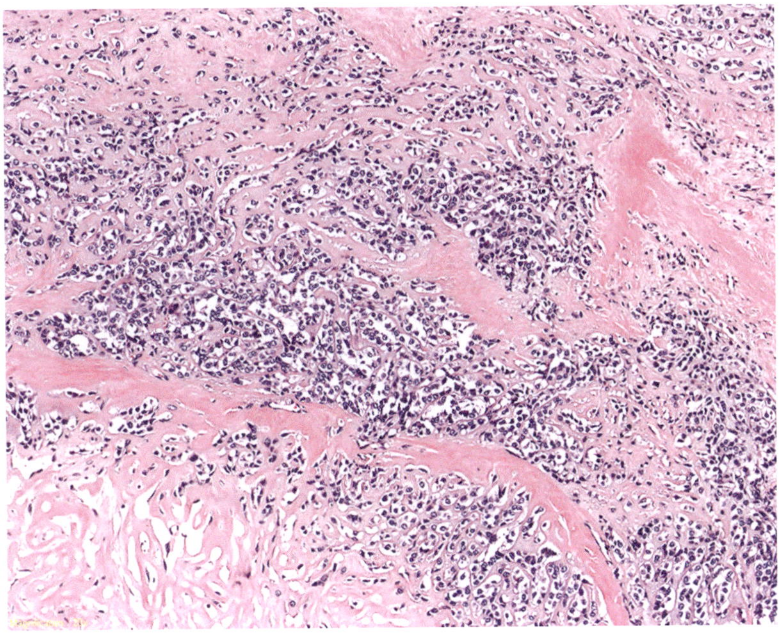

Fig. 13.65 Sertoli cell tumor. A sclerosing tumor pattern is characterized by dense hyalinized fibrous stroma penetrating between the tumor cells or subdividing the tumor into cellular nests

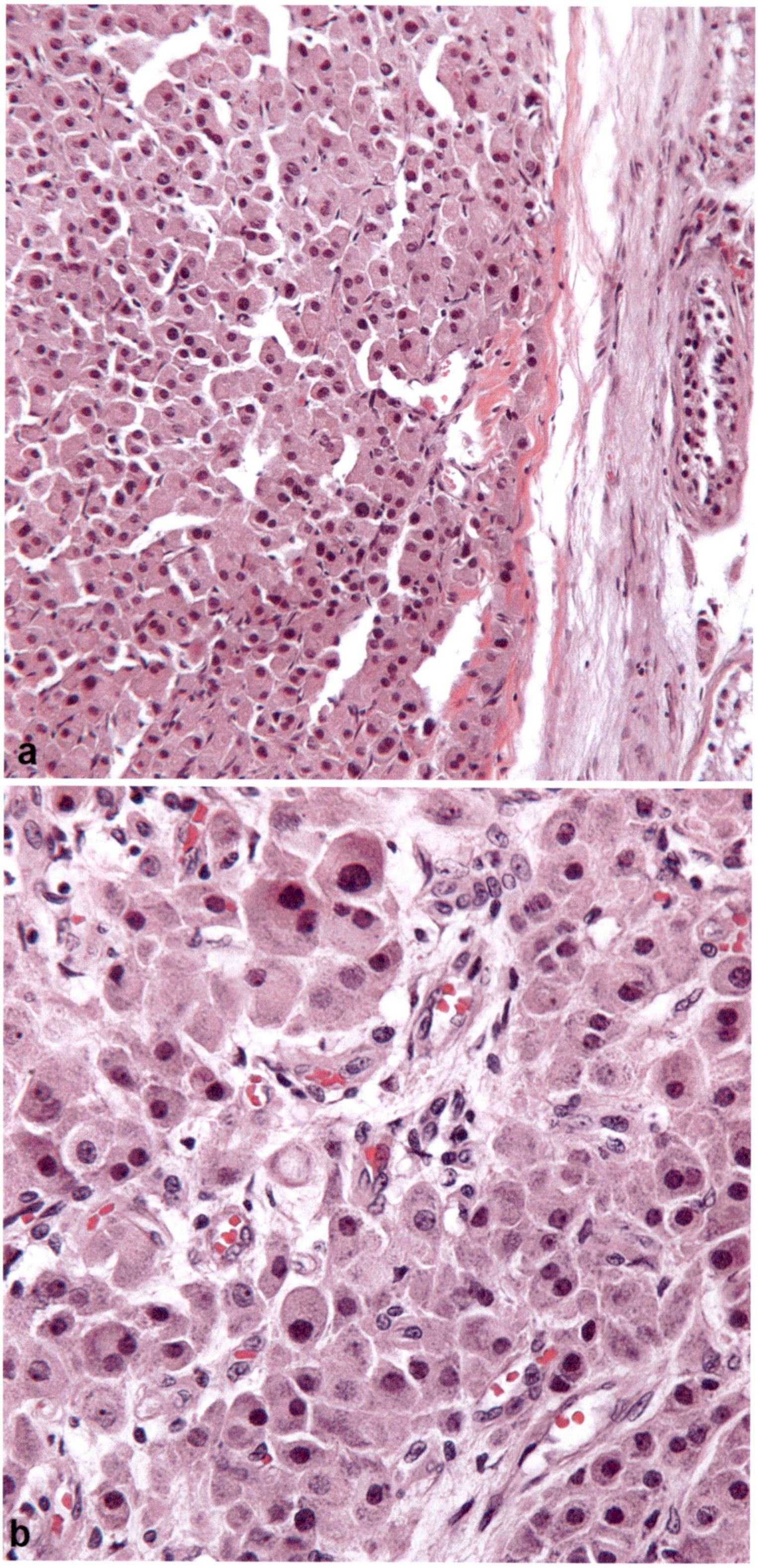

Fig. 13.66 Leydig cell tumor. (**a**) This tumor is composed of uniform polygonal cells with round nuclei and well-developed eosinophilic cytoplasm. A fibrous capsule separates the tumor from the testicular tissue on the right side. (**b**) Higher magnification of the tumor shows the solid tumor growth pattern and the uniformity of well-differentiated cells with abundant eosinophilic cytoplasm

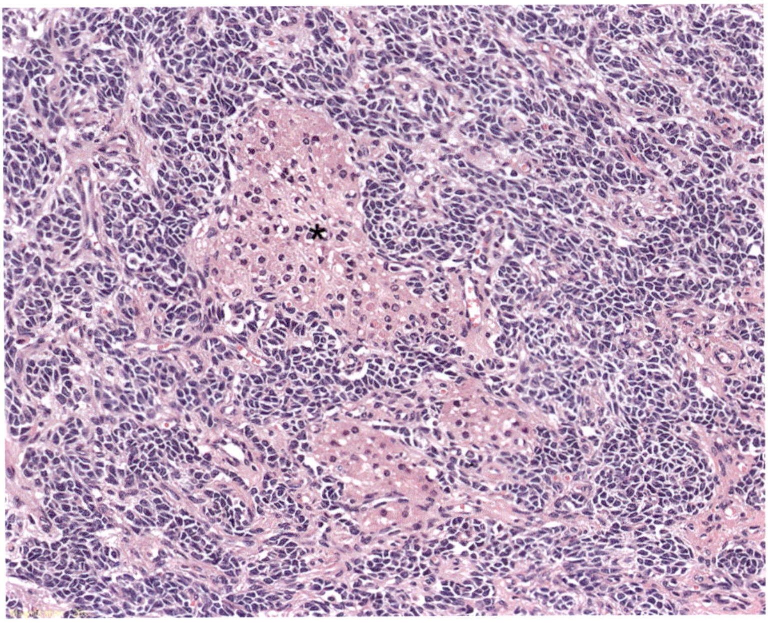

Fig. 13.67 Mixed sex cord cell tumor. This tumor consists of eosinophilic Leydig cells (*asterisk*) and Sertoli cells forming abortive tubules

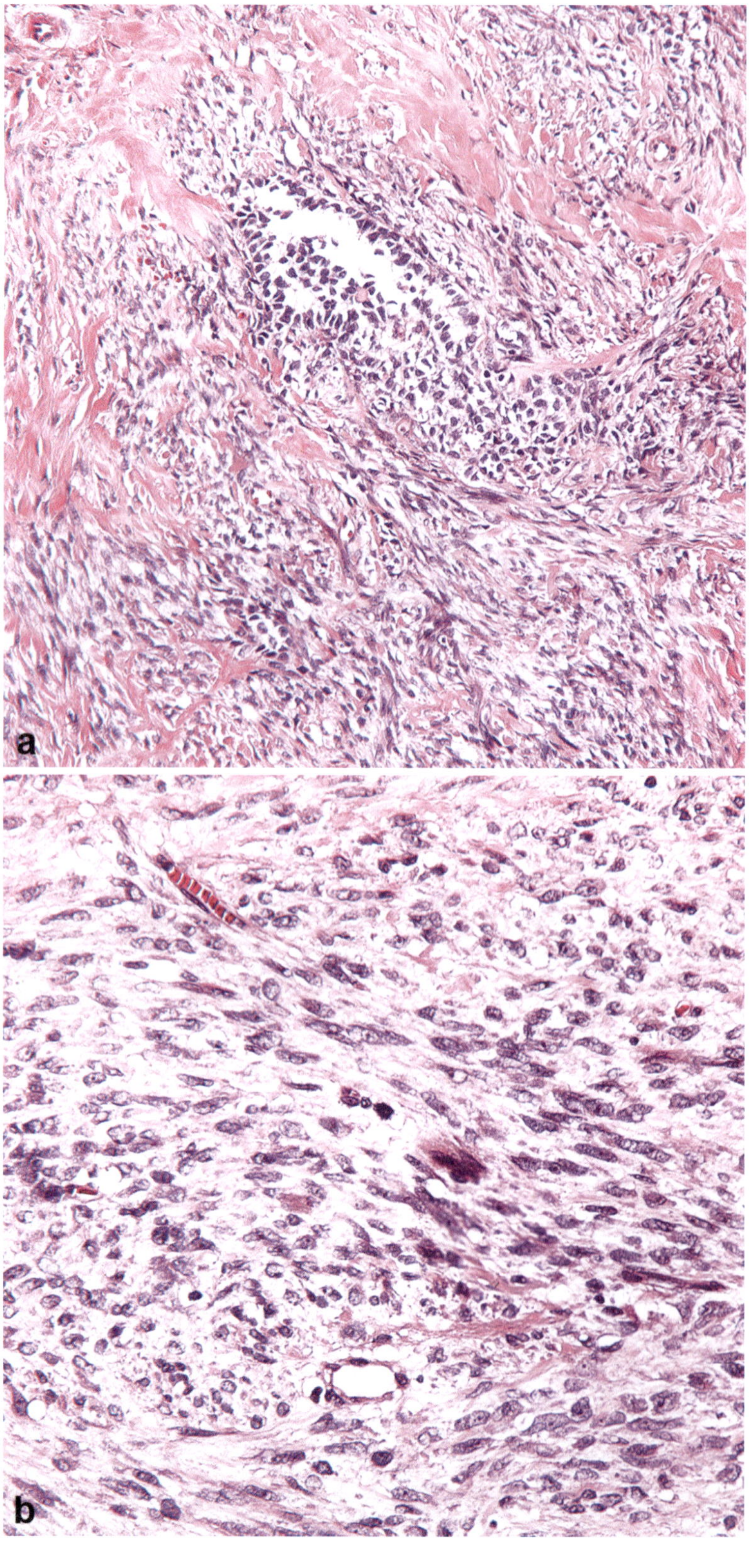

Fig. 13.68 Incompletely differentiated sex cord tumor. (**a**) This tumor consists of dense fibrous stroma with focal grouping of columnar cells around a central lumen. Tumor cells also form cords and abortive tubules. (**b**) Higher magnification of the spindle cell "undifferentiated" component of this tumor illustrates the fibroblastic appearance of tumor cells with only focal condensation but no tubule formation. The sex cord nature of this tumor can be demonstrated immunohistochemically with antibodies to inhibin

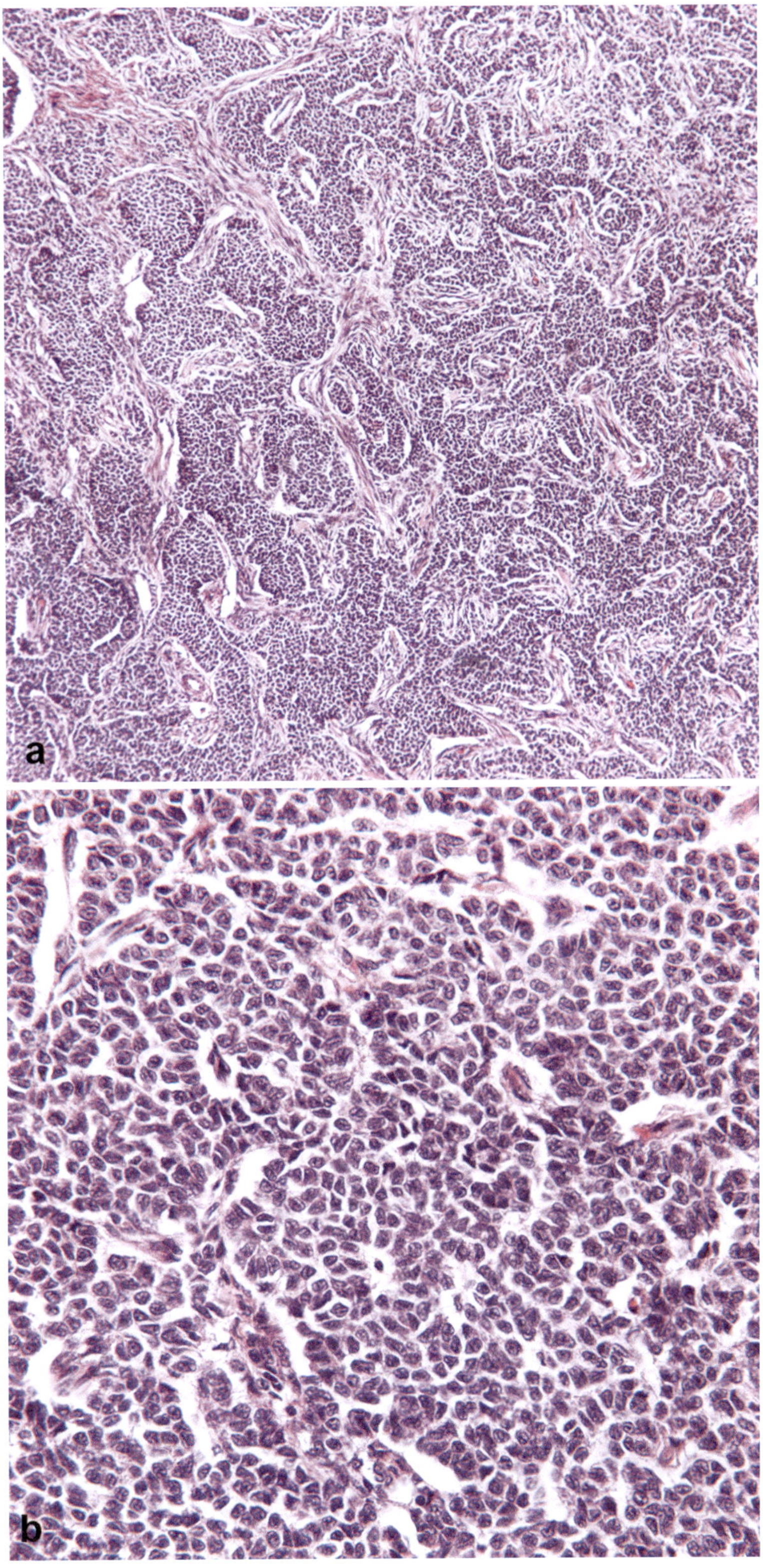

Fig. 13.69 Granulosa cell tumor. (**a**) This tumor forms a nest and cord composed of small cells with scant cytoplasm, resembling its ovarian counterpart. (**b**) Higher magnification of the tumor shows cords and follicular nesting of cells similar to those observed in granulosa cell tumor of the ovary

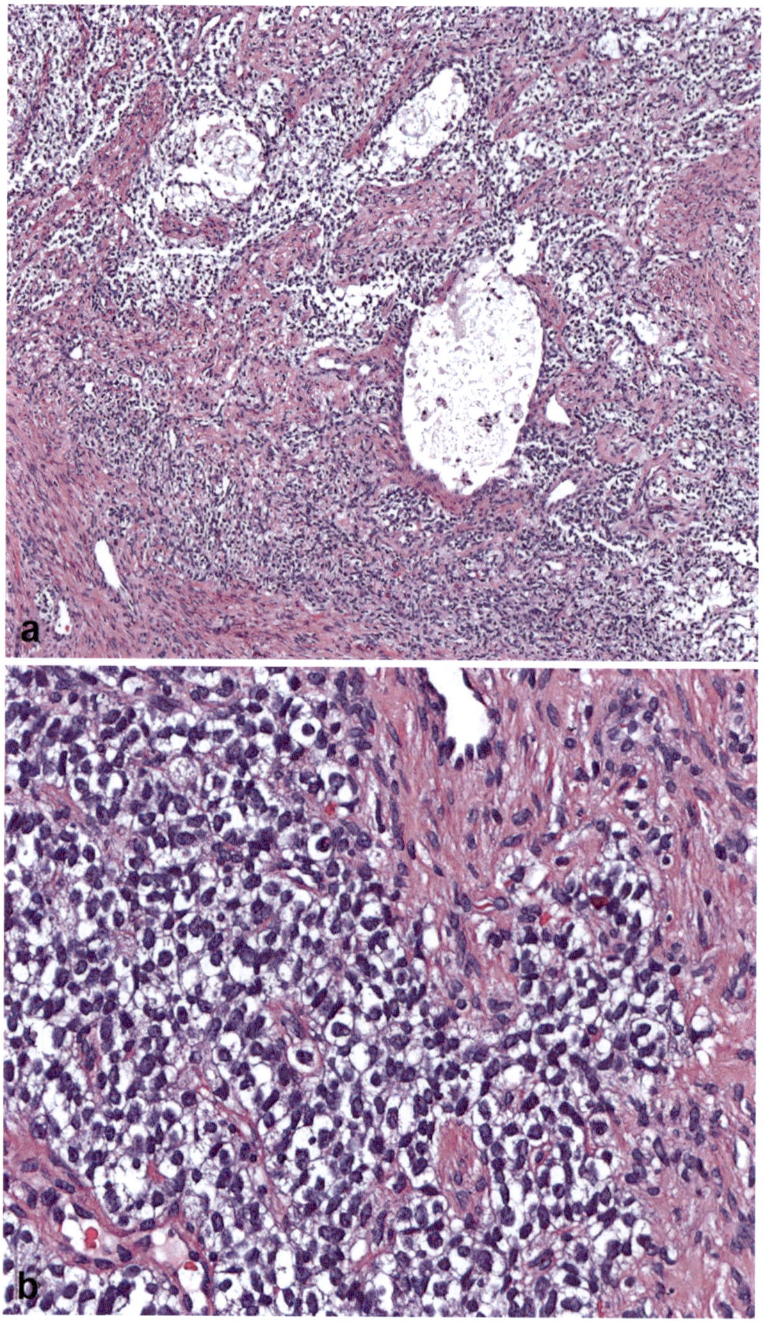

Fig. 13.70 Juvenile granulosa cell tumor. (**a**) This neonatal tumor is composed of cellular nests surrounding cystic structures. (**b**) Solid parts of the tumor are composed of uniform cells with oval or round nuclei and clear cytoplasm

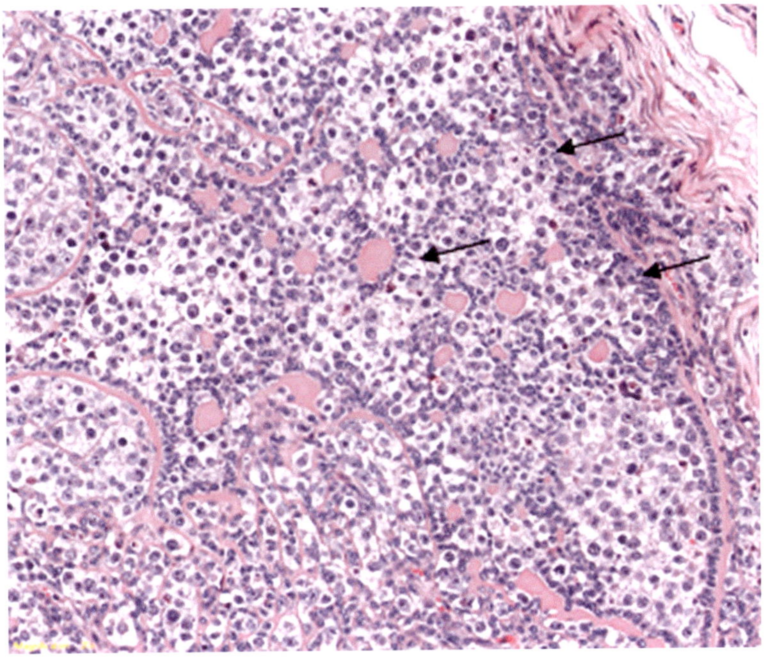

Fig. 13.71 Gonadoblastoma. This tumor is composed of round germ cells and sex cord cells surrounding round globules composed of basement membrane material (*arrows*)

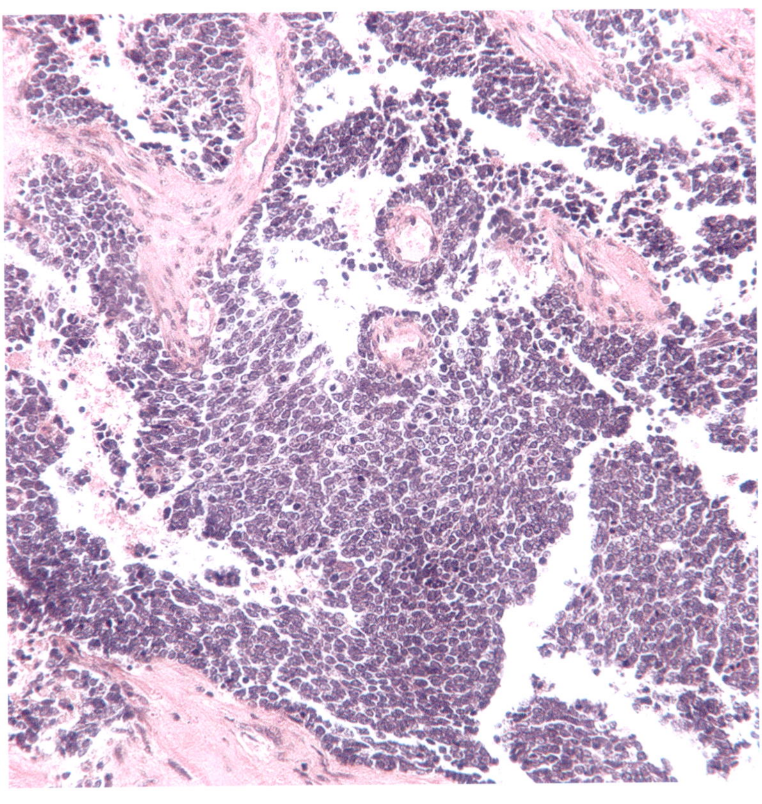

Fig. 13.72 Nephroblastoma. This tumor is composed of blastema-like undifferentiated cells, resembling renal Wilms' tumors

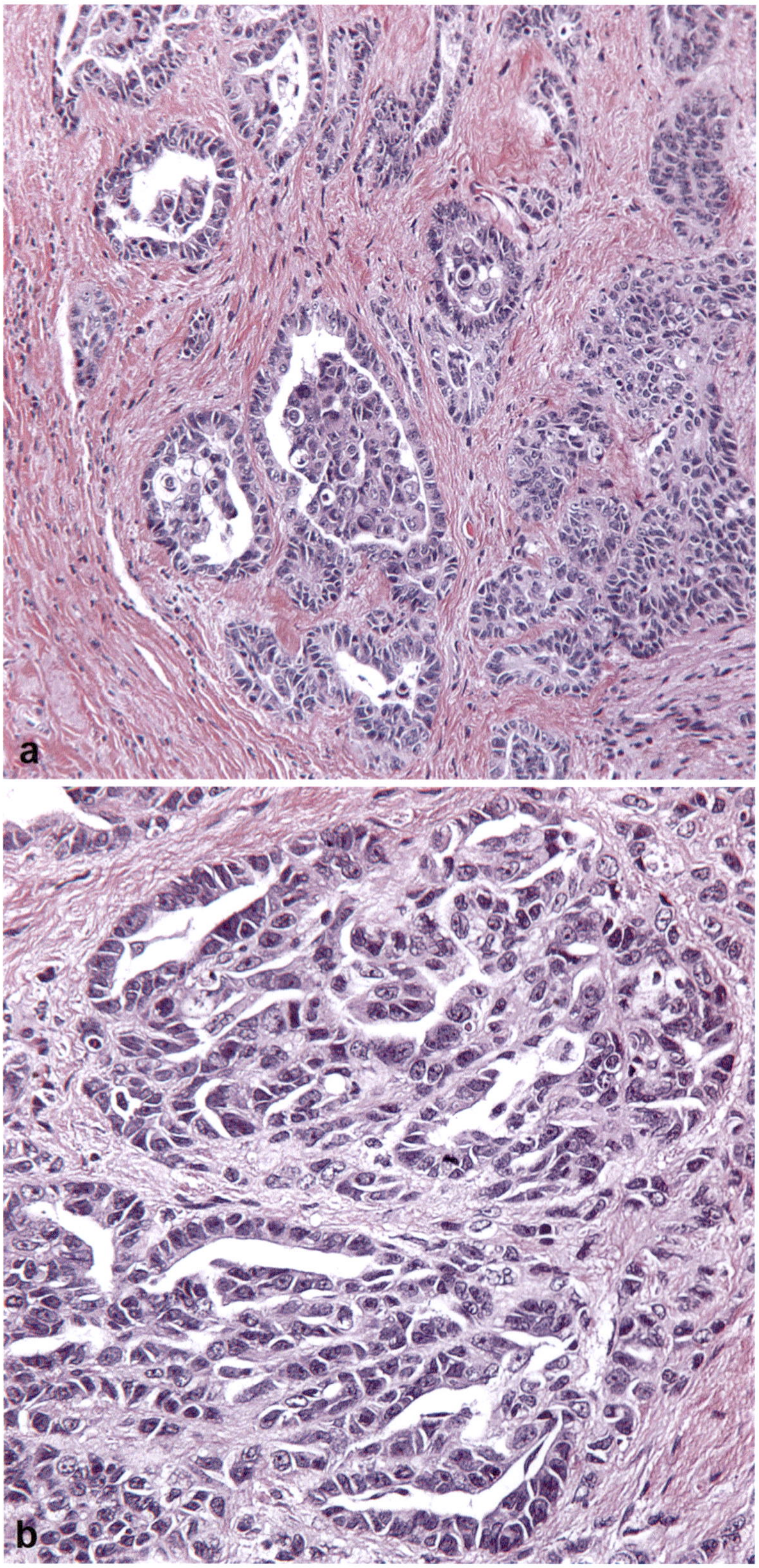

Fig. 13.73 Rete testis carcinoma. (**a**) This tumor has features of an adenocarcinoma, with intraluminal papillary projections and solid clumps of hyperchromatic cells on the right side of the figure. (**b**) Higher magnification illustrates the interanastomosing cords and papillae, which obliterate the larger glands, giving them a cribriform appearance

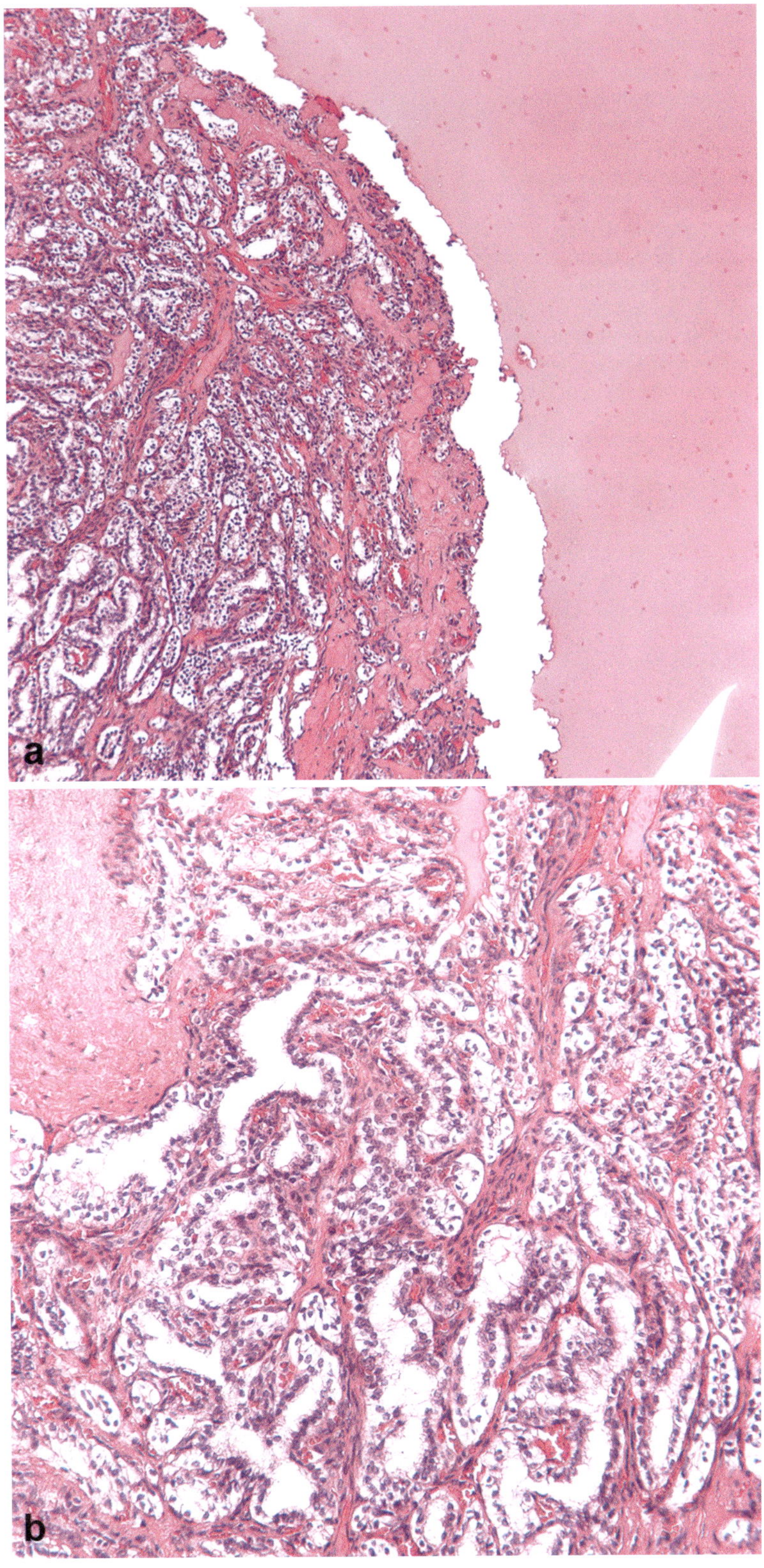

Fig. 13.74 Papillary cystadenoma of the epididymis. (**a**) These tumors typically occur in patients with von Hippel–Lindau syndrome and are composed of duct-like and papillary structures protruding into dilated cystic spaces partially filled with proteinaceous material. (**b**) Higher magnification shows that the tumor is composed of clear cells forming duct-like structures

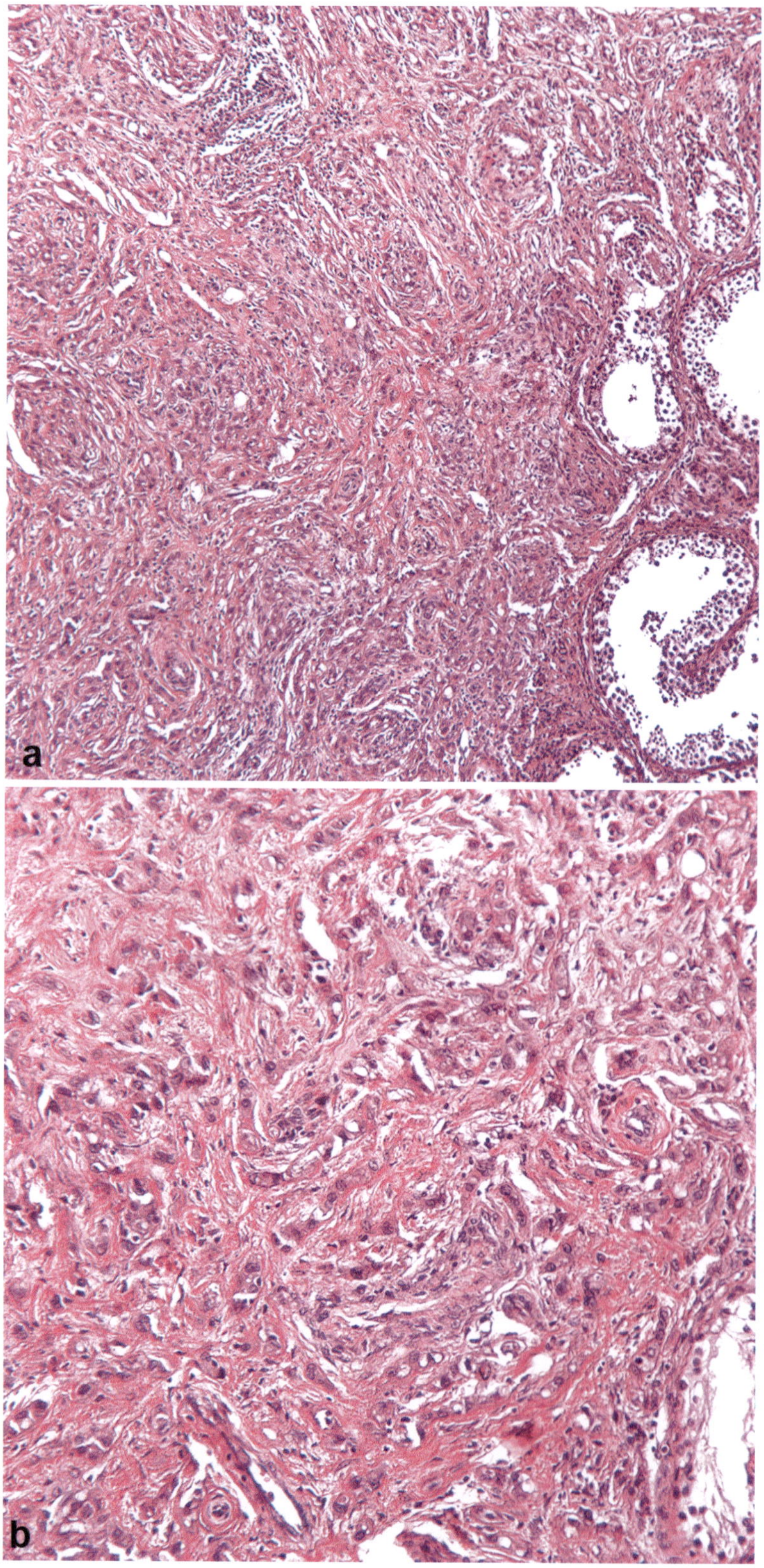

Fig. 13.75 Adenomatoid tumor of the testis. (**a**) This tumor is composed of cuboidal cells lining slit-like spaces in a fibromuscular stroma. (**b**) Higher magnification of the tumor illustrates the flat cuboidal cells lining the slit-like spaces. These cells react with antibodies to keratin and have been shown to be mesothelial in origin. Adenomatoid tumors represent the most common neoplasms of the testicular adnexa

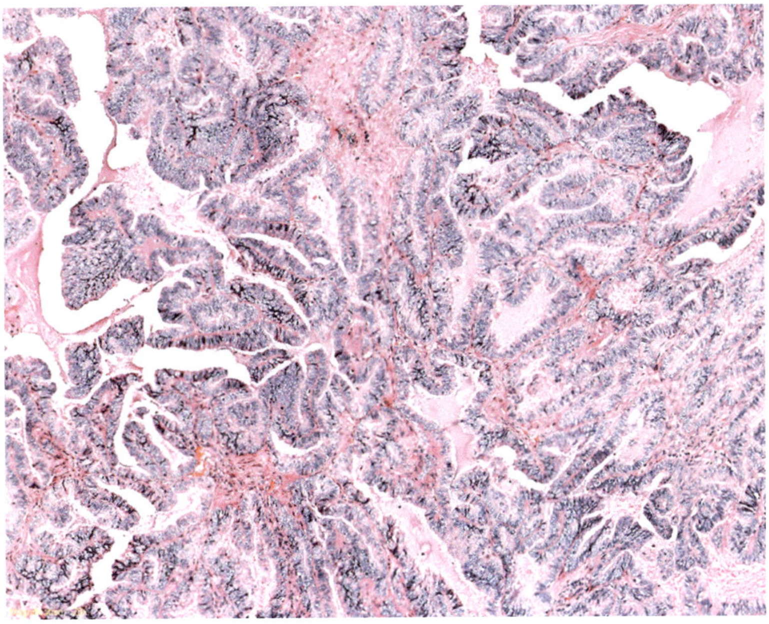

Fig. 13.76 Borderline paratesticular müllerian cystadenoma. This tumor, composed of atypical columnar cells lining papillae, resembles serous tumors of the ovary

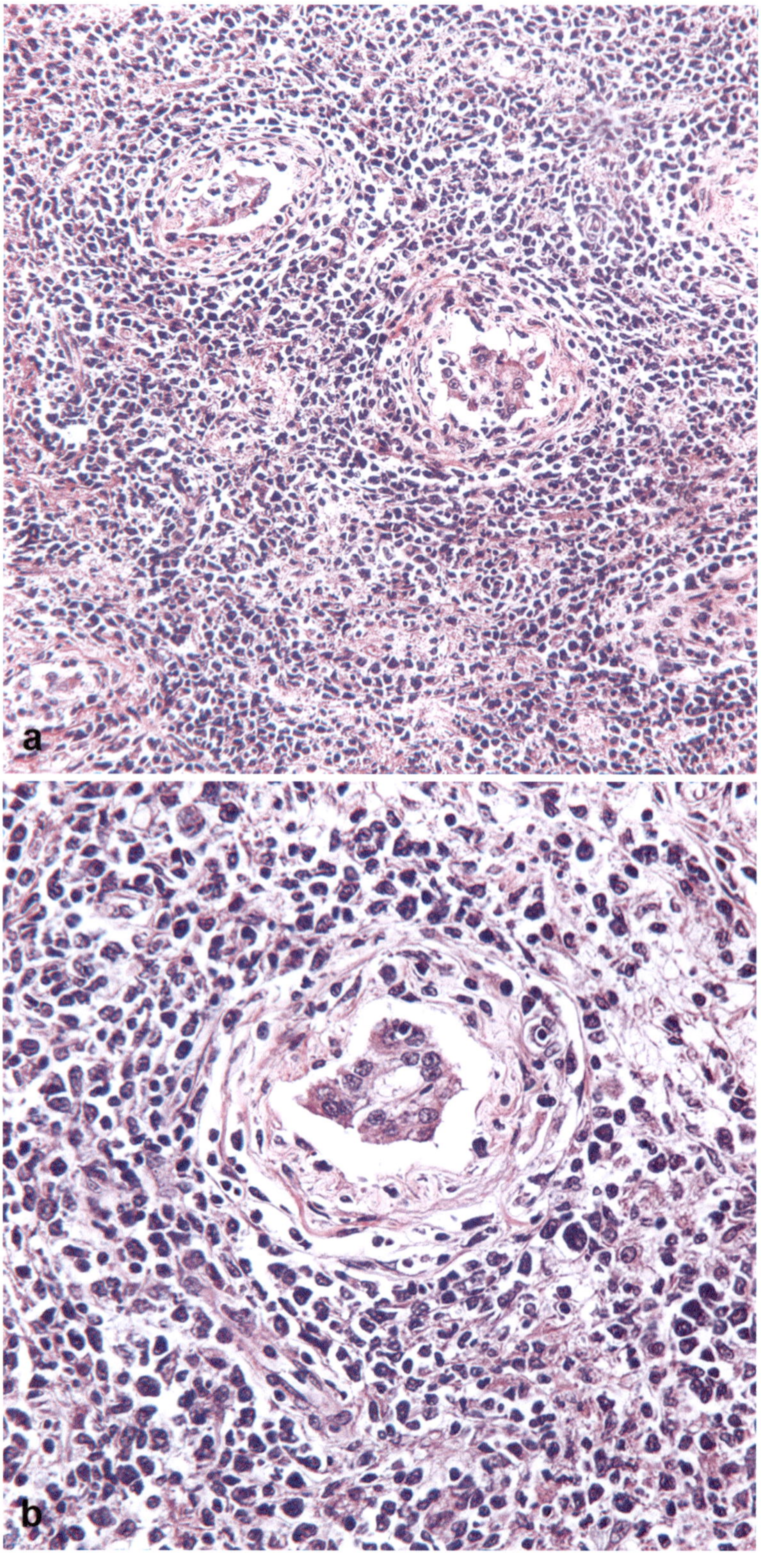

Fig. 13.77 Lymphoma of the testis. (**a**) This B cell lymphoma infiltrating the interstitial spaces of the testis has destroyed most of the seminiferous tubules; those that remain are damaged and show no signs of spermatogenesis. (**b**) Higher magnification shows atypical lymphocytes infiltrating the testis

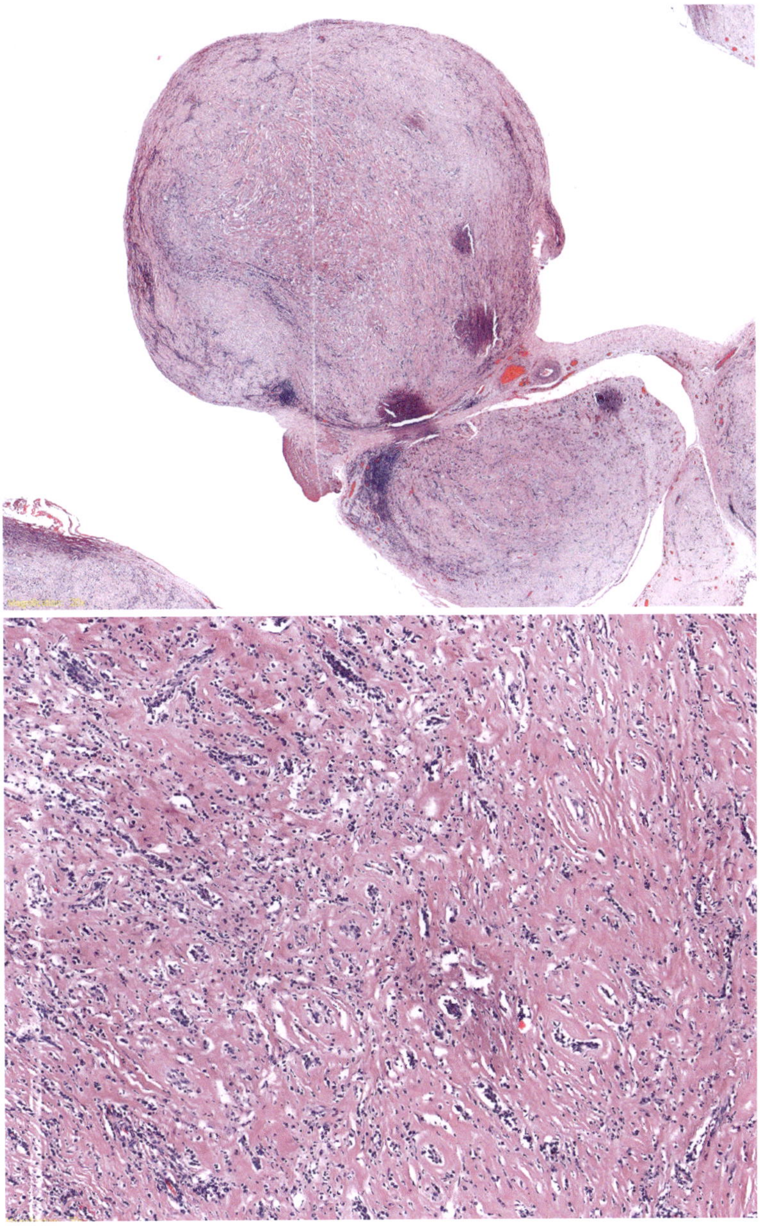

Fig. 13.78 Fibromatous (nodular) periorchitis. *Left*, Pedunculated nodule composed of dense fibrous tissue on the surface of the tunica albuginea. *Right*, Higher magnification shows that the nodule is composed of hyalinized collagen and scattered plasma cells

References

1. Eble J, Sauter G, Epstein JI, Sesterhen IA, editors. Pathology and genetics of tumours of the urinary system and male genital organs. World Health Organization classification of tumours. Lyon: IARC Press; 2004.
2. Mikuz G, editor. Clinical pathology of urologic tumors. London: Informa Healthcare; 2007.

Congenital Testis Pathology

14

Manuel Nistal and Pilar González-Peramato

This chapter groups several lesions that either are present at birth and clinically evident in the newborn or develop later in life. The lesions present in newborns are cystic dysplasia of the rete testis, fetal gonadoblastoid testicular dysplasia (FGTD), ectopic testicular parenchyma, adrenal cortical tissue ectopia, splenic ectopia, and microlithiasis. Other lesions become clinically evident later in life as a consequence of a primary pathology in the adrenal glands, such as congenital adrenal hyperplasia.

Cystic Dysplasia of the Rete Testis

Cystic dysplasia of the rete testis is a congenital lesion first reported in 1973 as *cystic dysplasia of the testis* by Leissring and Oppenheimer [1], who found it in a 4-year-old boy with right renal agenesis. This dysplasia consists of the cystic transformation of an excessively developed rete testis.

Cystic dysplasia of the rete testis has been reported in both cryptorchid and scrotal testes, from infancy to adulthood, and with a unilateral or bilateral presentation [2]. So far, about 50 cases have been reported [3], most of them in children, with an average age at presentation of 5.8 years (range, 0–23 years). The most remarkable clinical symptom is scrotal swelling, followed by scrotal pain. About 35 % of patients have abdominal cryptorchidism.

The extent of cystic transformation varies widely, from lesions that hardly exceed the mediastinum testis to generalized testicular involvement, and, as a result, the zone of preserved seminiferous tubules is reduced to a peripheral crescent (Fig. 14.1). The epithelial lining is cuboidal or flattened, and the cells are immunohistochemically (expressing both cytokeratin and vimentin) and ultrastructurally similar to those of normal rete testis epithelium (Figs. 14.2 and 14.3).

The epididymis shows alterations in most cases [4]. The head usually is small and contains a few ductuli efferentes showing a dilated lumen and abundant loose stroma. The ductus epididymidis appears dilated and has an atrophied epithelium consisting of cuboidal cells lacking stereocilia, as well as a thickened connective tissue wall (Fig. 14.4).

The ultrasound image of cystic dysplasia of the rete testis may help in making the diagnosis.

M. Nistal, M.D., Ph.D. (✉)
Department of Pathology, University Hospital La Paz, Madrid, Spain

Department of Anatomy, Histology, and Neuroscience, University Autonoma of Madrid, Madrid, Spain
e-mail: mnistal.hulp@salud.madrid.org

P. González-Peramato, M.D., Ph.D.
Department of Pathology, University Hospital La Paz, Madrid, Spain

Department of Pathology, University Autonoma of Madrid, Madrid, Spain

D. Ježek (ed.), *Atlas on the Human Testis*,
DOI 10.1007/978-1-4471-2763-5_14, © Springer-Verlag London 2013

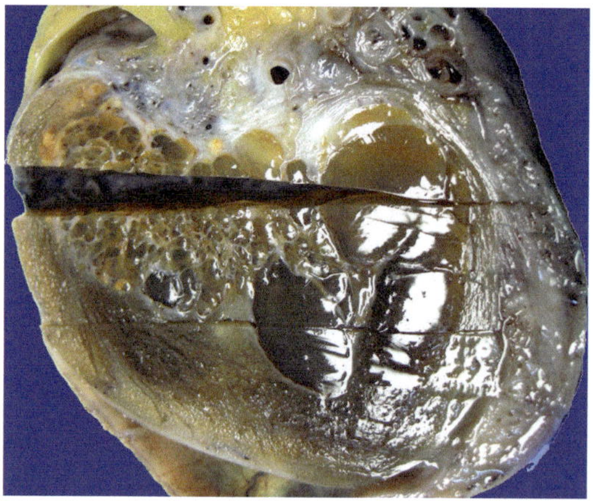

Fig. 14.1 Cystic dysplasia of the rete testis in an adult patient. This gross image of a longitudinal section of the testis shows many cystic formations of various sizes. The cysts occupy most of the testicular parenchyma. Many cystic formations are filled with a colloid-like material

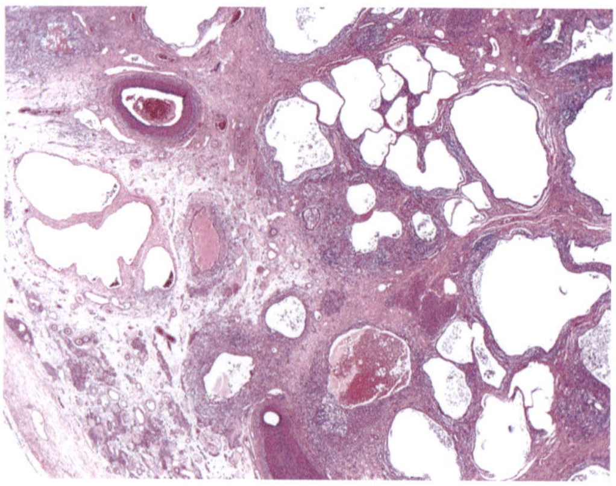

Fig. 14.2 Cystic dysplasia of the rete testis in an adult patient. Cystic formations reach the albuginea opposite the rete testis. Between the cavities and albuginea, a testicular parenchyma with hyalinized tubules and Leydig cell hyperplasia may be observed

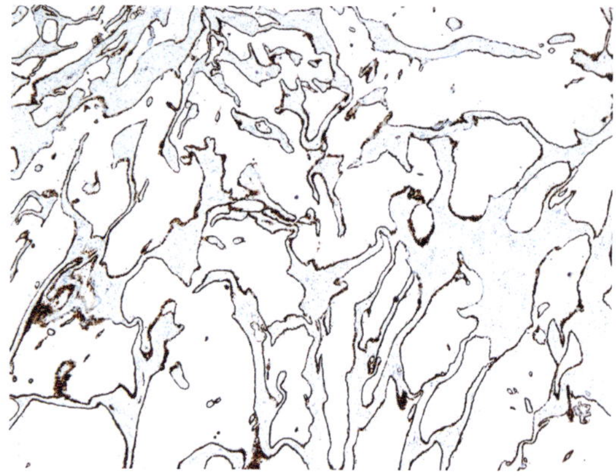

Fig. 14.3 Cystic dysplasia of the rete testis in an adult patient. Characteristic AE1/AE3 cytokeratin immunoexpression is visible in the epithelium lining the cavities

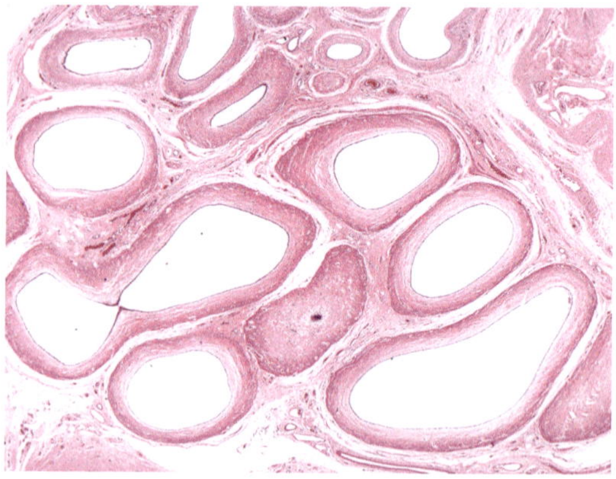

Fig. 14.4 Spermatic pathway in cystic dysplasia of the rete testis. These transverse and longitudinal sections of ductus epididymis show dilatation of the lumen, fibrosis of the wall, and poorly developed muscular layers

The major problem in patients with cystic dysplasia is its high degree of association with ipsilateral anomalies of the kidney or urinary system, rather than the lesion itself. The most frequent anomalies are renal agenesis (51 %) [1, 5], renal dysplasia (21 %), hydronephrosis (8 %), dilated or cystic epididymis (8 %), absence of an ipsilateral ureteric orifice (6 %), ureteral duplication (4 %), urethral stenosis, and an ectopic ureter draining into a seminal vesicle [6].

The etiology and pathogenesis of cystic dysplasia are unknown. However, considering that the rete testis is a mesonephric derivative and that most of the associated renal malformations may be the result of a failure in renal blastema induction by the mesonephros, one may conclude that in cystic dysplasias, the mesonephros, which gives rise to both the rete testis and renal structures, has been altered.

Classically, the treatment of choice for cystic dysplasia was orchidectomy, after considering the possibility of testis-sparing surgery [7]. Today, however, conservative therapy is recommended if the ultrasound diagnosis is highly

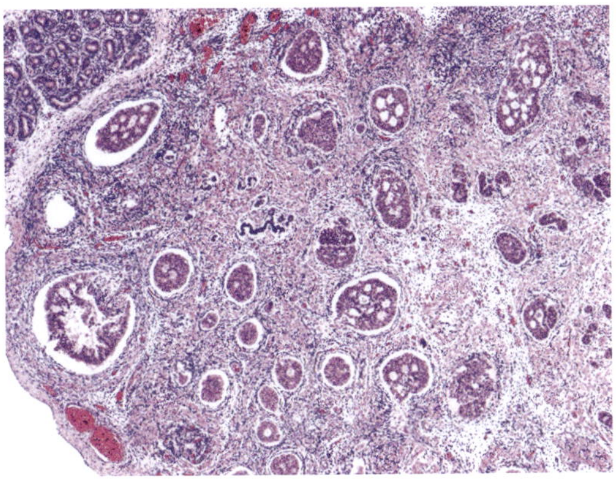

Fig. 14.5 Fetal gonadoblastoid testicular dysplasia in a newborn. Shown are the testis and part of the epididymis. The testicular architecture is distorted by multiple nodules showing a cribriform pattern. Among these nodules, normally developed seminiferous tubules may be observed

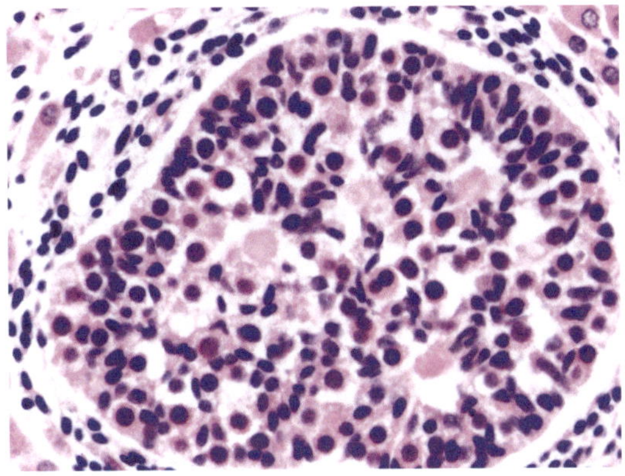

Fig. 14.6 Fetal gonadoblastoid testicular dysplasia in a newborn. Two types of cells (spindle and spherical) may be recognized in the nodular formations, which are surrounded by a layer of fusiform cells. Note the nodular eosinophilic bodies inside the nodular formations. Fetal Leydig cells may be observed in the interstitium

suggestive of the disease and tumor markers (α-fetoprotein and human chorionic gonadotropin β-subunit) are negative. Spontaneous regression has been reported in several cases [3, 8].

In revising a large collection of perinatal specimens, we observed a similar cystic dysplasia involving the epididymis. In 18 of 20 cases identified, renal and/or urinary tract malformations also were present [9].

Fetal Gonadoblastoid Testicular Dysplasia

Fetal gonadoblastoid testicular dysplasia is an exceptional disorder. In all cases of FGTD studied, the testes were grossly unremarkable and showed a similar histologic testicular pattern. The tunica albuginea was normal, and the seminiferous tubules were normally developed in the central zone of the testicular parenchyma. At the periphery, beneath the tunica albuginea, the testicular parenchyma showed a crescent-like zone with large malformed tubules, cords, and solid nodules of spherical or irregular shape embedded in a fibrous connective tissue resembling that of the ovarian stroma and organized into several concentric layers. Each of these structures was surrounded by a delimiting basement membrane (Fig. 14.5). The malformed structures comprised three cell types: those with a vesicular nucleus and vacuolated cytoplasm, those with a hyperchromatic nucleus, and those similar to germ cells. The former two types of cells were arranged in a pseudostratified pattern with the longest axis perpendicular to the periphery of the nodules. The third cell type was very similar to fetal spermatogonia (Fig. 14.6). Immunohistochemically, the cells in the nodular lesions expressed inhibin whereas the surrounding spindle stromal cells expressed vimentin and muscle-specific actin. Isolated morphologically, cells resembling germ cells expressed placental alkaline phosphatase (PLAP) (Figs. 14.7, 14.8, and 14.9). These formations showed continuity with the seminiferous tubules. The remaining testicular parenchyma showed normal development for patient age. The rest of the testicular parenchyma expressed vimentin in Sertoli cells. Inhibin was intensely and diffusely expressed in Sertoli and Leydig cells [10].

Because FGTD has been reported in patients with Walker-Warburg syndrome, there is a basis for the hypothesis that the genetic deficiency causing this syndrome also might be responsible for the abnormal testicular development [11]. Because no patient has survived beyond a few months of life, it is not possible to ascertain how the lesion would have progressed; therefore, we

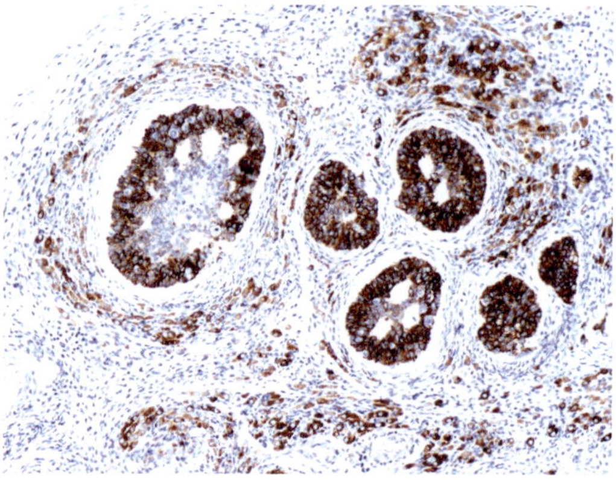

Fig. 14.7 Fetal gonadoblastoid testicular dysplasia in a newborn. Most of the nodular formations express inhibin, as well as Leydig cells located in the periphery of the nodules

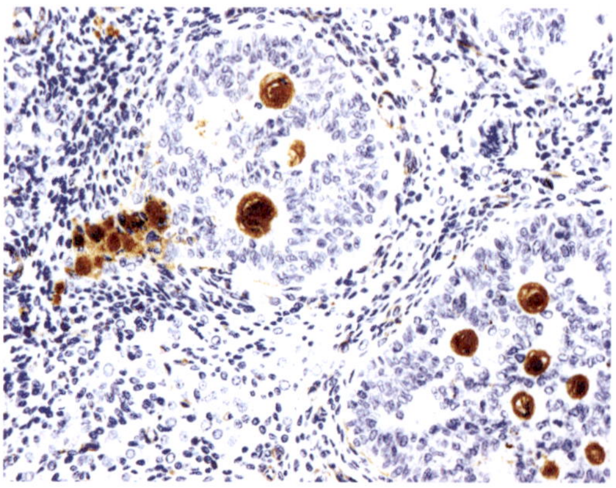

Fig. 14.8 Fetal gonadoblastoid testicular dysplasia in a newborn. Eosinophilic formations inside the nodules show strong immunoexpression of collagen IV

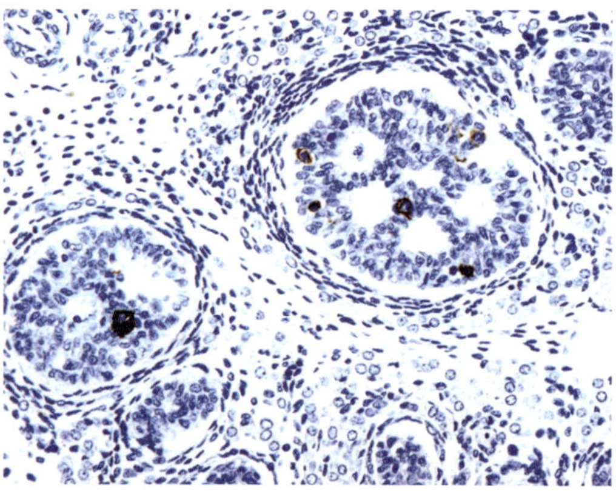

Fig. 14.9 Fetal gonadoblastoid testicular dysplasia. Morphologically similar germ cells inside the nodules express PLAP

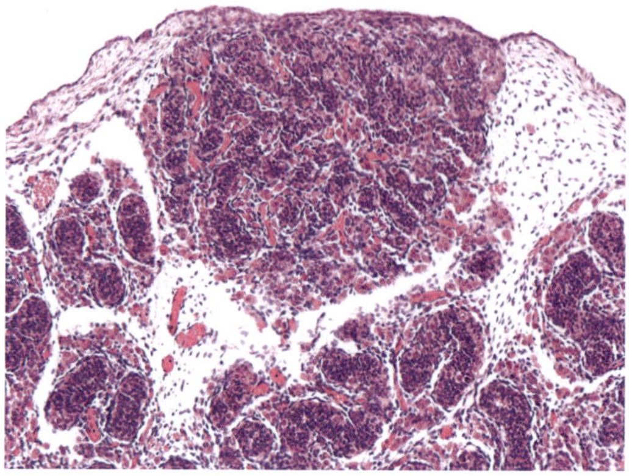

Fig. 14.10 Ectopic testicular parenchyma in albuginea. Ectopic seminiferous tubules and Leydig cells inside the albuginea show the same development as the adjacent testicular parenchyma

can only speculate about the potential progression pathway of FGTD. Incidentally, it is puzzling to recognize the vague resemblance between FGTD and granulose cell tumor of the juvenile type, another congenital testicular tumor. Further study is needed to determine whether FGTD is a precursor lesion of a tumor (e.g., full-blown gonadoblastoma, juvenile granulose cell tumor, or another neoplasm) or a tissue that will undergo atrophy (as it may be surmised by the observations of Weinberg) [12]. However, such elucidation will have to wait until FGTD is recognized in patients with longer survival.

The differential diagnosis of FGTD concerns two conditions: testicular dysgenesis and gonadoblastoma. Testicular dysgenesis shows abnormal tubular formations located within the tunica albuginea and the adjacent testicular parenchyma. The differential diagnosis with regard to gonadoblastoma is more difficult. It should be taken into account that most gonadoblastomas arise in the characteristic gonads of intersex stages (an ovarian streak gonad or a dysgenetic testis).

Ectopic Testicular Parenchyma

Ectopic testicular components, including germ cells, seminiferous tubules—either isolated or associated with Leydig cells—and Leydig cells alone, may be found in the tunica albuginea (Fig. 14.10). The presence of a well-collagenized

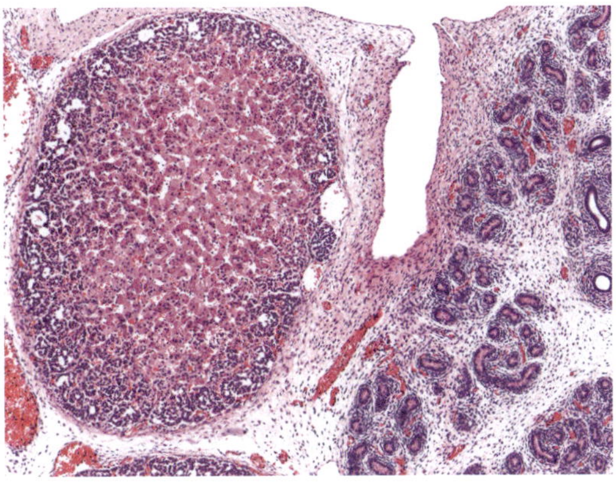

Fig. 14.11 Adrenal cortex tissue ectopia in the epididymis of a newborn. Note the absence of an adrenal medulla

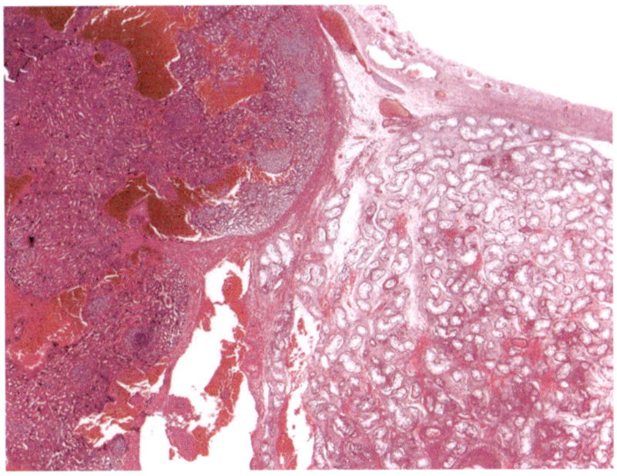

Fig. 14.12 Splenogonadal fusion. The splenic and testicular parenchymas are separated only by conjunctive tissue

tunica albuginea containing well-configured ectopic seminiferous tubules is frequent in cryptorchidism, and it is not associated with disorders of sexual development [13, 14]. During infancy and puberty, the ectopic testicular parenchyma is recognized because it forms one or several blue maculae instead of showing the normal white color. In adult men, the most striking (although not the most frequent) picture is the presence of a group of multiple small vesicles, which has been compared with a bunch of grapes, emerging through the tunica albuginea.

Adrenal Cortical Tissue Ectopia

The ectopic adrenal cortex is one of the most frequent incidental findings in male urologic surgery [15]. Its incidence is about 15 % in paratesticular structures in newborns [16]. Most cases have been reported in pediatric patients. It has been observed in the spermatic cord, tunica albuginea, epididymis (Fig. 14.11), testis, and hernial sacs [17]. These ectopies are more common on the right side. Macroscopically, ectopic adrenal formations are yellowish spherical or ovoid nodules measuring 1–4 mm in diameter.

The ectopic tissue may be functional and, under several circumstances, develops hyperplasia and tumors. Compensatory hyperplasias have been reported in patients with Addison's disease, and signs of hyperfunction have been found in those with Cushing's syndrome [18].

Adrenal ectopies have been observed between the rete testis and adjacent seminiferous tubules. Most cases are associated with adrenal cytomegaly and, in some patients, Wiedemann-Beckwith syndrome [19].

Splenic Ectopia (Splenogonadal Fusion)

The occurrence of splenic tissue near the testis, although infrequent, has great significance because this tissue may be misinterpreted clinically as a testicular tumor [20, 21]. At present, about 170 cases have been reported, 16.8 % of which are autopsy findings [20, 22, 23]. More than 70 % of the splenic ectopies reported occurred in males younger than 20 years (approximately 50 % of them were aged less than 10 years), although the reported ages vary from newborns up to 81-year-old men [24].

Splenogonadal fusions have been classified into two types, continuous and discontinuous, with the former being more frequent (55 %). In the continuous type, the ectopic splenic tissue is connected to the eutopic spleen by a cord, which is either splenic or fibrous tissue. In the discontinuous presentation, the splenic tissue is not joined to the eutopic spleen (Fig. 14.12).

Splenogonadal fusion usually occurs on the left side (98 % of cases). The anomaly appears to be associated with inguinal hernia in more than

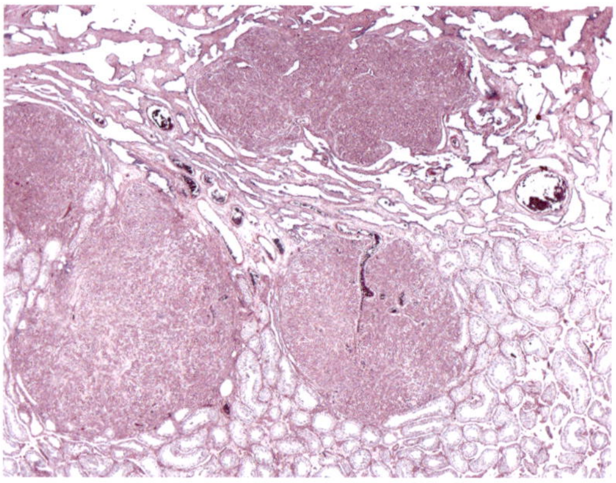

Fig. 14.13 Tumor in adrenogenital syndrome in an adult patient. Multiple nodules resembling an adrenal cortex are seen adjacent to the rete testis.

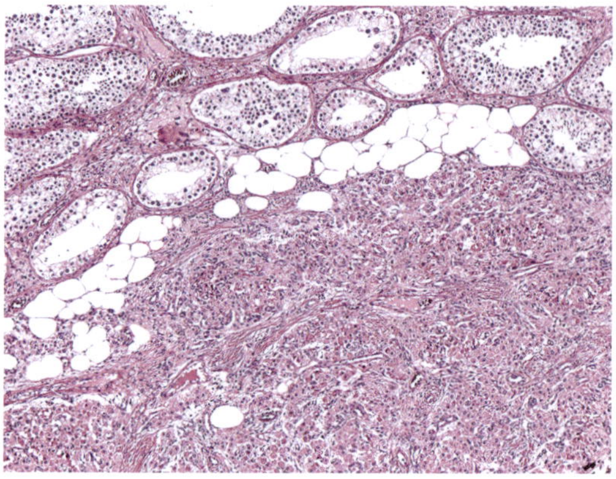

Fig. 14.14 Tumor in adrenogenital syndrome showing a ring of adipocytes between the tumor and the testicular parenchyma. Seminiferous tubules with reduced spermatogenesis are adjacent to the tumor

one-third of the cases and with cryptorchidism in 31 % of cases (bilateral in 59 % of them) [25]. The continuous type is associated with several malformations.

Tumors in Adrenogenital Syndrome

Among patients with congenital adrenal hyperplasia, such as that caused by 21β-hydroxylase deficiency, 63 % will show a large testicular size when studied during both infancy and adulthood [26]. This condition is a result of the development of "tumors" of adrenal-like cells when treatment has not been adequate. These congenital adrenogenital syndrome tumors may originate from adrenal cortical rest cells, or multipotent testicular cells from the hilar region of the testis (Fig. 14.13) [27]. They consist of polygonal cells with abundant eosinophilic and granular cytoplasm. The nuclei vary in size and occasionally show prominent nucleoli. The cells are arranged in small nests with intersecting fibrous bands (Figs. 14.14 and 14.15). The resemblance between these cells and those of a Leydig cell tumor is great; therefore, differential diagnosis is very difficult and may be made mainly on a clinical, hormonal, and immunohistologic basis.

Clinically, tumors in congenital adrenal hyperplasia are more frequently bilateral, and other symptoms of congenital adrenal hyperplasia

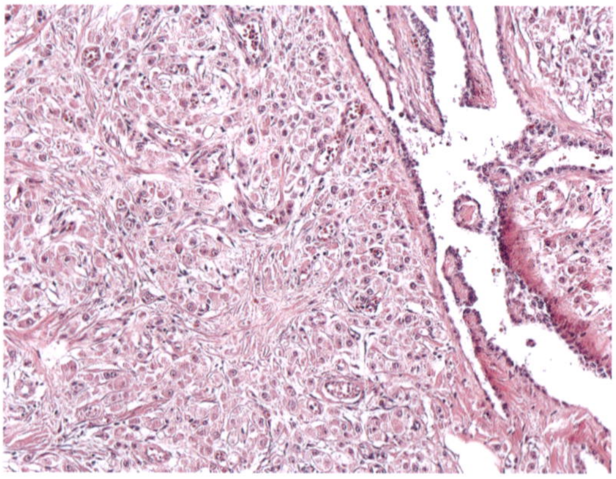

Fig. 14.15 Tumor of adrenogenital syndrome adjacent to the rete testis showing large eosinophilic polygonal cells separated by abundant collagen bands

should be demonstrated. From a hormonal perspective, serum levels of androgen and estrogens are not increased; immunohistologically, although both congenital adrenal hyperplasia and Leydig cell tumors express calretinin and inhibin, tumor cells in congenital adrenal hyperplasia show intense immunoexpression for CD56 and are negative for androgen receptor. In contrast, Leydig cell tumors show moderate or negative expression for CD56 but are positive for androgen receptor [28].

Tumors of congenital adrenal hyperplasia do not show Reinke's crystals; however, only 40 % of Leydig cell tumors do.

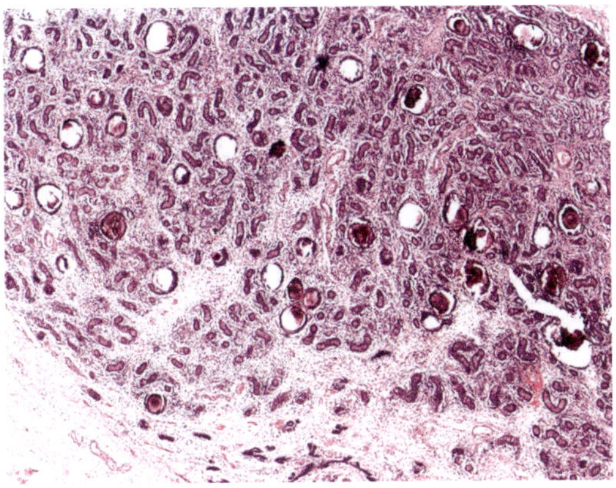

Fig. 14.16 Microlithiasis an infantile cryptorchidism. The testicular section shows numerous completely calcified microliths

Tumors in adrenogenital syndrome may become autonomous and secrete cortisol or androgens, although they usually become smaller after corticosteroid treatment [29]. In nonrespondent cases, tumor removal with maintenance of the normal testicular parenchyma is advised.

Testicular Microlithiasis

Testicular microlithiasis is characterized by the presence of concentric laminated calcifications inside the seminiferous tubules (Fig. 14.16). Most cases are identified incidentally on testicular ultrasound scanning. Ultrasonography characteristically shows multiple 1- to 3-mm foci of increased echogenicity distributed randomly within the testicular parenchyma. This disorder has been reported in both infancy and adulthood. In infants, testicular microliths have been found in those with normally descended testes, undescended testes [30], Klinefelter's syndrome, Down's syndrome [31], XY undermasculinization, polyorchidism [32], congenital adrenal hyperplasia [33], or secondary hypogonadism [34]. In adults, testicular microliths are frequent in cryptorchid testes, either descended or undescended; at the periphery of testicular germ cell tumors [35]; in patients with carcinoma in situ [36]; in those with testicular dysgenesis [37]; and in testes with ischemic injury [38].

The incidence of testicular microlithiasis varies from fewer than 2 cases per 100,000 in Asia and Africa to 8–10 cases per 100,000 in Denmark and Norway. The two types of testicular microlithiasis, defined according to the number of microliths, are *classic* (more than five microliths) and *limited* (fewer than five).

The major significance of testicular microlithiasis is that, together with other lesions, it is part of the *testicular dysgenetic syndrome*. Microliths are composed of hydroxyapatite crystals and are formed either by calcification of intraluminal cellular rests or by internalization of basal lamina material that becomes mineralized and migrates toward the lumen of the tubule [39].

The management and follow-up of patients with testicular microlithiasis depend on the associated pathology. In cases of asymptomatic testicular microlithiasis, no follow-up is needed, except for testicular self-examination to control eventual enlargement of the gonads or scrotal swelling [40]. For patients with cryptorchidism, testicular atrophy, subfertility, or disorders in sexual development, testicular biopsy is recommended to rule out intratubular germ cell neoplasia of unclassified type or testicular germ cell tumor. The contralateral testis of patients with microlithiasis associated with germ cell tumors should be followed up after surgery and chemotherapy [41].

References

1. Leissring JC, Oppenheimer ROF. Cystic dysplasia of the testis: a unique congenital anomaly studied by microdissection. J Urol. 1973;110:362–3.
2. Nistal M, Regadera J, Paniagua R. Cystic dysplasia of the testis: light and electron microscopic study of three cases. Arch Pathol Lab Med. 1984;108:579–83.
3. Jeyaratnam R, Bakalinova D. Cystic dysplasia of the rete testis: a case of spontaneous regression and review of published reports. Urology. 2010;7:687–90.
4. Robson WL, Thomason MA, Minette LJ. Cystic dysplasia of the testis associated with multicystic dysplasia of the kidney. Urology. 1998;51:477–9.
5. Glantz L, Hansen K, Caldamone A, Madeiros LJ. Cystic dysplasia of the testis. Hum Pathol. 1993;24:1142–5.
6. McGee SM, Hutcheson JC, Vandersteen DR, Reinberg Y, Wolpert JJ. Cystic dysplasia of testis associated with ectopic ureter causing chronic orchialgia. Urology. 2009;73(1423):e7–8.

7. Noh PH, Cooper CS, Snyder 3rd HM. Conservative management of cystic dysplasia of the testis. J Urol. 1999;162:2145.
8. Butler M, Servaes S, Bellah R. Cystic dysplasia of the testis: spontaneous regression. Pediatr Radiol. 2011; 41:1346–8.
9. Nistal M, González-Peramato P, Sousa G, García-Cabezas MA, Rodríguez JI, Cajaiba MM. Cystic dysplasia of the epididymis: a disorder of mesonephric differentiation associated with renal maldevelopment. Virchows Arch. 2010;456:695–702.
10. Nistal M, Rodríguez JI, García-Fernández E, Cajaiba MM, Reyes-Múgica M. Fetal gonadoblastoid testicular dysplasia: a focal failure of testicular development. Pediatr Dev Pathol. 2007;10:274–81.
11. Hung NA, Silver MM, Chitayat D, Provias J, Toi A, Jay V, Becker LE. Gonadoblastoid testicular dysplasia in Walker-Warburg syndrome. Pediatr Dev Pathol. 1998;1:393–404.
12. Weinberg AG. Walker-Warburg syndrome. Pediatr Pathol. 1989;9:749–55.
13. Nistal M, Paniagua R, Leon L, Regadera J. Ectopic seminiferous tubules in the tunica albuginea of normal and dysgenetic testes. Appl Pathol. 1985;3:123–8.
14. Cajaiba MM, García-Fernández E, Reyes-Múgica M, Nistal M. The spectrum of persistence of testicular blastema and ectopic testicular parenchyma: a possible result of focal delay in gonadal development. Virchows Arch. 2007;451:89–94.
15. Ventura L, Leocata P, Hind A, Greco L, Ventura T. Ectopic adrenal tissue in the spermatic cord. Case report and review of the literature. Arch Ital Urol Androl. 1998;70:15–8.
16. Dahl EV, Bahn RC. Aberrant adrenal cortical tissue near the testis in human infants. Am J Pathol. 1962;40:587–98.
17. Ketata S, Ketata H, Sahnoun A, FakhFakh H, Bahloul A, Mhiri MN. Ectopic adrenal cortex tissue: an incidental finding during inguinoscrotal operations in pediatric patients. Urol Int. 2008;81:316–9.
18. Rose EK, Enterline HT, Rhoads JE, Rose E. Adrenal cortical hyperfunction in childhood. Report of a case with adrenocortical hyperplasia and testicular adrenal rests. Pediatrics. 1952;9:475–84.
19. Nistal M, Paniagua R. Congenital anomalies of the testis and the epididymis. In: Testicular and epididymal pathology. New York: Thieme Stratton; 1984. p. 74–5.
20. Petrik PK. Spleen in the scrotum. Am J Surg Pathol. 1988;12:414–5.
21. Karaman MI, Gonzales Jr ET. Splenogonadal fusion: report of 2 cases and review of the literature. J Urol. 1996;155:309–11.
22. Alivizatos G, Skolarikos A, Sopilidis O, Ferakis N, Chorti M. Splenogonadal fusion: report of a case and review of the literature. Int J Urol. 2005;12:90–2.
23. Patafio FM, Dufton JA, Walker DR. Case of the month #173: splenogonadal fusion. Can Assoc Radiol J. 2011;62:302–4.
24. Shen XC, Du CJ, Chen JM, Zhang ZW, Qiu YQ. Splenogonadal fusion. Chin Med J. 2008;121:383–4.
25. Cortes D, Thorup JM, Visfeldt J. The pathogenesis of cryptorchidism and splenogonadal fusion: a new hypothesis. Br J Urol. 1996;77:285–90.
26. Unuvar T, Demir K, Abaci A, Atas A, Cakmakci H, Bober E. A 2-year-old boy with a testicular mass. Diagnosis: testicular tumor of adrenogenital syndrome due to 11-beta-hydroxylase deficiency. Pediatr Ann. 2010;39:471–4.
27. Paner GP, Kristiansen G, McKenney JK, Amin MB. Rete testis-associated nodular steroid cell nests: description of putative pluripotential testicular hilus steroid cells. Am J Surg Pathol. 2011;35:505–11.
28. Wang Z, Yang S, Shi H, Du H, Xue L, Wang L, et al. Histopathological and immunophenotypic features of testicular tumour of the adrenogenital syndrome. Histopathology. 2011;58:1013–8.
29. Cutfield RG, Bateman JM, Odell WD. Infertility caused by bilateral testicular masses secondary to congenital adrenal hyperplasia (21-hydroxylase deficiency). Fertil Steril. 1983;40:809–14.
30. Nistal M, Paniagua R, Diez-Pardo JA. Testicular microlithiasis in 2 children with bilateral cryptorchidism. J Urol. 1979;121:535–7.
31. Vachon L, Fareau GE, Wilson MG, Chan LS. Testicular microlithiasis in patients with Down syndrome. J Pediatr. 2006;149:233–6.
32. Rajbabu K, Morel JC, Thompson PM, Sidhu PS. Multi-cystic (rete testis) supernumerary testis in polyorchidism with underlying microlithiasis: ultrasound features. Australas Radiol. 2007;51(Spec):B56–8.
33. Poyrazoglu S, Saka N, Agayev A, Yekeler E. Prevalence of testicular microlithiasis in males with congenital adrenal hyperplasia and its association with testicular adrenal rest tumors. Horm Res Paediatr. 2010;73:443–8.
34. Wasniewska M, Raiola G, Teresa A, Galati MC, Zirilli G, Catena MA, et al. Gynecomastia disclosing diagnosis of Leydig cell tumour in a man with thalassemia, secondary hypogonadism and testis microlithiasis. Acta Biomed. 2009;80:286–8.
35. Ikinger U, Wuster K, Terwey B, Mohring K. Microcalcifications in testicular malignancy: diagnostic tool in occult tumor? Urology. 1982;19:525–8.
36. Wegner HE, Hubotter A, Andresen R, Miller K. Testicular microlithiasis and concomitant testicular intraepithelial neoplasia. Int Urol Nephrol. 1998;30: 313–5.
37. Hoei-Hansen CE, Sommer P, Meyts ER, Skakkebaek NE. A rare diagnosis: testicular dysgenesis with carcinoma in situ detected in a patient with ultrasonic microlithiasis. Asian J Androl. 2005;74:445–7.

38. Mazzilli F, Delfino M, Imbrogno N, Elia J, Spinosa V, Di Nardo R. Seminal profile of subjects with testicular microlithiasis and testicular calcifications. Fertil Steril. 2005;84:243–5.
39. Nistal M, Martinez-Garcia C, Paniagua R. The origin of testicular microliths. Int J Androl. 1995;18:221–9.
40. Silveri M, Bassani F, Colajacomo M, Orazi C, Adorisio O. Management and follow-up of pediatric asymptomatic testicular microlithiasis: are we doing it well? Urol J. 2011;8(4):287–90.
41. Tan MH, Eng C. Testicular microlithiasis: recent advances in understanding and management. Nat Rev Urol. 2011;8:153–63.

15 Vascular Testis Pathology

Manuel Nistal and Pilar González-Peramato

Testicular lesions secondary to vascular pathology, either arterial or venous, are characterized by their severity, their occurrence in patients of any age, and their frequently dramatic clinical presentation. At diagnosis, patients with testicular torsion usually have very evolved lesions, leading to removal of the testis. Varicocele is considered an important cause of male infertility. Testicular and epididymal vasculitis may simulate a tumoral process requiring orchio-epididectomy. Arteriosclerosis of testicular vessels plays an important role in hormonal dysfunction in older men.

Testicular Torsion

The most frequent cause of testicular infarct is spermatic cord torsion, followed by traumatism, incarcerated inguinal hernia, epididymitis, and vasculitis. Spermatic cord torsion is a surgical emergency, and if surgery is delayed more than 8 h, testicular viability is compromised.

There are two peaks of maximal incidence: the perinatal period and puberty [1]; however, torsion is not uncommon in adults [2]. Left-sided torsions predominate slightly (60 %); bilateral torsions are infrequent and more common in newborns than in adults. Only 1 % of torsions are bilateral and synchronous, and these rarely are observed in ages other than the neonatal period. Cases of familial presentation have been reported [3, 4].

Several factors predispose to testicular torsion:

- Anatomic anomalies in testicular suspension. A significant number of patients present an abnormally high reflection of the tunica vaginalis, so that part of the spermatic cord remains sheathed, giving rise to the so-called bell-clapper deformity.
- Abnormal position of the testis. Testicular torsion is more frequent in cryptorchid and retractile testes than in normal testes.
- Other factors, such as trauma.

Two classic anatomic forms of testicular torsion are known: the high, supravaginal, or extravaginal form, which is actually spermatic cord torsion, and low or intravaginal torsion. Each form of testicular torsion (extra- and intravaginal) appears at a different age. Extravaginal torsions are characteristic of infancy and childhood (92 %) [5]. In contrast, intravaginal torsions are more frequent during puberty and adulthood. Neonatal torsions comprise 12 % of all testicular

M. Nistal, M.D., Ph.D. (✉)
Department of Pathology, University Hospital La Paz, Madrid, Spain

Department of Anatomy, Histology, and Neuroscience, University Autonoma of Madrid, Madrid, Spain
e-mail: mnistal.hulp@salud.madrid.org

P. González-Peramato, M.D., Ph.D.
Department of Pathology, University Hospital La Paz, Madrid, Spain
Department of Pathology, University Autonoma of Madrid, Madrid, Spain

D. Ježek (ed.), *Atlas on the Human Testis*,
DOI 10.1007/978-1-4471-2763-5_15, © Springer-Verlag London 2013

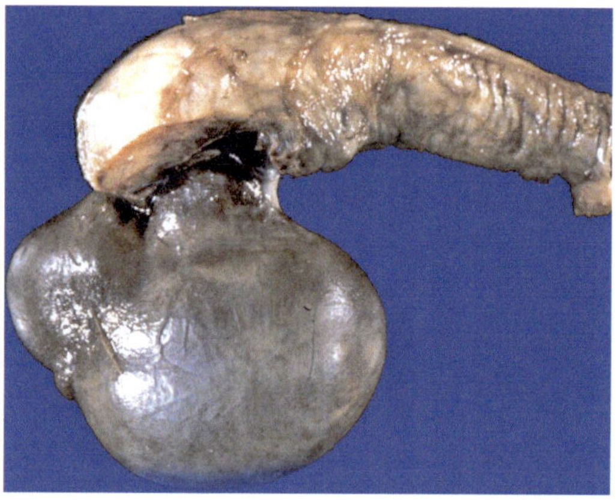

Fig. 15.1 Testicular torsion. Gross image of long-term intravaginal testicular torsion in a pubertal patient. The testis shows the so-called bell-clapper deformity

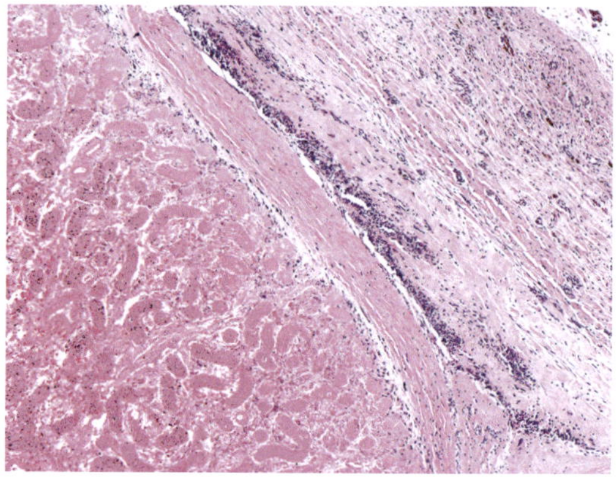

Fig. 15.2 Long-standing testicular torsion. Note the complete necrosis of the seminiferous tubules and interstitium. The testicular coverings show calcification and fibrosis

torsions in infancy. Most intrauterine torsions are asymptomatic [6]; however, pubertal and adult patients show many symptoms, the most frequent of which are ipsilateral testicular pain (89 % of cases), vomiting (39 %), and urinary symptoms (5 %). The differential diagnosis should comprise all causes of acute scrotum, including testicular appendix torsion, strangulated hernia, testicular tumor, traumatic hemorrhagic hydrocele, orchitis, epididymitis, testicular vasculitis, scrotal abscess, and idiopathic testicular infarct.

The testicular lesion resulting from spermatic cord torsion is hemorrhagic infarct of the testis [7]. Macroscopically, the testis is firm, red or bluish at first, and black afterward (Fig. 15.1). Histologic lesions are well developed in most neonatal torsions and range from necrosis to fibrosis with calcium and hemosiderin deposits (Fig. 15.2).

In testicular torsion in adults, the histologic lesions have been classified in three progressive categories [8]:

- Grade I lesions. These lesions are found mainly in the testicular interstitium and consist of edema, vascular congestion, and focal hemorrhages. They are present in 28.5 % of the testes from our series [9].
- Grade II lesions. In these lesions, the seminiferous epithelium shows sloughing of all germ cell types. The severity of these lesions is greater in the center than at the periphery of the testis. They are present in 26.5 % of the testes from our series [9].
- Grade III lesions. These lesions are the characteristic lesions of hemorrhagic infarct with necrosis of the seminiferous epithelium. They may be observed in 45 % of testicular torsion cases.

In general, there is a correlation between the duration of torsion and the degree of histologic lesions. Most testes with torsion of less than 4 h have grade I lesions, those with more than 12 h of torsion have grade III lesions, and grade II lesions are found in testes with between 4 and 8 h of torsion.

Testes with intravaginal torsion may be bearers of primary lesions such as spermatogenesis arrest, Sertoli cell-only (SCO) tubules, hypoplastic zones, and microliths.

Some patients with untreated spermatic cord torsion not only observed that the testis became atrophic and retracted to the inguinal region, but weeks or even months later, they developed a painful tumor in that region. The cause might be the development of a lesion known as *lipomembranous fat necrosis* (Figs. 15.3 and 15.4) [10].

Localized (segmental) infarction has been reported in patients with polycythemia, hypercoagulable states, S-protein or antithrombin deficiency, sick cell anemia, vasculitis, angiitis by hypersensibility, familial Mediterranean fever, nonbacterial thrombotic endocarditis, cholesterol embolism folate metabolism defects associated with epididymitis, intermittent testicular torsion, laparoscopic inguinal hernia repair, fibroplasia of

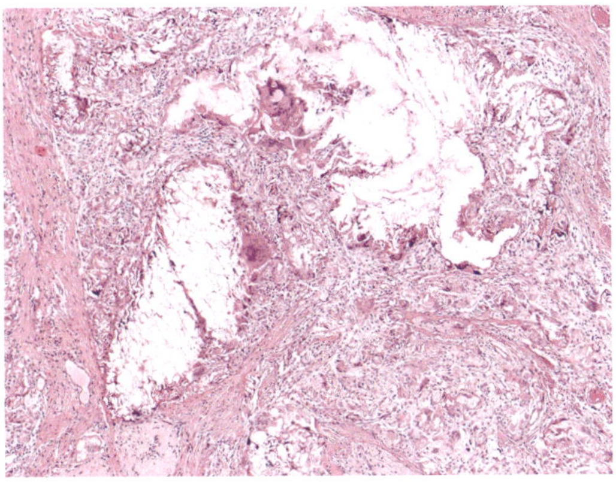

Fig. 15.3 Lipomembranous fat necrosis of a spermatic cord in testicular torsion. Necrotic tissue with cystic cavities limited by eosinophilic, wavy outlined, refringent, hyaline membranes is surrounded by a granulomatous reaction

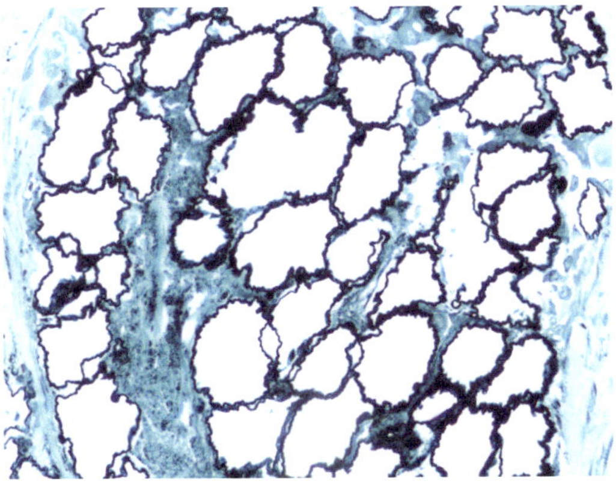

Fig. 15.4 Lipomembranous fat necrosis of a spermatic cord in testicular torsion. Note the spiculated pattern of membranes with Sudan black staining

spermatic artery intimae after intense physical exercise, herniorrhaphy varicocelectomy, vasectomy, testicular trauma, and obstetric trauma, although this disorder also may be spontaneous [11, 12]. Bilateral forms have been reported [13]. The extent of the infarct depends on the severity of the vascular lesion (Fig. 15.5).

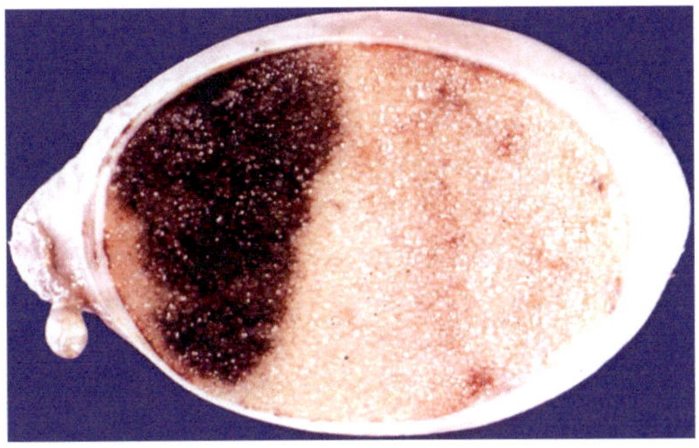

Fig. 15.5 Gross image of a testicular segmental infarct showing a well-delimited hemorrhagic area of parenchyma with a wedge shape

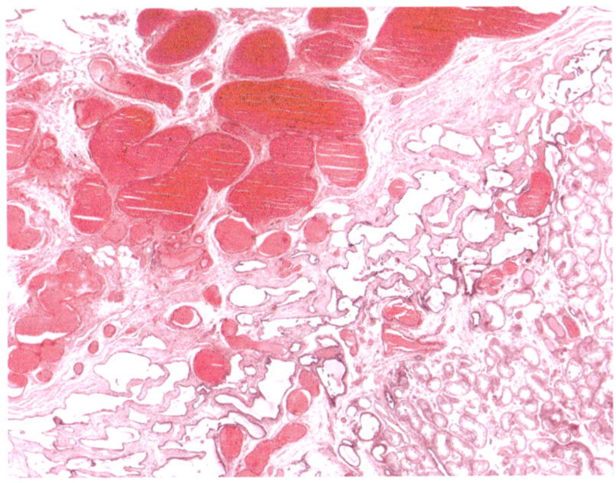

Fig. 15.6 Varicocele. The dilation of venous vessels of both the pampiniform plexus and centripetal testicular veins is evident. Note the rete testis ectasia

Varicocele

Varicocele is a pathologic disorder characterized by the elongation, dilatation, and winding of the veins that drain the testis and its coverings, resulting in blood reflux along the inner spermatic vein (Fig. 15.6). Varicocele is found in 15–20 % of the general adult male population and in 35 % of infertile men [14]. Most varicoceles affect the left testis. Only 1–2 % of varicoceles occur on the right side. The incidence of bilateral varicocele ranges from 15–70 % [15].

The diagnosis of varicocele should be based on physical examination. Based on size, three degrees of clinical varicocele have been established [16]:

- Grade 1: small varicoceles that are invisible and impalpable but may be identified on Valsalva maneuver
- Grade 2: midsized (1–2 cm) invisible but palpable varicoceles
- Grade 3: large (>2 cm) visible varicoceles described as a “bag of worms” or “squishy tube” on palpation

In patients with varicocele, the ipsilateral testis is softer and smaller on palpation and appears less descended than the contralateral testis. Scrotal thermography, ultrasonography, duplex Doppler ultrasound, and venous scintigraphy are the noninvasive techniques used most often for diagnosis.

Because of the high incidence of left-sided varicocele, most hypotheses proposed to explain the origin of varicocele refer to the peculiar features of the left spermatic vein. Several of the following elements are thought to be involved in varicocele development: (1) congenital absence and incompetence of internal spermatic vein valves, (2) congenital alterations in the vascular wall connective tissue, (3) elevated hydrostatic pressure in the left internal spermatic vein, (4) drainage difficulty in the left internal spermatic vein, and (5) alterations in the fasciomuscular pump of the spermatic cord.

The most outstanding macroscopic finding of varicocele is the arrest of ipsilateral testicular growth at puberty (testicular hypotrophy) or an ipsilateral decrease in testicular size (atrophy) in adulthood. This atrophy is present in half of varicocele patients. This is a very important finding because patients with varicocele and ipsilateral testicular hypotrophy have lower total motile sperm counts than those without hypotrophy or atrophy, and the degree of varicocele is correlated with the degree of testicular hypotrophy. Subclinical varicocele also may result in a decrease in testicular size.

Histologic testicular lesions in varicocele involve both the seminiferous tubules and the testicular interstitial tissue. Most testes affected by varicocele have diffuse lesions in the seminiferous tubules, although these lesions are more severe in focal areas. The wide variability in tubular diameter is outstanding. The most altered tubules show marked lumen dilatation, which usually is inversely correlated with the degree of seminiferous epithelium atrophy. The tubules with a more reduced diameter present an enlarged tunica propria [17, 18] and also may show sperm accumulation in the lumen, tubules with Sertoli cells only or Sertoli cells and isolated spermatogonia, peritubular inflammatory infiltrates, and tubular hyalinization (Fig. 15.7).

The most common alteration in the seminiferous epithelium is the presence of an adluminal compartment lesion consisting of young spermatids sloughing in some cases or primary spermatocytes sloughing in others [19]. Germ cell loss produces disorganization of the adluminal compartment (Fig. 15.8). In the lumen, in addition to sloughed cells, one may see some maturative anomalies of adult spermatids consisting of spermatids with an elongated head and a thin base. Sertoli cells show apical cytoplasm vacuolization, dilatation of smooth endoplasmic reticulum, and alterations in the Sertoli cell–germ cell specialized junctions [19]. Spermatids are disoriented in the apical cytoplasm of Sertoli cells.

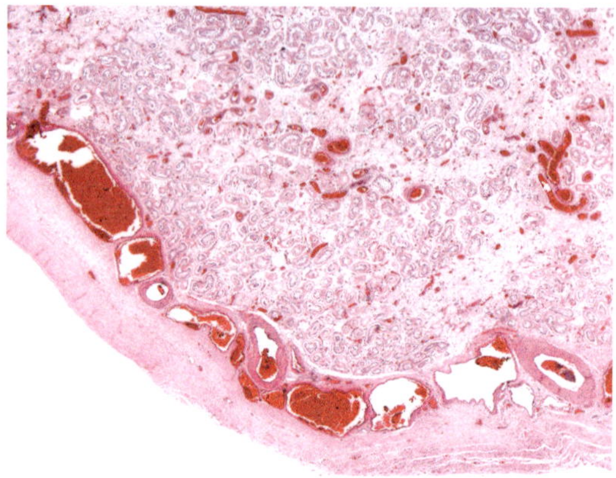

Fig. 15.7 Long-term varicocele. The testicular parenchyma shows varicose veins in the tunica vasculosa and inside the parenchyma. Note the decreased size of the seminiferous tubules

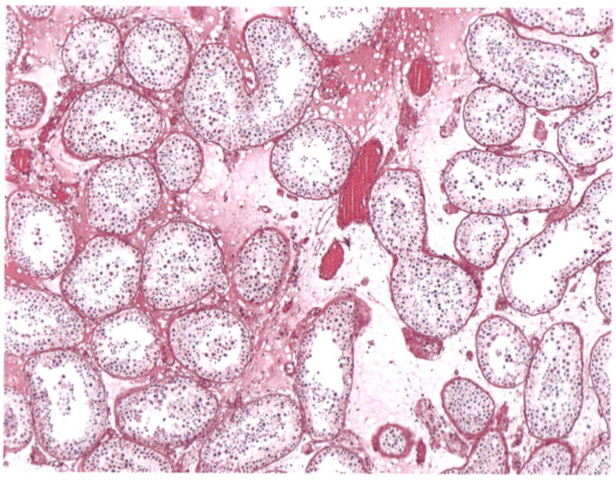

Fig. 15.8 Patient with varicocele who underwent consultation for infertility with oligozoospermia. Testicular biopsy showed congestive testicular veins and ectatic seminiferous tubules with epithelium of variable height

The tunica propria undergoes alterations in all its structures. The basal membrane appears thickened [20], and the numbers of myoid cells and collagen fibers increase.

The testicular interstitium appears enlarged by diffuse edema and an increased proliferation of collagen fibers. Dilatation of lymphatic vessels and blood stagnation in small vessels also may be observed. Leydig cells have been reported to be atrophic, hypoplastic, or hyperplasic [17, 18].

Biopsy of the contralateral testis in varicocele patients shows histologic alterations in many cases [20]. Comparison of the biopsy specimens from both testes reveals that the most severe lesions are in the ipsilateral testis [17].

Semen quality is markedly altered in half of varicocele patients. There is a decrease in sperm and an increase in tapering and amorphous sperm forms, similar to the so-called stress pattern. The following anomalies have been found in spermiograms of varicocele patients: a high proportion of abnormal spermatozoa; an alteration in the middle piece, with insufficient elimination of the cytoplasmic droplet [21]; low progressive motility; a possible decrease in longevity; increased production of reactive oxygen species; and depressed total antioxidant capacity levels [22].

The main pathogenetic theories postulated to be involved in adult varicocele are: alteration in testicular thermoregulation, hypoxia, toxic effects of renal and adrenal metabolites, endocrine alterations, such as intrinsic deficiency in testicular steroidogenesis or defects in Sertoli cell function, partial obstruction of the proximal sperm excretory ducts and alteration in epididymal vascularization and function.

Although it generally is assumed that varicocele contributes to male infertility, there is no general agreement about the efficacy of surgical treatment in improving fertility.

Adolescent varicocele deserves special mention. Most varicoceles in adults start at a pubertal age; they are rare in preadolescents [23]. The incidence of varicocele increases suddenly at puberty, with a weighted average of 16.3 % [24]. This incidence is similar to that of adult varicocele. The most outstanding feature of adolescent varicocele is testicular hypotrophy in 77 % of patients [25]. Histologic testicular lesions in adolescent varicoceles involve the seminiferous tubules and testicular interstitium, although these lesions are not as severe as those of adult varicoceles.

As a general rule, control of testicular growth during puberty is necessary, especially in patients with grade II or III varicocele, to enable the early detection of an arrest in testicular growth.

Vasculitis

The term *vasculitis* refers to processes characterized by necrosis and inflammation of the vascular wall that may occur in any type of blood vessel in any organ.

Panarteritis Nodosa

The alteration in panarteritis nodosa may involve both the testis and epididymis [26]. The frequency of testicular alterations in panarteritis, calculated from autopsy series, is estimated at 60–86 %, although clinical diagnosis is infrequent (2–18 %) [27]. Only rarely is testicular or epididymal alteration the first symptom of the disease. In these cases, the clinical manifestation may be orchitis, epididymitis, testicular torsion, or tumor [28].

The testis and epididymis usually show arterial lesions in different evolutionary stages (fibrinoid necrosis, inflammatory reaction, thrombosis, or aneurysm); (Fig. 15.9). The testicular parenchyma initially shows areas of hemorrhagic infarct then, later, tubular sclerosis with interstitial fibrosis.

The same testicular symptomatology has been reported in a necrotizing arteritis that affects only the testis and/or epididymis (isolated arteritis of the testis and epididymis); (Fig. 15.10). In the presence of histologic evidence of necrotizing arteritis involving the testis or epididymis, all clinical, hematologic, and biochemical studies necessary to rule out systemic disease should be performed.

Henoch–Schönlein Purpura

Henoch–Schönlein purpura is a disease of unknown etiology characterized by a diffuse vasculitis

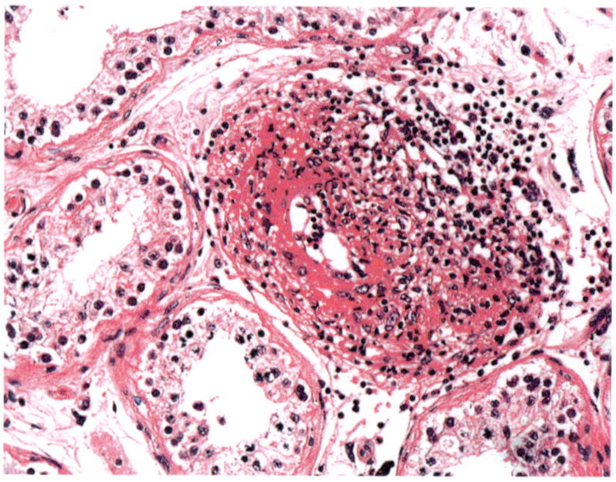

Fig. 15.9 Testicular involvement in systemic panarteritis nodosa. An arteriole with fibrinoid necrosis of its wall surrounded by inflammatory infiltrate may be observed

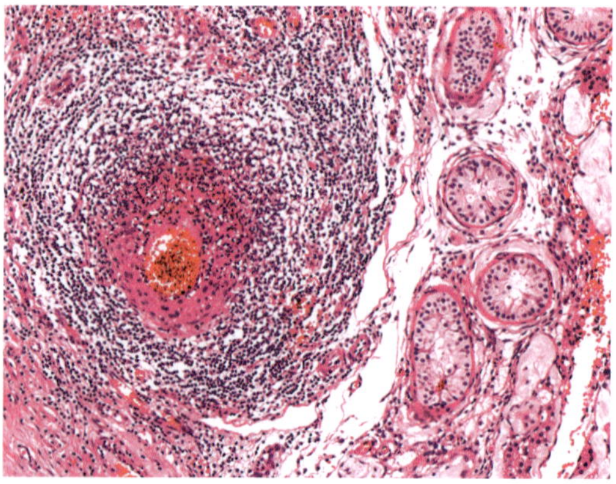

Fig. 15.10 Isolated arteritis of the testis. Shown is segmental necrosis of an artery wall in a cryptorchid testis. The testicular parenchyma shows hyalinized and SCO tubules lined by dysgenetic Sertoli cells

predominantly involving the skin, gastrointestinal tract, joints, and kidneys. Scrotal pain or swelling is present in 10–15 % of affected boys. The disorder causes alterations in blood capillaries and veins and, subsequently, increased permeability, exudation of polymorphonuclear leukocytes, and hemorrhage. The alteration may occur in the testis, epididymis, spermatic cord, or scrotal skin.

Wegener's Granulomatosis

Wegener's granulomatosis is characterized by glomerulonephritis, necrotizing vasculitis of mid-sized and small blood vessels, and granulomas that appear mainly in the upper and lower respiratory tracts. Testicular damage is estimated to occur in 9 % of patients. The epididymis may be the sole location of the disease. Histologic lesions are characterized by extravascular granulomas and either necrotizing or granulomatous vasculitis in small arteries and veins [29]. The consequence is development of numerous testicular infarcts.

Thromboangiitis Obliterans

Thromboangiitis obliterans is a disease that principally affects middle-aged men who are heavy smokers. It may appear in spermatic blood vessels as part of the systemic development of the disease [30], may be an isolated presentation, or may simulate a spermatic cord tumor [31]. Lesions may appear macroscopically as nodular indurations along the spermatic cord.

Giant-Cell Vasculitis

A vasculitis similar to that affecting the temporal artery has been observed in other locations. Spermatic cord and testicular vessels may be affected in the context of a generalized disease. The clinical symptomatology is that of a malignant tumor.

Vasculitis Associated with Germ Cell Tumors

Vasculitis associated with germ cell tumors may be either necrotizing or granulomatous and may be observed in the testis and epididymis (Fig. 15.11). Lesions may even involve the contralateral nontumoral testis.

Arteriosclerosis

In general, testicular lesions are not directly related to the degree of aortic arteriosclerosis, but they are related to the degree of cerebral artery

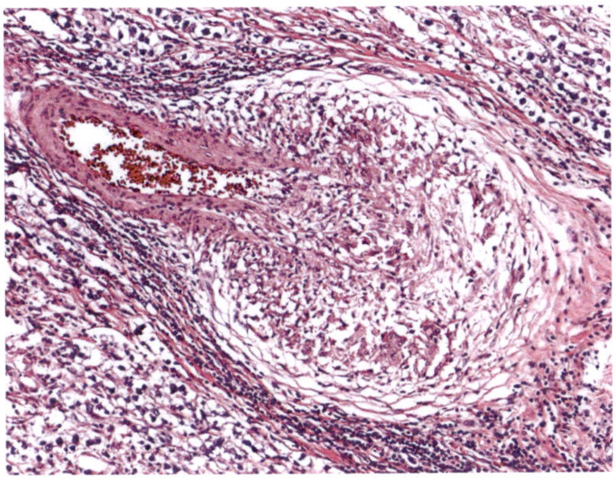

Fig. 15.11 Granulomatous arteritis associated with seminoma. Shown is an inflammatory lesion affecting all layers of the vessel wall. In the vessel, note the numerous multinucleated giant cells, which are surrounded by tumoral cells

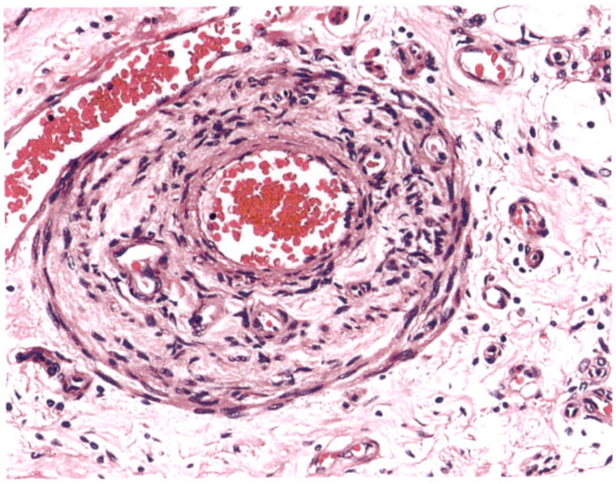

Fig. 15.13 Arteriosclerosis obliterans. Shown are small intratesticular arteries of the tunica vasculosa with a marked prominence of vasa vasorum, mainly in the medial and adventitial layers. Note the hyalinized seminiferous tubules in the right upper corner

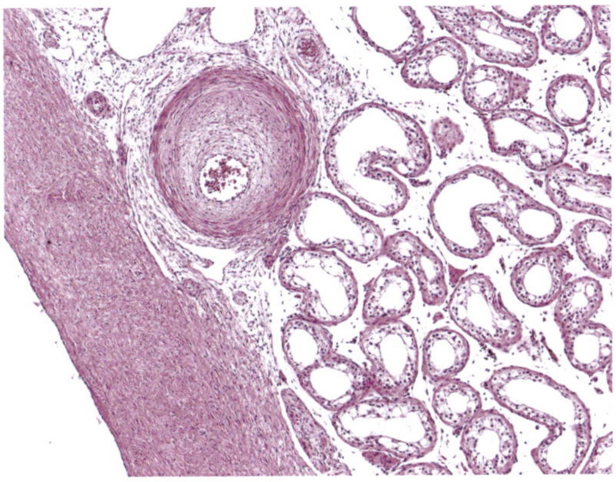

Fig. 15.12 Arteriosclerosis. An artery in the tunica vasculosa shows thickening of the intimal layer. There is marked ectasia of the seminiferous tubules, which have an atrophic epithelium

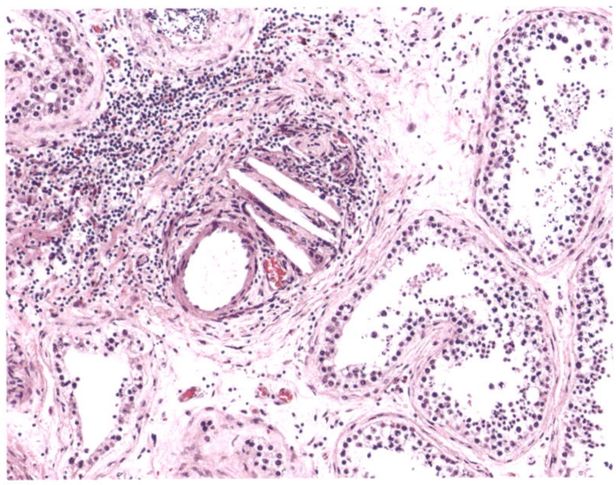

Fig. 15.14 Cholesterol thromboembolism with material from an abdominal atheroma. A granulomatous reaction with multinucleate giant cells engulfing cholesterol crystals may be observed

arteriosclerosis. Ischemia secondary to an arterial lesion involves changes in both the epididymis and testis (Fig. 15.12).

Lesions of the epididymis usually are more severe than those of the testis and may contribute, through an obstructive mechanism, to the worsening of testicular lesions. The caput epididymis is the most frequent location of epididymal lesions. In all epididymal regions, the lesions may be either necrotic or atrophic. Necrotic lesions involve several ductuli efferentes that become fully necrotic. The best-preserved ductuli show inflammatory infiltrates, consisting mainly of macrophages that destroy the epithelial lining of the ductuli and are eliminated in the lumen. The evolved lesions show a central zone consisting of cell debris, cholesterol needles, and ceroid granulomas (Figs. 15.13 and 15.14). The adjacent ductuli efferentes may undergo epidermoid metaplasia. This histologic pattern has been called *ischemic epididymitis* [32]. Atrophic lesions are more frequent in the caput epididymis and have either a diffuse or a focal pattern. Macrophages containing ceroid pigment may be seen among

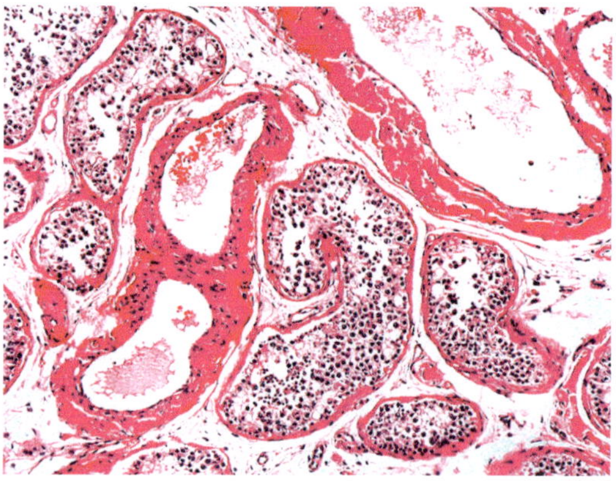

Fig. 15.15 Amyloidosis. Note the abundant deposit of amyloid material in the walls of the arteries, veins, and seminiferous tubules stained with Congo red

the ductuli, and when the number of macrophages is high, they give rise to ceroid granulomas.

Epididymal lesions, which hinder testicular drainage and fluid reabsorption, cause a cystic transformation of the rete testis and a seminiferous tubule atrophy displaying a lobular pattern. The Leydig cells maintain their normal appearance.

Testicular lesions in acute ischemia consist of an ischemic infarct. Necrotic testicular parenchyma (ischemic orchitis) may be secondarily infected and eliminated through a scrotal fistula.

Amyloidosis

The male genital system is damaged mainly in reactive systemic amyloidosis, some dyscrasias of immunocytes with amyloidosis, long-term hemodialysis, some inherited diseases, and senile amyloidosis [33]. Isolated testicular amyloidosis is rare (Fig. 15.15).

References

1. Nistal M, Alcoba M, Contreras F. Torsión del cordón espermático. A propósito de 87 nuevos casos. Revisión de la literatura. Arch Esp Urol. 1971;24:385–98.
2. Davol P, Simmons J. Testicular torsion in a 68-year-old man. Urology. 2005;66:195.
3. Cubillos J, Palmer JS, Friedman SC, Freyle J, Lowe FC, Palmer LS. Familial testicular torsion. J Urol. 2011;185:2469–72.
4. Shteynshlyuger A, Freyle J. Familial testicular torsion in three consecutive generations of first-degree relatives. J Pediatr Urol. 2011;7:86–91.
5. Das S, Singer D. Controversies of perinatal torsion of the spermatic cord: a review, survey and recommendations. J Urol. 1990;143:231–3.
6. Melekos MD, Asbach HW, Markou SA. Etiology of acute scrotum in 100 boys with regard to age distribution. J Urol. 1988;139:1023–5.
7. Skoglund RW, McRoberts JW, Ragde H. Torsion of the spermatic cord: a review of the literature and an analysis of 70 new cases. J Urol. 1970;104:604–7.
8. Mikuz G. Testicular torsion: simple grading for histological evaluation of tissues damage. Appl Pathol. 1985;3:134–9.
9. Nistal M, Martinez C, Paniagua R. Primary testicular lesions in the twisted testis. Fertil Steril. 1992;57:381–6.
10. Nistal M, Gonzalez-Peramato P, Paniagua R. Lipomembranous fat necrosis in three cases of testicular torsion. Histopathology. 2001;38:443–7.
11. Nistal M, Palacios J, Regadera J, Paniagua R. Postsurgical focal testicular infarct. Urol Int. 1986;41:149–51.
12. Calcagno C, Gastaldi F. Segmental testicular infarction following herniorrhaphy and varicocelectomy. Urol Int. 2007;79:273–5.
13. Bayne A, Koslin B, Daneshmand S. Bilateral segmental testicular infarction. Scientific World Journal. 2007;7:779–83.
14. Jarow JP, Coburn M, Sigman M. Incidence of varicoceles in men with primary and secondary infertility. Urology. 1996;47:73–6.
15. Preziosi P, Miano R, Bitelli M, Ciolfi MG, Micali S, Micali F. Right varicocele associated with inferior vena cava malformation in situs inversus: percutaneous treatment with retrograde sclerotherapy. J Endourol. 2001;15:1001–3.
16. Dubin L, Amelar RD. Varicocele size and results of varicocelectomy in selected subfertile men with varicocele. Fertil Steril. 1970;21:606–9.
17. Etriby A, Girgis SM, Hefnawy H, Ibrahim AA. Testicular changes in subfertile males with varicocele. Fertil Steril. 1967;18:666–71.
18. Dubin L, Hotchkiss RS. Testis biopsy in subfertile men with varicocele. Fertil Steril. 1969;20:51–7.
19. Cameron DF, Snydle FE, Ross MH, Drylie DM. Ultrastructural alterations in the adluminal testicular compartment in men with varicocele. Fertil Steril. 1980;33:526–33.
20. Spera G, Medolago-Albani L, Coia L, Morgia C, Gonnelli S, Ghilardi C. Histological, histochemical, and ultrastructural aspects of interstitial tissue from the contralateral testis in infertile men with monolateral varicocele. Arch Androl. 1983;10:73–8.
21. Zini A, Defreitas G, Freeman M, Hechter S, Jarvi K. Varicocele is associated with abnormal retention of cytoplasmic droplets by human spermatozoa. Fertil Steril. 2000;74:461–4.
22. Hendin BN, Kolettis PN, Sharma RK, Thomas Jr AJ, Agarwal A. Varicocele is associated with elevated

spermatozoal reactive oxygen species production and diminished seminal plasma antioxidant capacity. J Urol. 1999;161:1831–4.
23. Vasavada S, Ross J, Nasrallah P, Kay R. Prepubertal varicoceles. Urology. 1997;50:774–7.
24. Skoog SJ, Roberts KP, Goldstein M, Pryor JL. The adolescent varicocele: what's new with an old problem in young patients? Pediatrics. 1997;100:112–21.
25. Lyon RP, Marshall S, Scott MP. Varicocele in childhood and adolescence: implication in adulthood infertility? Urology. 1982;19:641–4.
26. Azar N, Guillevin L, Du Huong LT, Herreman G, Meyrier A, Godeau P. Symptomatic urogenital manifestations of polyarteritis nodosa and Churg-Strauss angiitis: analysis of 8 of 165 patients. J Urol. 1989;142:136–8.
27. Nigro KG, Abdul-Karim FW, Maclennan GT. Testicular vasculitis. J Urol. 2006;176:2682.
28. Eilber KS, Freedland SJ, Rajfer J. Polyarteritis nodosa presenting as hematuria and a testicular mass. J Urol. 2001;166:624.
29. Agraharkar M, Gokhale S, Gupta R. Wegener's granulomatosis diagnosed by testicular biopsy. Int Urol Nephrol. 2002–2003;34:559–64.
30. Cope EMB. Thrombo-angiitis obliterans of the spermatic cord. S Afr Med J. 1968;42:872–3.
31. Kuwahara M, Matsushita K, Nakamura K, Yoshinaga H, Aki M, Fujisaki N, et al. Thromboangitis obliterans of the spermatic cord: a case report. Hinyokika Kiyo. 1993;39:369–72.
32. Nistal M, Mate A, Paniagua R. Granulomatous epididymal lesion of possible ischemic origin. Am J Surg Pathol. 1997;21:951–6.
33. Esslimani M, Serre I, Granier M, Robert M, Baldet P, Costes V. Urogenital amyloidosis: clinico-pathological study of 8 cases. Ann Pathol. 1999;19:487–91.

16 New Approaches to Diagnosing Male Infertility, Part I: Microarray Analysis of Testicular Biopsy Specimens from Men with Intact and Impaired Spermatogenesis

Andrej-Nikolai Spiess and Wolfgang Schulze

In recent years, there have been major advances in microarray technology in terms of the number of genes under investigation (features) and the optimization of platforms. Starting with only 48 spotted cDNAs in 1995 [1], the past two decades have seen the development of many microarray platforms that can monitor the expression of thousands of genes simultaneously. Combined with the now standard method of cRNA amplification from small amounts of mRNA [2], as routinely encountered in biopsies, this offers the possibility of investigating global expression changes in small samples obtained from tumors or other clinical investigations.

With respect to human spermatogenesis, microarray technology has the potential to unveil the dramatic changes in gene expression that correlate with an altered composition of germ cells within the tubular compartment. This paradigm may be viewed from two sides: (1) From the viewpoint of a sample with severe spermatogenic (e.g., Sertoli cell-only [SCO]) impairment, comparing that sample with one that has less or no impairment (resulting in the presence of more differentiated germ cells) is actually a monitoring of gene expression during the progression of spermatogenesis [3]. (2) Investigating gene expression changes from low to high impairment describes the molecular effects of germ cell loss. In general, most of the expression changes from both viewpoints are dominated by germ cell-specific transcripts. Only a small proportion (about 4 %) of the genes changes their expression inversely to the germ cell content—that is, they are downregulated with an increase in germ cells and upregulated during germ cell loss [4]. This set of genes usually pertains to the somatic compartment and may be attributed to Sertoli, Leydig, peritubular myoid, or mast cells.

One of the most crucial aspects for obtaining high-quality microarray data for testicular gene expression is to use samples with an extremely homogeneous germ cell composition. Microarray technology has proven to be so sensitive that even a little "contamination" from foci with germ cell types other than those of the morphologic classification may result in the misclassification of the biopsy sample as a different spermatogenic type. This prerequisite can be met through morphologic classification of the biopsy specimen by semithin sectioning, using several sections from different regions of the sample to eliminate any foci of spermatogenesis that significantly deviate from the rest of the sample. Typical histologic sections with a homogeneous appearance are shown in Fig. 16.1; all six phenotypes of seminiferous epithelium—tubular atrophy (TA), SCO, spermatogonial presence (SGO), meiotic arrest (MA), hypospermatogenesis (HYS), and full spermatogenesis (FS)—are defined in Table 16.1. We routinely classify testicular biopsy samples by the modified Johnsen score (mJS; *see also*

A.-N. Spiess, Ph.D. (✉) • W. Schulze, M.D.
Department of Andrology, University Hospital Hamburg-Eppendorf, Hamburg, Germany
e-mail: a.spiess@uke.de

D. Ježek (ed.), *Atlas on the Human Testis*,
DOI 10.1007/978-1-4471-2763-5_16, © Springer-Verlag London 2013

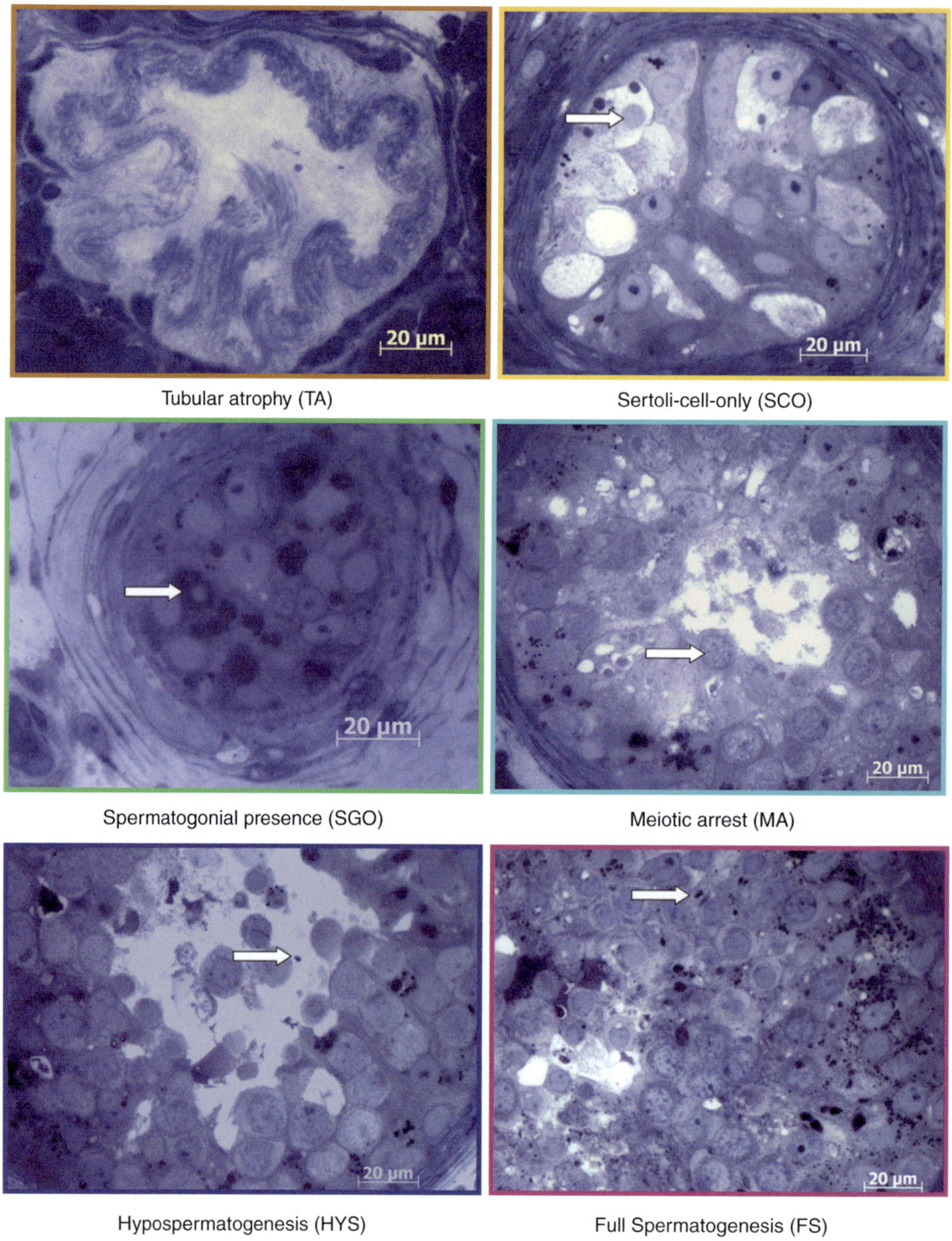

Fig. 16.1 Histologic images of the six different phenotypes of seminiferous epithelium discussed in this chapter. The biopsy specimens were obtained from men with TA, SCO syndrome (the *white arrow* indicates a Sertoli cell nucleus), SGO (the *white arrow* indicates a spermatogonial nucleus), MA (the *white arrow* indicates a primary spermatocyte), HYS (the *white arrow* indicates the sperm head, and FS (the *white arrow* indicates the sperm head). These samples were fixed with 5.5 % glutaraldehyde/1 % osmium, stained with toluidine blue, embedded in Epon, and cut into semithin (1–1.5-μm) sections using an ultramicrotome (Reichert-Jung Supercut 2065; Leica Microsystems, Buffalo Grove, IL). The sections were photographed with the Axiovert 100 microscope (Carl Zeiss MicroImaging GmbH, Jena, Germany). The *scale bar* indicates 20 μm. The colored margins of the micrographs correspond to the sample coloring in Table 16.1 and Fig. 16.2

Table 16.1 Phenotypes of the seminiferous epithelium

Abbreviation	Color code	Samples, *n*	Full nomenclature	Cellular composition	mJS
TA	(orange)	1	Tubular atrophy	Absence of any cells in seminiferous tubules	1
SCO	(yellow)	5	Sertoli cell only	Only Sertoli cells	2
SGO	(green)	2	Spermatogonial presence	Sertoli cells + spermatogonia	3
MA	(light blue)	5	Meiotic arrest	Sertoli cells + spermatogonia + prim./sec. spermatocytes	5
HYS	(dark blue)	8	Hypospermatogenesis	Sertoli cells + spermatogonia + prim./sec. spermatocytes + elongated spermatids (quantitatively reduced)	8
FS	(purple)	8	Full spermatogenesis	Sertoli cells + spermatogonia + prim./sec. spermatocytes + elongated spermatids (quantitatively normal)	10

Table 16.1), which quantifies the cellular composition per tubule [5, 6] as opposed to the original Johnsen score, which is calculated as an average over many tubules [7]. With this classification, histologically homogeneous biopsies will result in a single score (e.g., mJS 5 for "meiotic arrest," *see* Table 16.1), whereas samples with a mixed histologic appearance ("mixed atrophy" [8]), such as MA with a few foci-containing SCO tubules, will result in mJS 5, 2. From the viewpoint of microarray analysis, it is this relative proportion of different germ cell types in the samples that will drive a subsequent clustering outcome.

In principle, all the aforementioned investigations are based on the hybridization of cRNA (usually 20–30 μg/biopsy) to high-content microarrays (in this case, Affymetrix Whole Genome U133 Plus 2.0 Gene Chips [Affymetrix, Santa Clara, CA]). After hybridization and washing, the microarrays are scanned and then the data are extensively preprocessed (normalization, low-intensity cutoff, missing value imputation). In this way, we obtained expression values for 54,613 transcripts in 29 biopsy samples classified by morphologic criteria and numbers per group, as defined in Table 16.1. From this large set of genes, we filtered the top 2,000 genes with respect to their highest variance σ^2 through all the samples. This is a common approach to obtain differentially expressed genes without imposing a particular grouping structure. These genes were clustered by hierarchic clustering of samples and genes, as displayed in the heat map representation in Fig. 16.2. Two major observations can be made from this analysis. First, all 29 samples cluster according to their corresponding morphologic classification. Hence, it can be deduced that each phenotype of the seminiferous epithelium has a clear and distinct molecular correlate. The sample cluster also separates into a premeiotic/postmeiotic branch, which suggests significantly different gene expression in these two groups. Gene clustering results in five individual clusters (C1–C5), four of which (C1–C4) exhibit a gradient expression pattern with increasing germ cell differentiation (i.e., C4 has significant differential expression with the appearance of spermatogonia, whereas C1 has the most differential expression in the late postmeiotic phase). C5 is a cluster that is unlike clusters C1 through C4 in that it contains genes that are downregulated with increasing germ cell differentiation. The five clusters C1 through C5 can be visualized by means of principle component analysis (PCA), in which the largest variance of the 2,000 genes is reduced to three principle components (PCA1–PCA3) that can be displayed on a

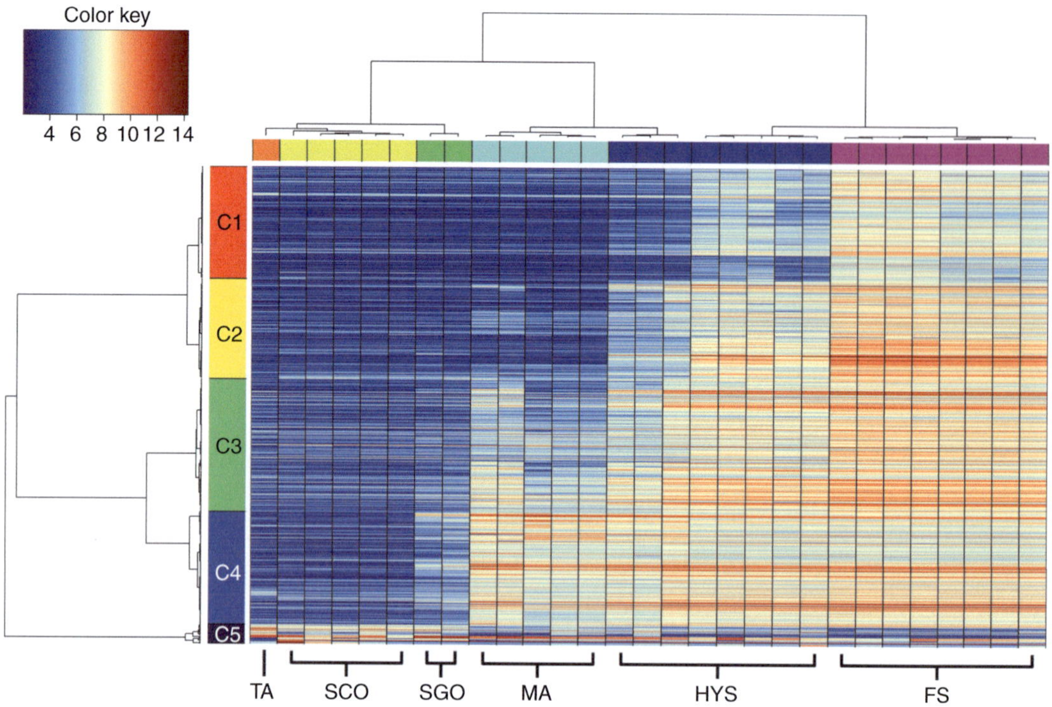

Fig. 16.2 Heat map representation of the top 2,000 genes with highly variable expression in testicular biopsy samples with different phenotypes of seminiferous epithelium. RNA from 29 testicular biopsies was hybridized to Human Genome U133 Plus 2.0 GeneChip (Affymetrix). After appropriate data transformation (quantile normalization, intensity cutoff filtering), the top 2,000 genes based on their variance across all samples were selected for hierarchic sample/gene co-clustering (Pearson correlation, average distance). The sample clustering (column dendrogram) results in a clear grouping of the different biopsy samples based on their gene expression profile correlating with the morphologic classification. The gene clustering (row dendrogram) results in the grouping of gene expression profiles into five distinct clusters (C1–C5), four of which (C1–C4) show different patterns of upregulation with increasing differentiation of the germ cells (TA → SCO → SGO → MA → HYS → FS) and one of which (C5) exhibits downregulation. The central heat map is color coded according to expression levels, as shown in the *color key*. For more details on the characteristics of the biopsy samples, see Table 16.1

three-dimensional plot (Fig. 16.3). Similar to Fig. 16.2, the main variance is defined by the separation of C1 through C4 from C5 on the first principle component PCA1, whereas the different patterns of differential expression throughout increasing germ cell differentiation are separated on PCA2 and PCA3.

Extending this, interesting interpretations of the different gene clusters can be derived from the functional analysis of the genes in clusters C1 through C5. In general, this is accomplished by using gene ontology (GO) term overrepresentation analysis of the genes in the individual clusters. Figure 16.4 depicts the averaged expression from all samples with increasing germ cell differentiation. The different clusters C1 through C5 are noted on the left. Concomitant with the pattern of largest differential expression in each cluster, the overrepresented GO terms correspond extremely well. For example, in C1, the largest expression changes are encountered in the late postmeiotic stages HYS+FS, with GO terms pointing to a significant enrichment of genes responsible for fertilization, sperm-to-egg fusion, and cell recognition. In C2, containing genes for an earlier postmeiotic phase, GO terms for spermatid development and sperm motility correlate well with this time point of germ cell differentiation. C3 and C4 are earlier premeiotic/meiotic clusters that contain genes involved in M phase, meiosis I, and germ cell development. Noteworthy again is C5, which has an inverse gene expression pattern and consists

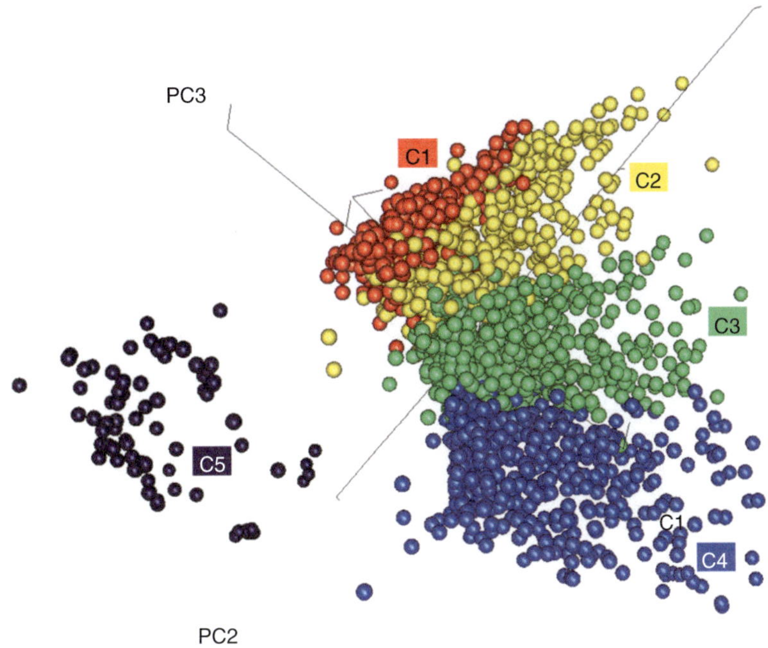

Fig. 16.3 Principle component analysis of the top 2,000 variant genes shown in Fig. 16.2. The gene expression of the top 2,000 variant genes is reduced to a three-dimensional display of the first three "principle components" (PC1–PC3) that describe the maximal variability in the dataset. The color coding of the genes (*spheres*) is taken from C1 through C5 in Fig. 16.2. Whereas PC1 clearly separates upregulated clusters C1 through C4 from downregulated cluster C5, PC2 and PC3 separate C1 through C4 based on their different expression patterns, with increasing differentiation of the germ cells. For a more thorough description of these clusters, see the text

of genes enriched for somatic transcripts involved in male sex differentiation, acute inflammatory response, and complement activation. In a more detailed study, we could show that many of the genes in C5 correspond to the interstitial mast cell numbers of the biopsy samples [4].

Finally, substantial insight may be gained by identifying important regulatory (protein) networks within the different clusters. In our example, we used specialized network software (Cytoscape 2.8.2 [Cytoscape Consortium], including the MiMI plugin [9]) to extract the most important protein interactions (Fig. 16.5). Similar to the GO terms in Fig. 16.4, the networks correlate significantly with the appearance of specific germ cell types in the different clusters: late postmeiotic cluster C1 contains a small network involved in genetic risk/organogenesis (PROX1, RORA); early postmeiotic clusters C2+C3 contain relevant networks for cell polarity (SMURF1), sperm tail function (AKAP3, ROPN1, DYNLL1), and protein ligation (CBL); (pre)meiotic cluster C4 has its network centered largely around M phase (CDC2), prophase I (CHEK1), and G2/M phase (ECT2); and C5 is dominated by networks correlating with complement activation.

In summary, human spermatogenesis or its failure may be characterized and categorized astonishingly well by means of microarray technology. Morphologic/histologic features are highly correlated with specific gene expression patterns driven largely by the germ cell composition of the seminiferous epithelium.

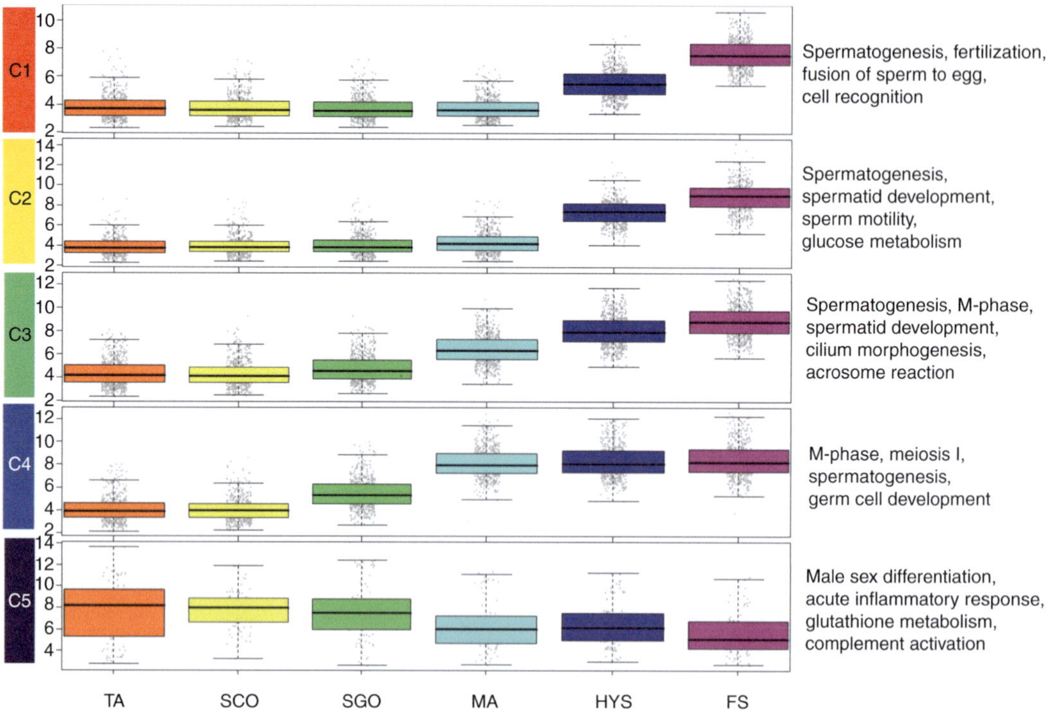

Fig. 16.4 Gene ontology term analysis of the five different gene clusters shown in Figs. 16.2 and 16.3. The five clusters (C1–C5) containing different numbers of genes (C1, $n=469$; C2, $n=419$; C3, $n=558$; C4, $n=473$; C5, $n=81$) were subjected to GO term overrepresentation analysis using the DAVID 6.7 functional annotation tools (http://david.abcc.ncifcrf.gov/). The five different clusters are denoted by the *color bars* on the left. For each cluster, the *box plots* display the distribution of all genes from the averaged expression values of each level of spermatogenic impairment. Overrepresented GO terms for each cluster are displayed on the right in ascending order of *p*-values obtained from Fisher's exact test with Benjamini-Hochberg correction

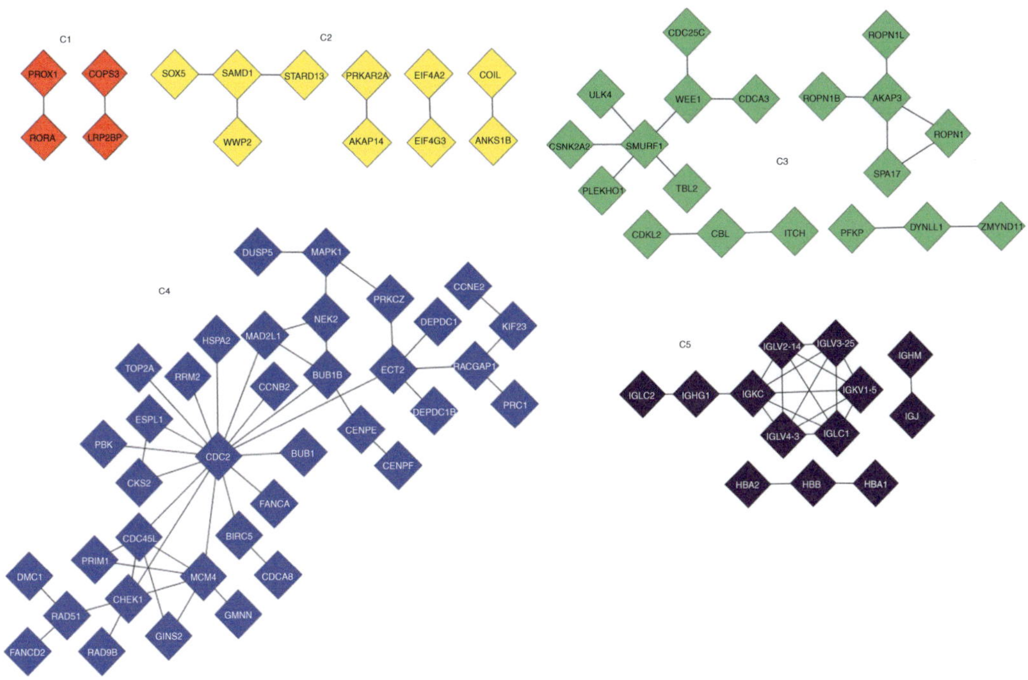

Fig. 16.5 Network analysis of the five different gene clusters shown in Figs. 16.2 through 16.4. The human genome nomenclature (HUGO) names were used to query several interaction databases (BIND, HPRD, IntAct, KEGG, MINT, Reactome) using the Cytoscape 2.8.2 software (www.cytoscape.org) and the MiMI plugin. Each cluster (C1–C5) is defined by dominating and individual interaction networks that do not overlap those of the other clusters. For more details on the networks' functions, see the text

References

1. Schena M, Shalon D, Davis RW, Brown PO. Quantitative monitoring of gene expression patterns with a complementary DNA microarray. Science. 1995;270:467–70.
2. Van Gelder RN, von Zastrow ME, Yool A, Dement WC, Barchas JD, Eberwine JH. Amplified RNA synthesized from limited quantities of heterogeneous cDNA. Proc Natl Acad Sci USA. 1990;87:1663–7.
3. Feig C, Kirchhoff C, Ivell R, Naether O, Schulze W, Spiess AN. A new paradigm for profiling testicular gene expression during normal and disturbed human spermatogenesis. Mol Hum Reprod. 2007;13:33–43.
4. Spiess AN, Feig C, Schulze W, Chalmel F, Cappallo-Obermann H, Primig M, Kirchhoff C. Cross-platform gene expression signature of human spermatogenic failure reveals inflammatory-like response. Hum Reprod. 2007;22:2936–46.
5. De Kretser DM, Holstein AF. Testicular biopsy and abnormal germ cells. In: Hafez ESE, editor. Human semen and fertility regulation in men. St. Louis: Mosby; 1976. p. 332–3.
6. Schulze W, Thoms F, Knuth UA. Testicular sperm extraction: comprehensive analysis with simultaneously performed histology in 1418 biopsies from 766 subfertile men. Hum Reprod. 1999;Suppl 1:82–96.
7. Johnsen SG. Testicular biopsy score count – a method for registration of spermatogenesis in human testes: normal values and results in 335 hypogonadal males. Hormones. 1970;1:2–25.
8. Sigg C, Hedinger C. Quantitative and ultrastructural study of germinal epithelium in testicular biopsies with "mixed atrophy. Andrologia. 1981;13:412–24.
9. Gao J, Ade AS, Tarcea VG, Weymouth TE, Mirel BR, Jagadish HV, States DJ. Integrating and annotating the interactome using the MiMI plugin for cytoscape. Bioinformatics. 2009;25:137–8.

17 New Approaches to Diagnosing Male Infertility, Part II: The Role of Confocal Microscopy and Three-Dimensional Reconstruction in Visualization of Reinke's Crystals

Viviana Kozina and Davor Ježek

The aim of this chapter is to illustrate the use of confocal microscopy and three-dimensional (3D) reconstruction in the visualization of various testicular structures. As an example, Reinke's crystals are pointed out. Reinke's crystals resemble hexagonal prisms of various sizes (mostly 2–3 μm). They are composed of parallel 5- to 10-nm filaments and do not have a unit membrane. Sometimes, a single crystal may be found in the cytoplasm and nucleus of Leydig cells. However, very often, two or even more crystals are present that may cause a deformity of the nucleus and infolding of the nuclear membrane. Occasionally, one may note electron-dense rod-shaped structures inside the crystal. Besides the remarkable Reinke crystals, other paracrystalline inclusions may be found within human Leydig cells.

Images of test specimens were acquired by a Leica SP2 AOSB confocal laser scanning microscope (Leica Microsystems, Buffalo Grove, IL) using an excitation wavelength of 488 nm yielded by an argon laser (20 mW). The images (with voxel dimensions of 37 × 37 × 81 nm) were acquired by HCX PL APO CS 100× oil immersion objective (NA = 1.40) using an emission wavelength ranging from 512 to 583 nm. A 3D reconstruction of a Reinke crystal was based on a series of confocal microscopic images, which were deconvoluted in the Huygens Professional software package (scientific volume imaging, Hilversum, The Netherlands) by the quick maximum likelihood estimation algorithm using the experimental point spread function. Subsequently, the 3D images were smoothed by a Gaussian filter and segmented by thresholding of the image connected with manual corrections as necessary. Finally, triangulated isosurfaces of segmented crystals were generated and surface rendering was applied. The image-processing algorithms and 3D visualizations were performed in the Ellipse (ViDiTo, Kosice, Slovakia; www.ellipse.sk) software environment using our custom-made plug-in modules (Smooth3D, Thresh, Surface).

The images presented (Figs. 17.1–17.5) clearly demonstrate the advantage of new imaging techniques such as confocal microscopy and 3D reconstruction in visualizing certain structures of interest within the testicular biopsy specimen.

V. Kozina, MEd • D. Ježek, M.D., Ph.D. (✉)
Department of Histology and Embryology,
School of Medicine, University of Zagreb,
Šalata 3, 10000 Zagreb, Croatia
e-mail: davor@mef.hr, davorjezek@yahoo.com

D. Ježek (ed.), *Atlas on the Human Testis*,
DOI 10.1007/978-1-4471-2763-5_17, © Springer-Verlag London 2013

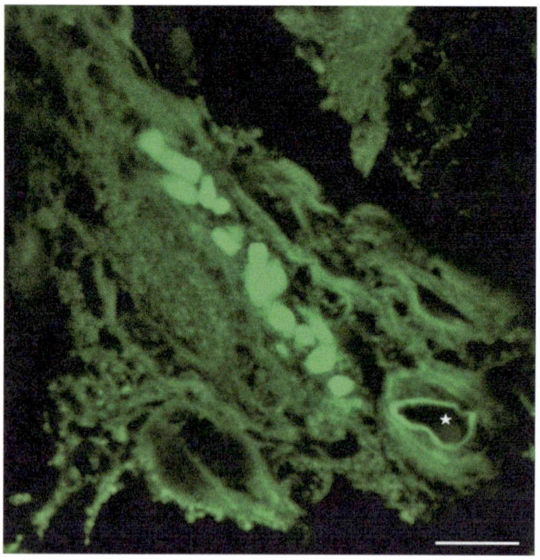

Fig. 17.1 Testicular biopsy specimen: part of the interstitium visualized on confocal microscopy. In one Leydig cell, numerous Reinke crystals are seen (*star*: small blood vessel) (Confocal microscopy; *scale bar*, 10 μm)

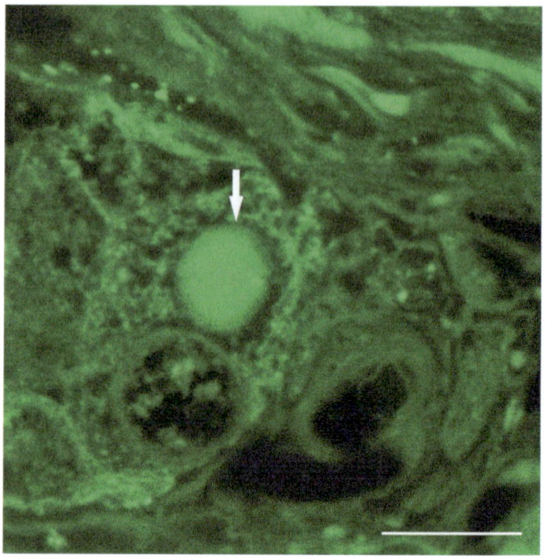

Fig. 17.3 Reinke crystal (*arrow*) near the nucleus of a Leydig cell (Confocal microscopy; *scale bar*, 15 μm)

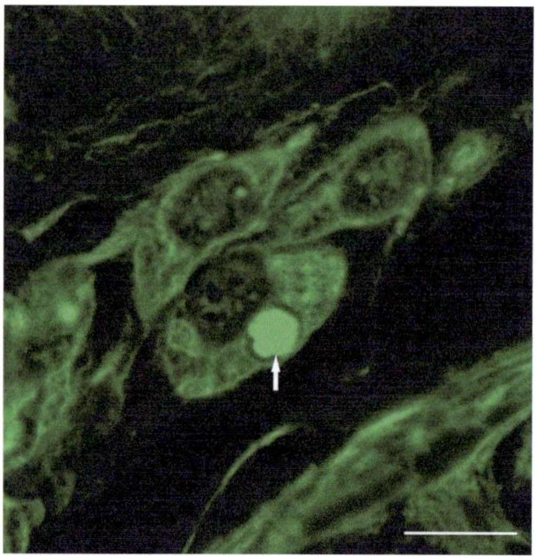

Fig. 17.2 Group of Leydig cells within the interstitial compartment. After the standard eosin staining, a Reinke crystal (*arrow*) becomes fluorescent (Confocal microscopy; *scale bar*, 15 μm)

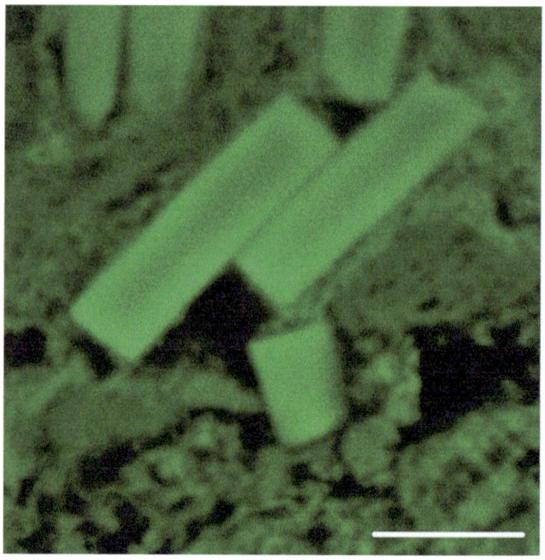

Fig. 17.4 At higher magnification, Reinke's crystals demonstrate a regular contour. Sometimes, several crystals are located within the same Leydig cell cytoplasm, reaching an impressive length of 7–10 μm (Confocal microscopy; *scale bar*, 10 μm)

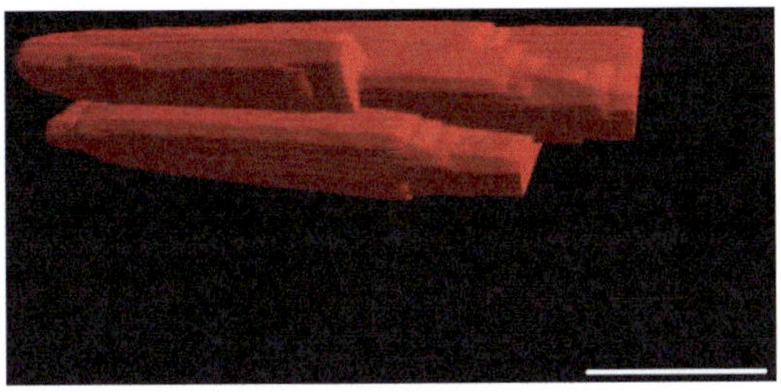

Fig. 17.5 3D reconstruction of Reinke's crystals based on confocal microscopic images (*Scale bar*, 5 μm)

Recommended Literature

de Kretser DM. Crystals of reinke in the nuclei of human testicular interstitial cells. Experientia. 1968;15:587–8.

Kozina V, Banek LJ, Knežević N, Geist D, Kubinová L, Kosović M, et al. Reinke's crystals in perivascular and peritubular Leydig cells. Cro Chem Acta. 2011a;84: 159–67.

Kozina V, Geist D, Kubinová L, Bilić E, Karnthaler HP, Waitz T, et al. Visualization of Reinke's crystals in normal and cryptorchid testis. Histochem Cell Biol. 2011b;135(2):215–28.

Lobo MV, Arenas MI, Alonso FJ, Gomez G, Bazán E, Paíno CL, et al. Nestin, a neuroectodermal stem cell marker molecule, is expressed in Leydig cells of the human testis and in some specific cell types from human testicular tumours. Cell Tissue Res. 2004;316: 369–76.

Nagano T, Ohtsuki I. Reinvestigation of the fine structure of Reinke's crystal in the human testicular interstitial cell. J Cell Biol. 1971;51:148–61.

Reinke F. Beiträge zur Histologie des Menschen. I. Über Krystalloidbildungen in den interstitiellen Zellen des menschlichen Hodens. Arch Mikr Anat. 1896;47: 34–44.

Schulze C. Sertoli cells and Leydig cells in man. Adv Anat Embryol Cell Biol. 1984;88:1–104.

Disorders of Sex Development

18

Manuel Nistal and Pilar González-Peramato

The term *disorders in sexual differentiation* (DSD) to designate all intersex disorders was introduced in the consensus classification of intersex disorders emanating from the Chicago Meeting in 2005 [1]. This classification considers the use of some terms, such as *pseudohermaphroditism*, *hermaphroditism*, *sex reversal*, and *intersex situation*, obsolete; however, it has given rise to many critical comments [2] because it was made on the basis of chromosomal constitution (peripheral karyotype of leukocytes). Specifically, in cases of gonadal dysgenesis, entities with a completely different histopathology, such as 45,X0 Turner's syndrome and its variants and mixed gonadal dysgenesis (45,X0/45,XY), have been included with Klinefelter's syndrome and its variants under the term *disorders in sex development*. On the other hand, other entities with a similar histopathology, such as 46,XY gonadal dysgenesis and 46,XX gonadal dysgenesis, have been classified into different groups. Two facts have prevented a solution to the problem in classifying these disorders: (1) the same karyotype may produce different phenotypes; and (2) it has been progressively understood that gonadal dysgenesis, instead of showing one cellular line, has mosaicisms or more complex chromosomal constitutions. Therefore, the classification of these disorders considering different viewpoints, including genetic, endocrinologic, clinical, and histopathologic, has not yet been resolved.

Pathologists count on a fact that is no less real than karyotype: the nature of the gonad that is in front of them. This advantage places them at the center of the problem. On the one hand, they can see what chromosomal or genetic mutations may cause the nature of the gonad; on the other hand, they can explain the variability of clinical symptoms through different parts of the gonad.

This chapter includes images of gonadal dysgenesis, ovotesticular disorder (true hermaphroditism), and male undermasculinization.

M. Nistal, M.D., Ph.D. (✉)
Department of Pathology, University Hospital La Paz, Madrid, Spain

Department of Anatomy, Histology, and Neuroscience, University Autonoma of Madrid, Madrid, Spain
e-mail: mnistal.hulp@salud.madrid.org

P. González-Peramato, M.D., Ph.D.
Department of Pathology, University Hospital La Paz, Madrid, Spain

Department of Pathology, University Autonoma of Madrid, Madrid, Spain

Gonadal Dysgenesis

Gonadal dysgenesis may be defined as an incomplete or defective gonadal differentiation as a result of disturbance in germ cell migration and/or its correct organization in the gonadal ridge. This anomaly is morphologically expressed as several grades of differentiation toward ovary, testis, or both. The term *gonadal dysgenesis* comprises a broad spectrum of lesions, ranging from absence of the gonad, to the classical streak

D. Ježek (ed.), *Atlas on the Human Testis*,
DOI 10.1007/978-1-4471-2763-5_18, © Springer-Verlag London 2013

gonad, to a testis surrounded by a stroma similar to the ovarian cortex. That is, the term includes all kinds of gonads until the architecture of the ovary or testis is fully defined.

Types of Gonads and Their Correlation with Clinical Syndromes

To better understand the different patterns, a pathologist may realize that it is very useful to consider that several histologic patterns may arise from an undifferentiated gonad. When the differentiation is toward ovary, two different stages may be seen: streak gonad with or without ovarian follicles and hypoplastic ovary. If the differentiation is toward testis, streak gonad with epithelial cords and dysgenetic testis may be observed. If the gonad has incomplete differentiation toward both testis and ovary without forming ovotestes, a streak-testis is formed, or the undifferentiated gonad may even disappear (Fig. 18.1).

At the beginning of its differentiation, the gonad may present an "undifferentiated gonadal tissue" (UGT) arrested at the fetal stage. In the literature, these gonads have been considered either streak gonad or ovary. In fact, it may present as a streak gonad with epithelial cord–like structures or focally in several types of gonads, such as dysgenetic testis, streak-testis, and ovotestis. UGT recognition is of great interest because of the high incidence of its transformation into gonadoblastoma [3]. The most significant characteristics of this peculiar gonad are discussed later, in "Streak Gonad with Epithelial Cord–Like Structures." The abundance of germ cells arranged in neither follicles nor seminiferous cords but enclosed in an ovarian-like stroma or cords of immature Sertoli-like/granulosa cells is outstanding (Fig. 18.2).

Grossly, and regardless of whether the gonad expresses differentiation toward testis or ovary, the streak gonad is an elongated, fibrous, pearly formation located in the natural place of the

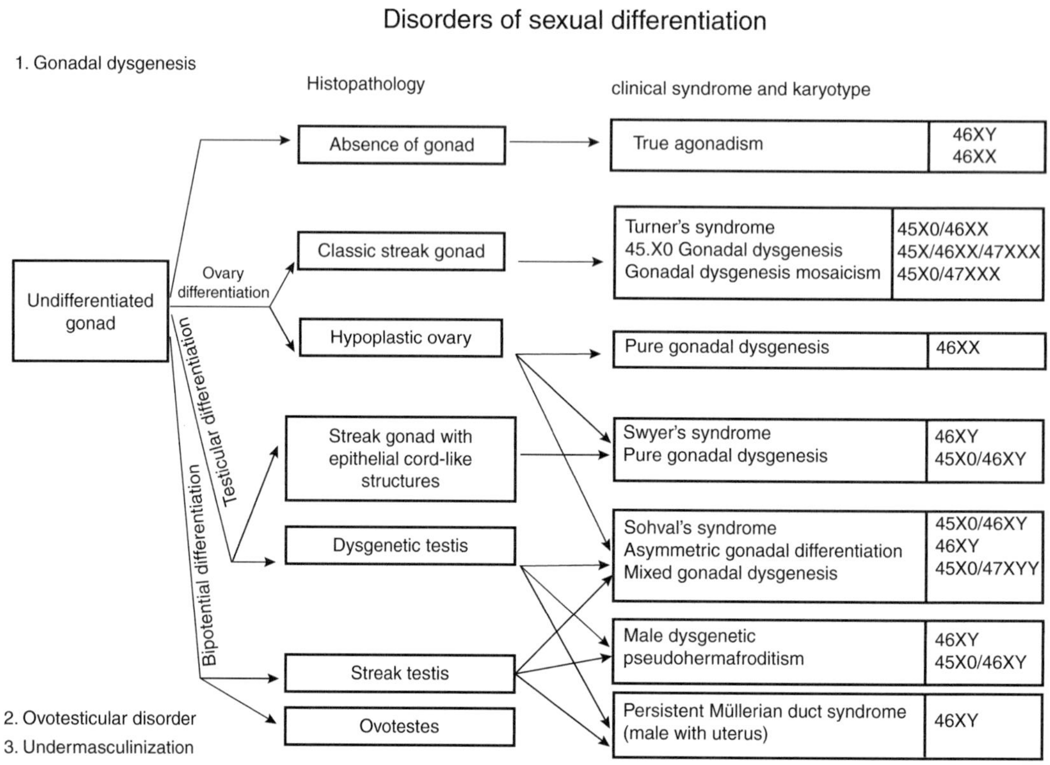

Fig. 18.1 Disorders of sexual differentiation. The diagram represents, from a histopathologic perspective, the different types of gonads that may be found in DSD and the corresponding karyotype and clinical syndrome

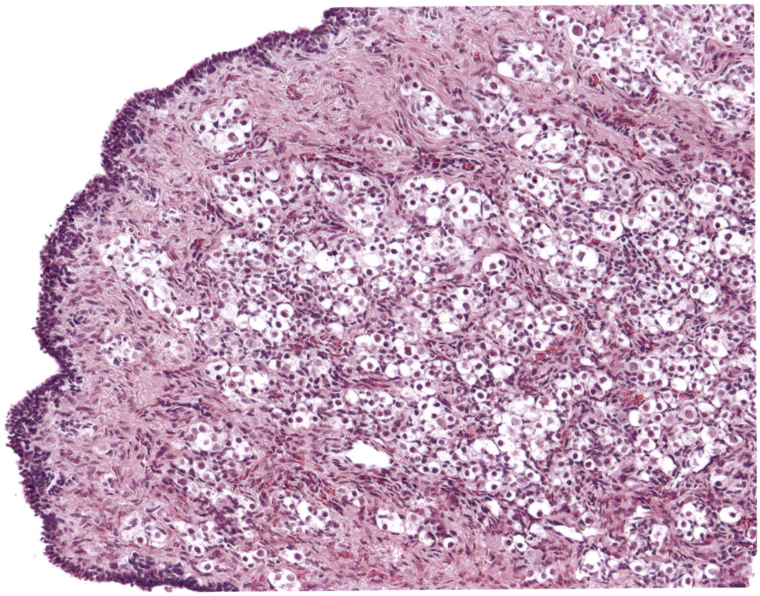

Fig. 18.2 Undifferentiated gonadal tissue. This image from a 1-year-old child shows the gross morphology of streak gonad, which consists of nests of large germ cells surrounded by small undifferentiated cells. Small undifferentiated cells also may be observed inside the nests. Among the nests, bundles of ovarian-like stroma may be recognized (H&E stain)

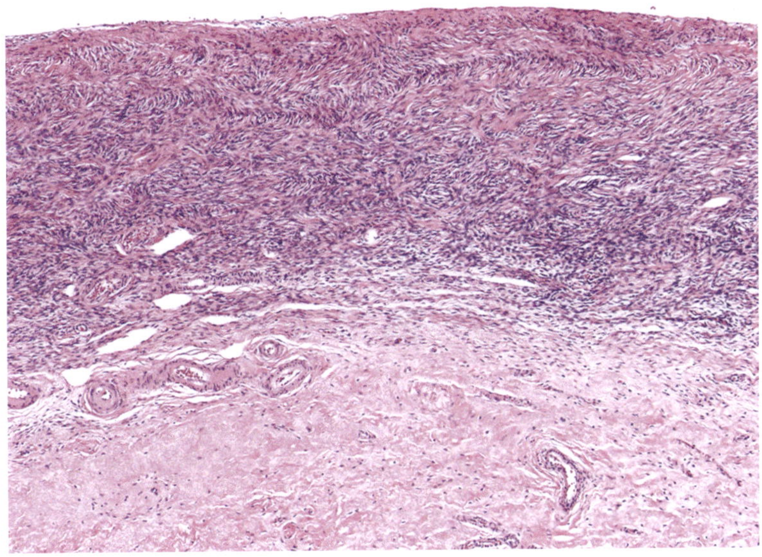

Fig. 18.3 Classical streak gonad. In this specimen from a 14-year-old patient with Turner's syndrome, bundles of fusiform cells are arranged parallel to the surface or are intertwined. No ovarian follicle is observed (H&E stain)

normal ovary. The streak gonad is 2–3 cm long and 0.5 cm wide in the adult, and may contain hilar cells or rete ovarii.

Classical Streak Gonad

Microscopically, classical streak gonad sometimes is made up of bundles of fibrous connective tissue and at other times by a whirled ovarian-like stroma. This is the classical gonad seen in 45,X0 gonadal dysgenesis or Turner's syndrome (Fig. 18.3).

Streak Gonad with Primordial Follicles

Inside an ovarian-like stroma, isolated ovarian follicles may be found. These germ cells have positive immunohistochemical expression for VASA and c-Kit and are negative for OCT3/4, PLAP, and TSPY. In adult patients, atretic follicles, multiple cortical cysts, abundant hilar cells, or glandular-like groups of clear cells in the hilus may be observed. This gonad is observed in 46,XX pure gonadal dysgenesis and mosaicisms 45,X0/46,XX, 45,X0/46,XX/47,XXX, and 45,X0/47,XXX (Fig. 18.4).

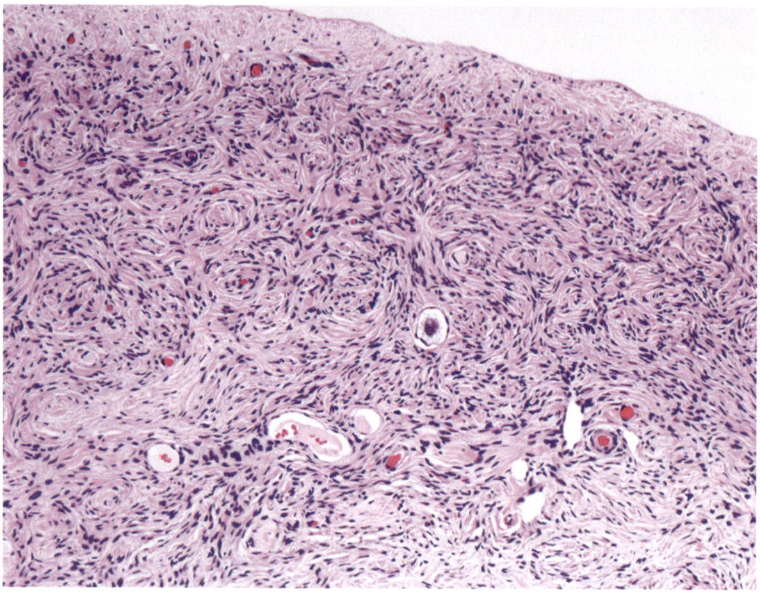

Fig. 18.4 Streak gonad with ovarian follicle. This image from a 20-year-old patient with Turner's syndrome (45,X0) shows classical streak gonad with a primordial follicle inside an ovarian-like stroma. Neither atresic follicles nor corpus albicans are observed (H&E stain)

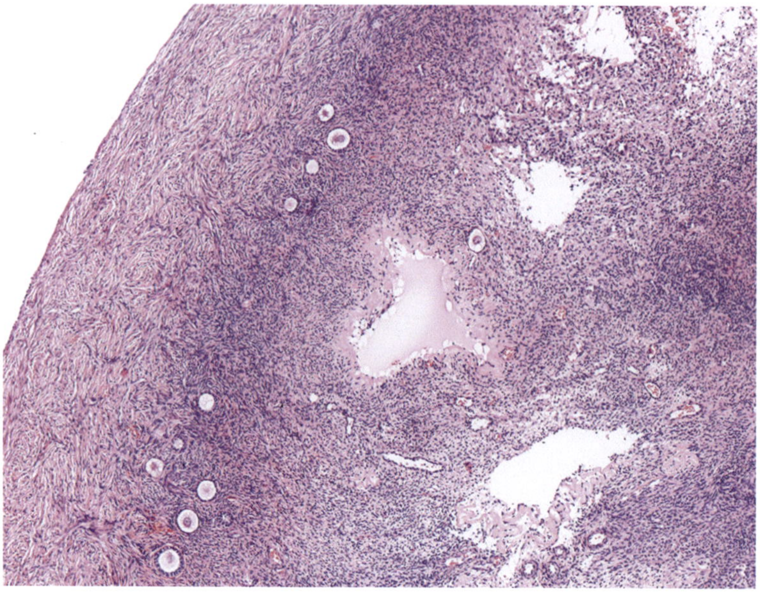

Fig. 18.5 Hypoplastic ovary. In this specimen from an 18-year-old patient with 46,XX pure gonadal dysgenesis, a row of primordial follicles can be observed parallel to the surface of the gonad. Note some atretic follicles in the center of the gonad. Maturation is not present in any of the follicles (H&E stain)

Hypoplastic Ovary

Hypoplastic ovary does not have the form of a streak gonad; it is small and ovoid with a smooth surface. Primary follicles, and occasionally follicles in development, may be conserved in the cortex with a regular pattern of distribution. This gonad is characteristic of 46,XX pure gonadal dysgenesis and some ovotesticular dysgenesis (true hermaphrodism) (Fig. 18.5).

Streak Gonad with Epithelial Cord–Like Structures

Streak gonad with epithelial cord–like structures consists of an ovarian cortex–like stroma crossed in different directions by epithelial cord–like structures (sex cords) that may form an anatomizing network. Sometimes, the cords are very thin; at other times, they are thick, with coarse outlines. These cord-like structures contain two types of cells. The

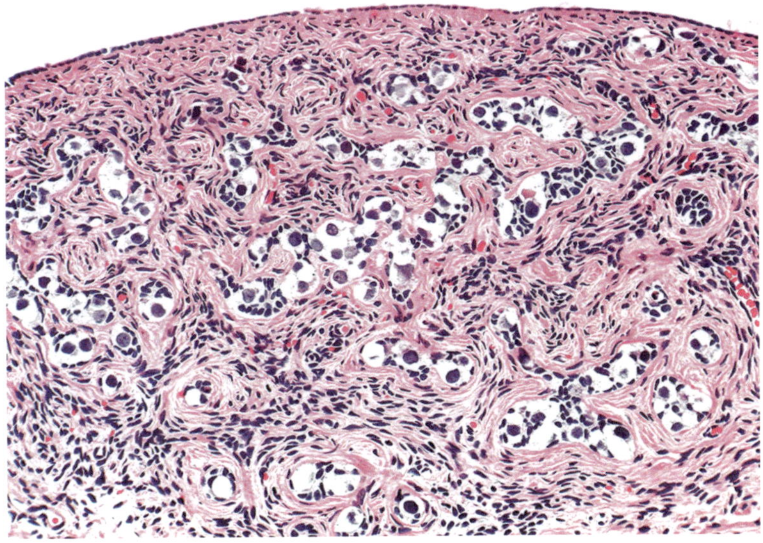

Fig. 18.6 Streak gonad with epithelial cord–like structures. This image from a 1-month-old child with 45,X0/46,XY gonadal dysgenesis shows cord-like structures formed by one or several germ cells surrounded by ovarian-like stroma (H&E stain)

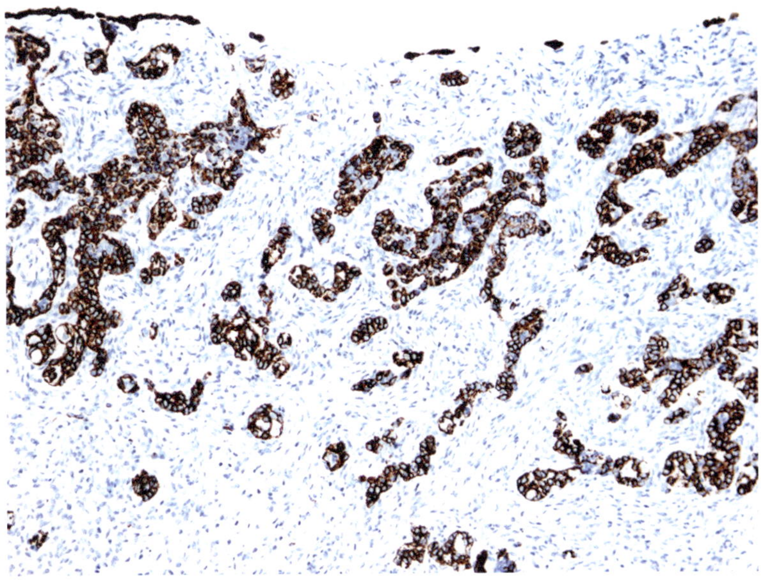

Fig. 18.7 Streak gonad with epithelial cord–like structures. In this sample from a 1-month-old child with 45,X0/46,XY gonadal dysgenesis, intense cytokeratin immunoexpression is seen in most cells in the cord-like, trabecular, and solid structures. Note the negative expression in germ cells (Cytokeratin AE1/AE3 stain)

majority are small cells with hyperchromatic nuclei, which are considered pre-Sertoli or immature Sertoli/granulosa cells (Fig. 18.6). These cells are intensely positive for AE1/AE3 (Fig. 18.7) and moderately positive for inhibin (Fig. 18.8) and D2-40. Antimüllerian hormone (AMH) is weakly expressed. The epithelial cord–like structures lean on a basal membrane of variable thickness. The second type of cells are germ cells, which may be isolated inside the stroma, forming cords, or located between Sertoli/granulosa cells. The number of these cells varies, and in some cords, they even form small nests. The immunohistochemical profile of germ cells is as follows: OCT3/4+, c-Kit+(Fig. 18.9), PLAP+, TSPY+(testis-specific protein Y-encoded), and VASA+. Overall, the histologic pattern of this gonad is similar to that of UGT. Although some streak gonads with epithelial cord–like structures are devoid of germ cells, they probably contained germ cells at some point during their development but the cells were eliminated by an apoptosis mechanism.

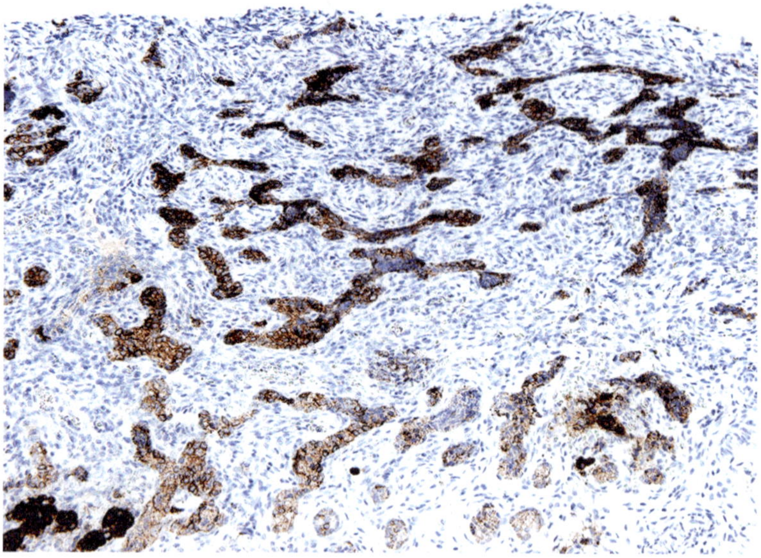

Fig. 18.8 Streak gonad with epithelial cord–like structures. This image from a 1-month-old child with 45,X0/46,XY gonadal dysgenesis shows intense expression of inhibin in epithelial cord–like structures. Expression is greater in superficial than deep cords and is absent in stromal and germ cells (Inhibin immunohistochemical stain)

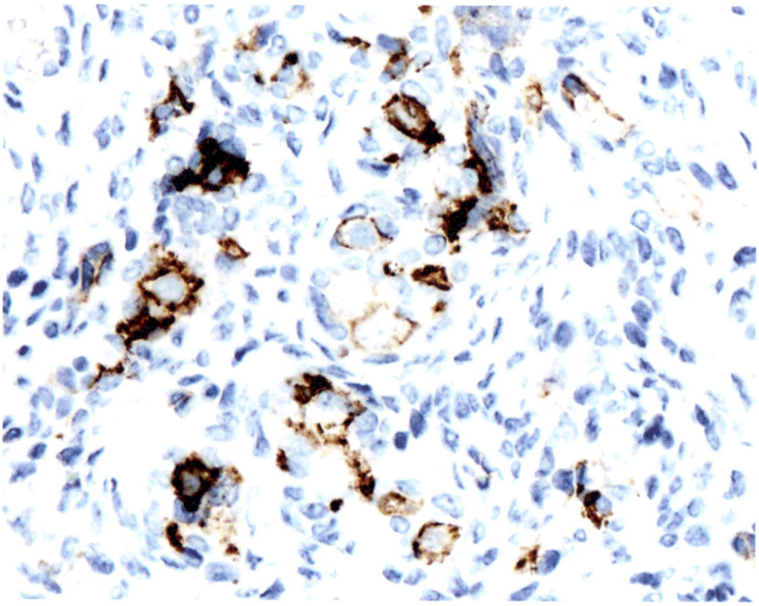

Fig. 18.9 Streak gonad with epithelial cord–like structures. Intense expression of c-Kit is seen in the membrane of most of the germ cells in the epithelial cord–like structures in this specimen from a 1-month-old child with 45,X0/46,XY gonadal dysgenesis (c-Kit immunohistochemical stain)

Dysgenetic Testis

Dysgenetic testis usually is small and cryptorchid, with a compact center containing seminiferous tubules with or without germ cells. The peripheral zone, instead of showing a well-collagenized albuginea as occurs in normal testes, shows an ovarian stroma with or without malformed seminiferous cords. These seminiferous cords have a strange anastomosis, and their diameter varies from superficial to deep areas without an exact topography. Seminiferous tubules or cords are separated by a wide and loose stroma (Figs. 18.10, 18.11, and 18.12). Sertoli cells express inhibin and AMH, and, like Leydig cells, they also express calretinin, especially in the more superficial seminiferous cords. However, in the central areas of the testis, calretinin is expressed only in the smaller tubules, which usually are devoid of germ cells, and in Leydig cells. The image of seminiferous tubules formed by calretinin-positive Sertoli cells forming a half-moon around

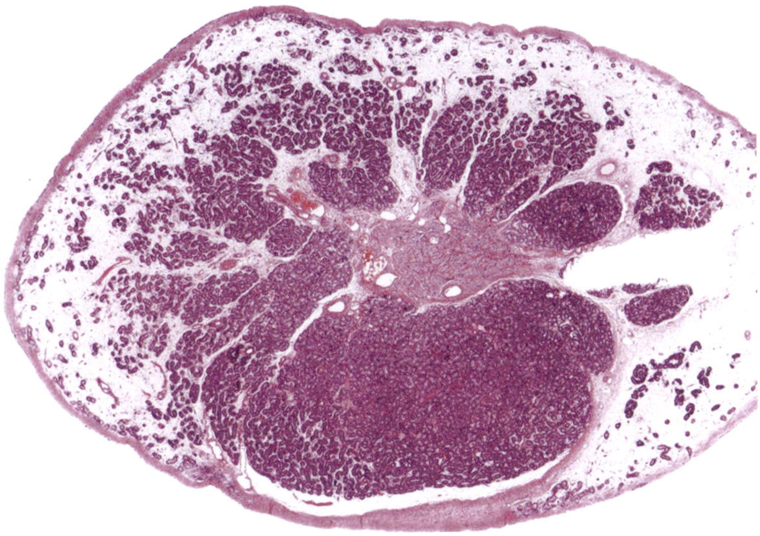

Fig. 18.10 Dysgenetic testis. Shown is a specimen from a 1-year-old child with 45,X0/46,XY gonadal dysgenesis. The gonad has a compact center containing seminiferous tubules and a crown-shaped peripheral zone with seminiferous tubules separated by wide myxoid stroma. Some of the tubules reach the albuginea (H&E stain)

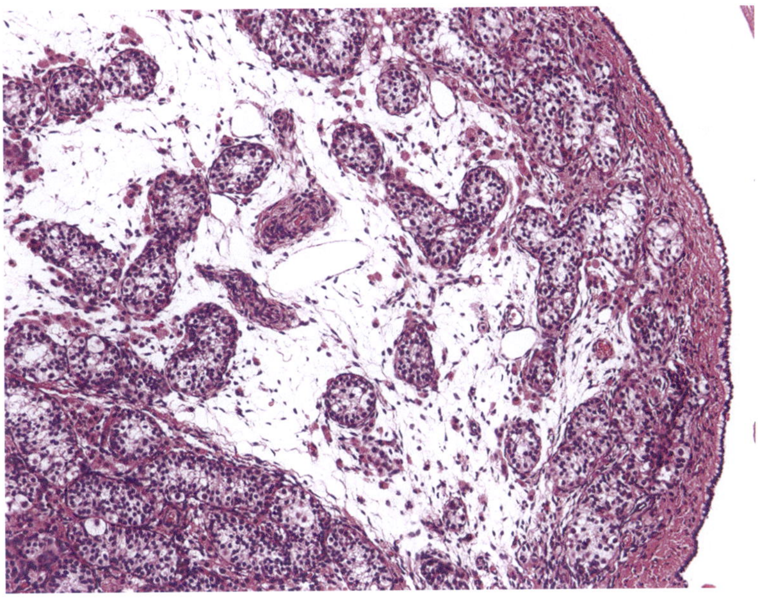

Fig. 18.11 Dysgenetic testis. This detail of the peripheral zone of the gonad shown in Fig. 18.10 shows seminiferous tubules with frequent malformations (bifurcations and tubules with irregular contours) arranged inside the albuginea. The germ cell number is quite variable from one tubule to another (H&E stain)

other negative cells is a frequent finding. Germ cells are found both in the center of seminiferous tubule and over the basal membrane. The germ cell phenotype is VASA+, TSPY+, D2-40+, and a subpopulation is OCT3/4+, c-Kit+, and PLAP+.

Studies of dysgenetic testes in pubertal and adult patients are not abundant. The whirled stroma in the albuginea and the irregular contour with strangely shaped seminiferous tubules are conserved. Sertoli cell maturation is incomplete, expressed by the absence of a tubular lumen and the nuclear characteristics of dysgenetic cells (round or ovoid nuclei rather than triangular nuclei with deep grooves of adult Sertoli cells). Most tubules are devoid of germ cells. The interstitium shows an apparent Leydig cell hyperplasia, both diffuse and peritubular. Immunohistochemical techniques allow a better understanding of the peculiar relationship among Leydig, peritubular, and Sertoli cells. With anti-calretinin and anti-inhibin antibodies, one can observe how many Leydig cells form concentric circles in the tubular wall. These Leydig cells are located either among peritubular cells or between

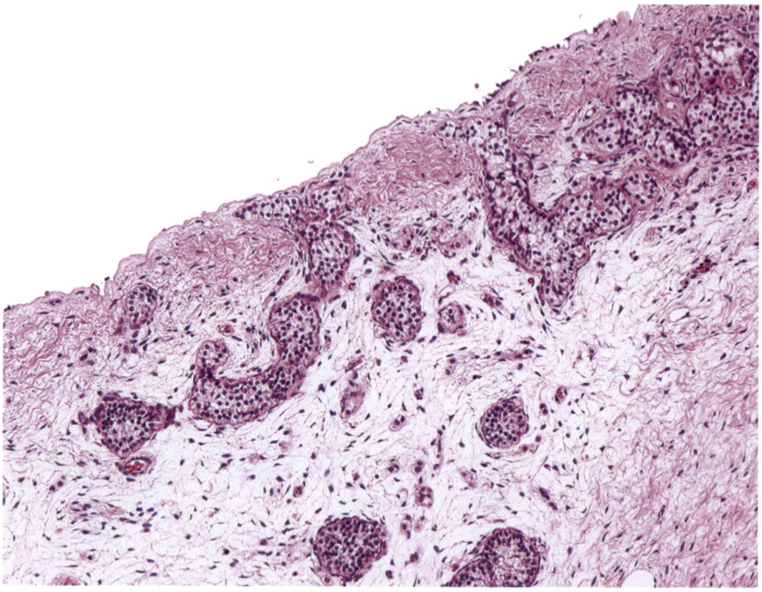

Fig. 18.12 Dysgenetic testis. This detail of the peripheral zone of the gonad shown in Fig. 18.10 reveals a thin, poorly collagenized albuginea without layers parallel to the surface and seminiferous tubules (epithelial cords) in its thickness (H&E stain)

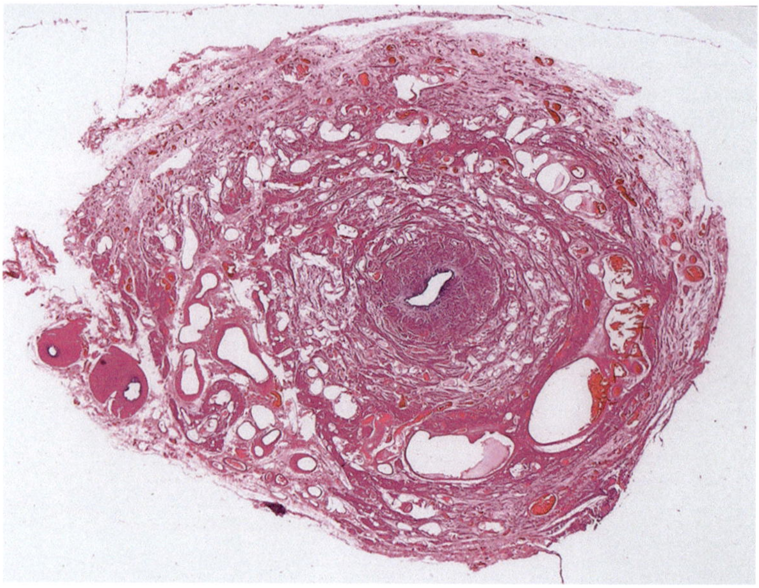

Fig. 18.13 Male with a uterus (46,XY). This adult has a normal testis on one side and a dysgenetic testis on the other associated with a hypoplastic uterus and two tubular structures corresponding to a vas deferens adjacent to a vascular plexus (H&E stain)

the peritubular cells and the basal membrane. Although most of the tubules are devoid of germ cells, as previously mentioned, some groups of tubules with undifferentiated germ cell neoplasia may be observed.

This gonad corresponds to that described by Jirasek [4] as "testicular dysgenesis" and is observed in the three syndromes classically included in male pseudohermaphroditism with müllerian rests: (a) mixed gonadal dysgenesis (Sohval syndrome or asymmetric gonadal differentiation), whose more frequent karyotypes are 45,X0/46,XY, 46,XY, and 45,X0/47,XYY; (b) male dysgenetic pseudohermaphroditism, which usually reveals karyotype 46,XY or 45,X0/46,XY; and (c) müllerian duct persistence syndrome (male with uterus), which has the 46,XY karyotype (Fig. 18.13).

Streak-Testis

Streak-testis has two intimately joined zones. The most voluminous part is formed by a dysgenetic testis with an ovarian cortex–like stroma close to the tunica albuginea. This stroma is in continuity with a streak gonad, which may or may not contain epithelial cords or some ovarian follicles. This is the gonad observed in some of the three previously mentioned syndromes: mixed gonadal dysgenesis, dysgenetic male pseudohermaphroditism, and müllerian duct persistence syndrome.

Basically, gonadal dysgenesis includes the following entities: 45,X0 gonadal dysgenesis, 46,XX gonadal dysgenesis, mixed gonadal dysgenesis, dysgenetic male pseudohermaphroditism, and persistent müllerian duct syndrome.

Ovotesticular Disorder (True Hermaphroditism)

Ovotesticular disorder of sexual development is an abnormal gonadal differentiation characterized by the presence of testicular parenchyma and ovary in the same individual (Figs. 18.14, 18.15, and 18.16). This disorder comprises fewer than 10 % of all DSDs [5]. The most frequent karyotype of true hermaphrodites is 46,XX (60 %), followed—in decreasing order of frequency—by the following chromosomal mosaicisms: 46,XX/46,XY, 46,XY/47,XXY, 45,X/46,XY, and 46,XX/47,XXY (33 %), and the 46,XY karyotype (7 %). These figures do not express the actual incidence of some karyotypes worldwide. Whereas in Europe a chromosomal mosaicism is found in 40.5 % of cases, in North America mosaicism appears in only 21 % of cases. Most true hermaphrodites in Africa (96.6 %) have the 46,XX karyotype. The phenotype of true hermaphroditism varies from female to normal male (10 %) [6]. The following gonadal types may be observed in hermaphrodites: ovotestis (44 %), ovary (33 %), and testis (22 %). Depending on the patient's gonads, true hermaphroditism may be classified as one of three types: *lateral hermaphroditism* (one gonad is a testis and the other an ovary [30 % of cases]), *bilateral hermaphroditism* (both gonads are ovotestes [20 %]), and *unilateral hermaphroditism* (one gonad is an ovotestis, the other is either a testis or an ovary [50 %]).

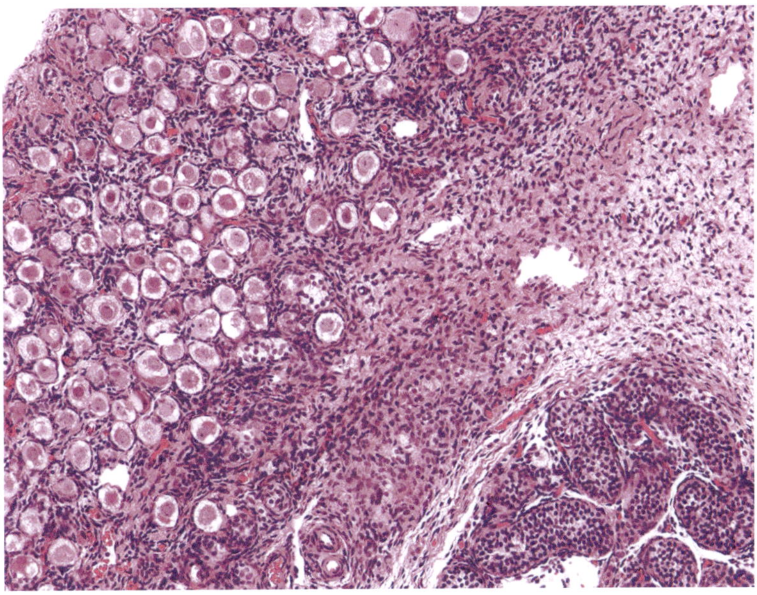

Fig. 18.14 Ovotestes. Shown is a specimen from a 1-year-old child with ovotesticular disorder (true hermaphroditism; 46,XY). Most of the gonad consists of ovarian parenchyma with many primordial ovarian follicles that cover some seminiferous tubules containing immature Sertoli cells. Germ cells are absent. Both parenchymas are fused (H&E stain)

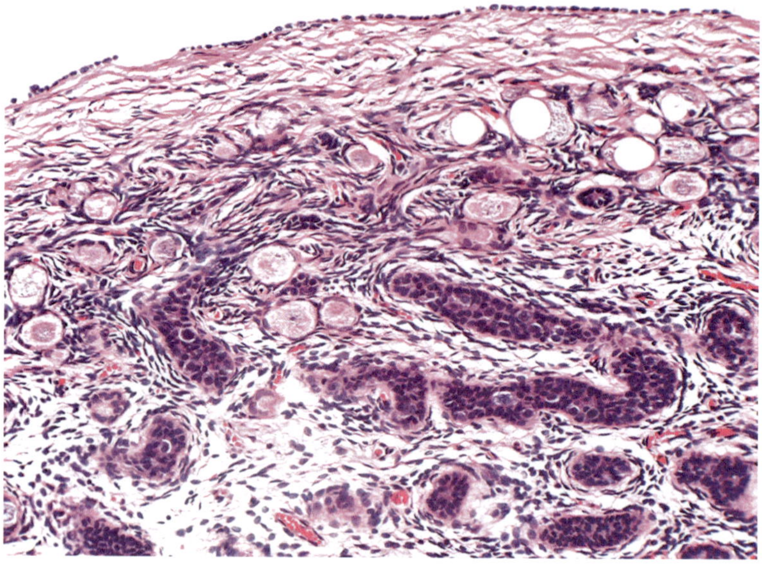

Fig. 18.15 Ovotestes. Shown is a specimen from a 2-year-old child with ovotesticular disorder (true hermaphroditism; 46,XY). The proportions of ovarian and testicular parenchyma vary greatly. In this particular case, only a small crown-shaped area of ovarian parenchyma with primordial follicles can be observed in an otherwise grossly normal testis (H&E stain)

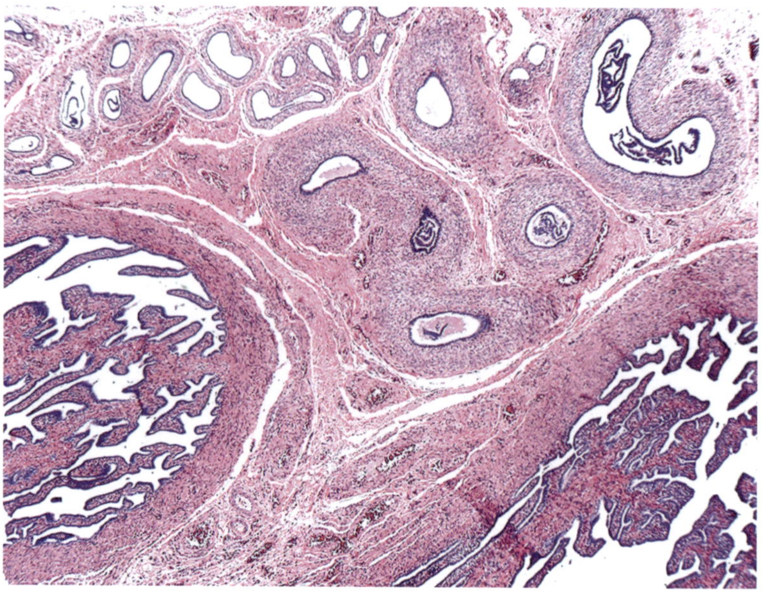

Fig. 18.16 Ovotesticular disorder. This section of müllerian and wolffian structures shows uterine fallopian tube and epididymal duct (H&E stain)

Male Undermasculinization

Normal male development requires adequate differentiation of the fetal testis, the synthesis and secretion of testicular hormones, and the correct action of these hormones on the target organs. AMH, secreted by Sertoli cells, inhibits the development of müllerian ducts toward uterus and Fallopian tubes. The testosterone produced by Leydig cells virilizes wolffian ducts. Transformation of testosterone into dihydrotestosterone by the enzyme 5α-reductase ensures the development of external genitalia. Alteration in some of these processes may lead to a syndrome of male undermasculinization.

Defective Leydig Cell Function

Defective Leydig cell function may present as one of two conditions: enzymatic defects

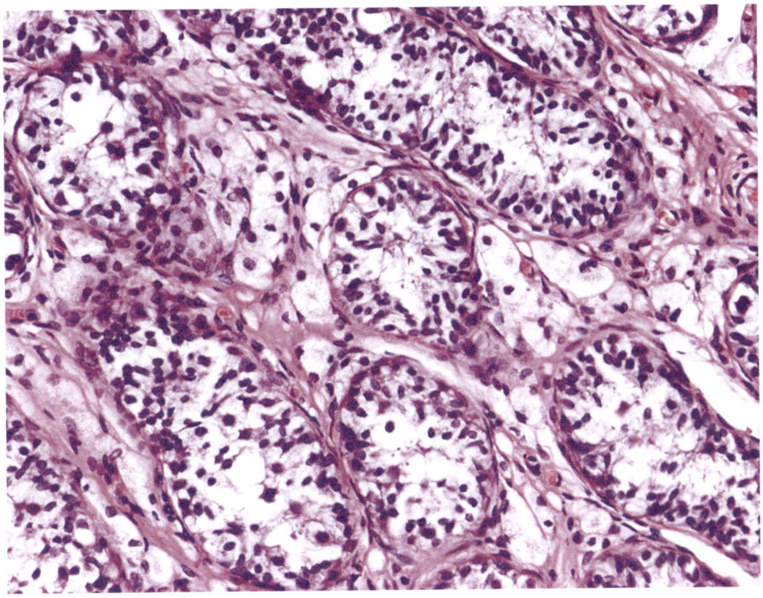

Fig. 18.17 Defective androgen synthesis (46,XY). This testis from a 3-year-old patient has seminiferous tubules, most of which are devoid of germ cells. Hypertrophic and hyperplastic Leydig cells stand out among seminiferous tubules. Leydig cells show clear and vacuolated cytoplasm and eccentric nuclei (H&E stain)

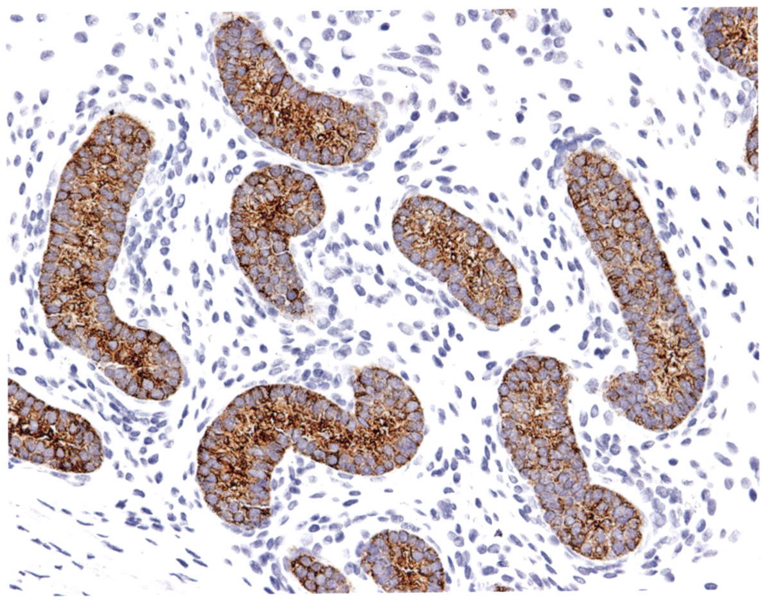

Fig. 18.18 Leydig cell hypoplasia (46,XY). This 6-month-old child has female external genitalia and undescended testes with no Leydig cells (Inhibin immunohistochemical stain)

leading to *insufficient androgen synthesis*, which comprises several inherited, autosomal recessive syndromes secondary to an error in testosterone biosynthesis, causing incomplete or absent virilization (Fig. 18.17); or *Leydig cell hypoplasia*, which is characterized by (a) female external genitalia in most cases; (b) absence of development of secondary sex characteristics (male or female) at puberty; (c) absence of a uterus and Fallopian tube while an epididymis and a vas deferens are present; (d) 46,XY karyotype; (e) low serum testosterone levels despite increased gonadotropin levels; (f) absence of response to human chorionic gonadotropin; (g) absence of an enzymatic defect in testosterone biosynthesis; and (h) small, undescended testes that appear grayish and mucous on section (Fig. 18.18).

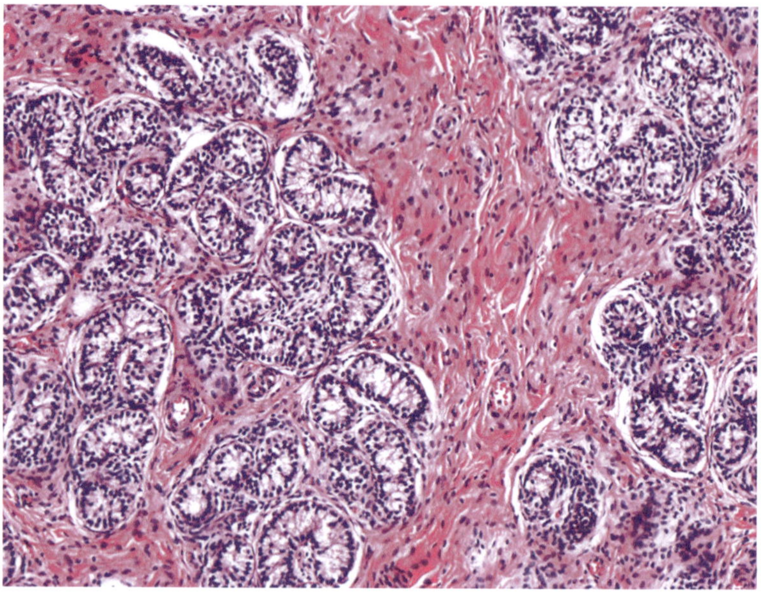

Fig. 18.19 Complete androgen insensitivity syndrome (46,XY). This 6-year-old boy has female external genitalia and undescended testes. Seminiferous tubules with scanty germ cells separated by abundant fibrous and cellular interstitium are present (H&E stain)

Defective Peripheral Action of Androgens

Androgens exert their function on differentiation and development of normal male phenotype via a single receptor protein, the androgen receptor (AR). Deficiencies in the peripheral action of androgens comprise the syndromes of androgen insensitivity (complete and incomplete testicular feminization syndrome) and a defect in 5α-reductase.

Syndromes of Androgen Insensitivity

Patients with syndromes of resistance to androgens comprise a group with a male genotype (46,XY) whose phenotype varies from female (testicular feminization) to normal male (the only reason for consultation is infertility), with intermediate degrees of undermasculinization between both phenotypes [7]. Based on phenotype, androgen insensitivity comprises three main forms: complete testicular feminization (CAIS; Figs. 18.19, 18.20, 18.21, 18.22, 18.23, and 18.24), partial testicular feminization (PAIS; Fig. 18.25), and slight or minimal androgen insensitivity (MAIS; Fig. 18.26). Androgen insensitivity syndromes are the effect of a total or partial resistance of target organs to androgen action [8]. This resistance is a result of the absence, decrease, or abnormality of the intracellular AR, or of postreceptor anomalies. In both complete and partial forms in adults, the testes vary in size from small to voluminous. These testes contain small seminiferous tubules, which are devoid of lumen and lined by Sertoli cells only. One third of patients show isolated germ cells. The testicular interstitium shows a fibrous tissue that resembles ovarian stroma. Two thirds of testes from CAIS patients show several bilateral whitish or yellowish nodules. These nodules are well delimited from the remaining testicular parenchyma and consist of small seminiferous tubules surrounded by a tunica propria that usually presents a certain degree of hyalinization and lacks elastic fibers. The Sertoli cells of these nodules are immature, and the interstitial Leydig cells are numerous. These nodules are considered tubulointerstitial hamartomas. About one fourth of testes from CAIS patients present Sertoli cell adenomas consisting of fetal- or infantile-like seminiferous tubules, although germ cells and peritubular myofibroblasts are lacking. About 80 % of patients have thick smooth muscle cell bundles in a paratesticular location (smoth muscle hamartoma).

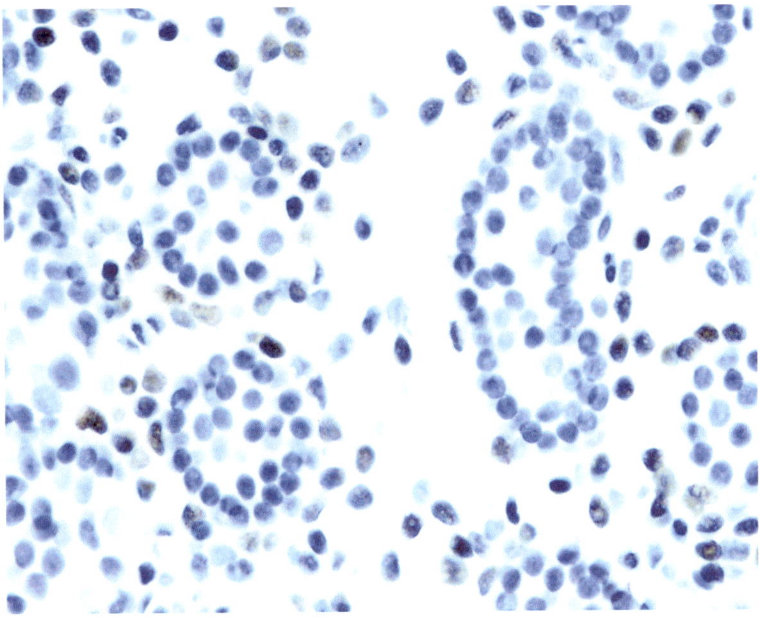

Fig. 18.20 Complete androgen insensitivity syndrome. The same case depicted in Fig. 18.19 shows positive expression in the nuclei of peritubular and immature interstitial cells and negative expression in all Sertoli cells (AR immunostain)

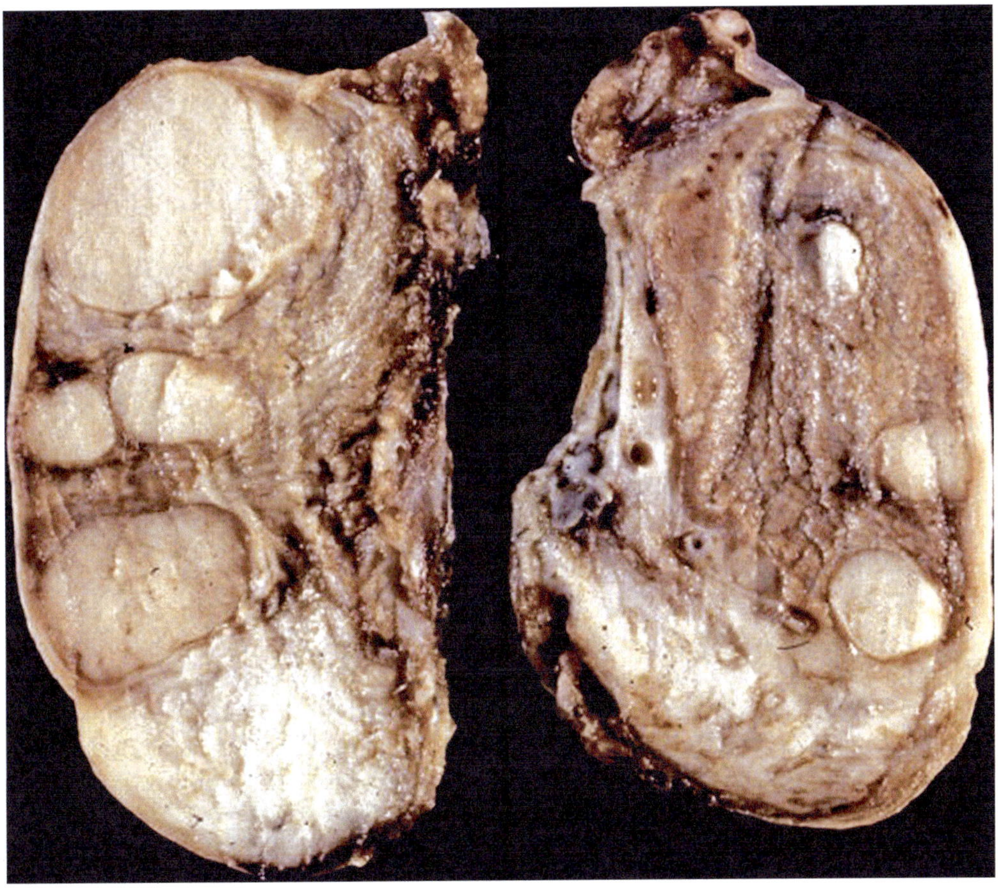

Fig. 18.21 Complete androgen insensitivity syndrome (46,XY), adult testis; transverse section of an intra-abdominal gonad. The testis has several whitish well-delimited nodules and a cap of tissue different from testicular parenchyma in the lower pole (smoth muscle hamartoma)

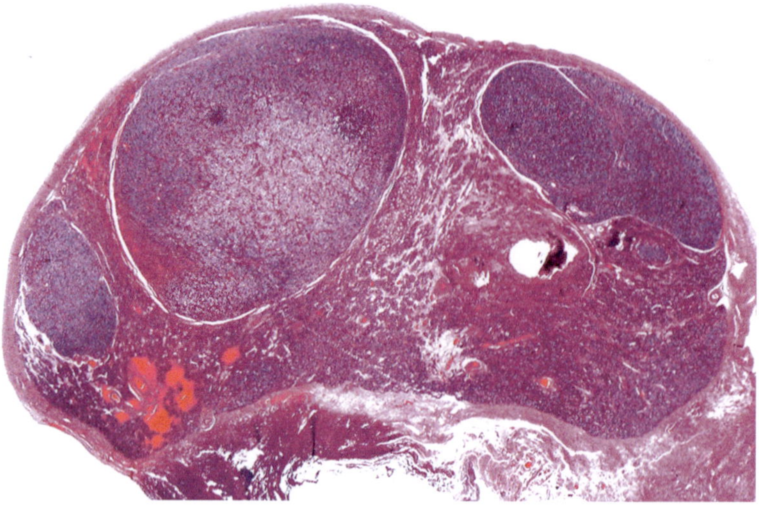

Fig. 18.22 Complete androgen insensitivity syndrome (46,XY), adult. The testicular parenchyma has three well-delimited unencapsulated nodules corresponding to tubulointerstitial hamartomas (H&E stain)

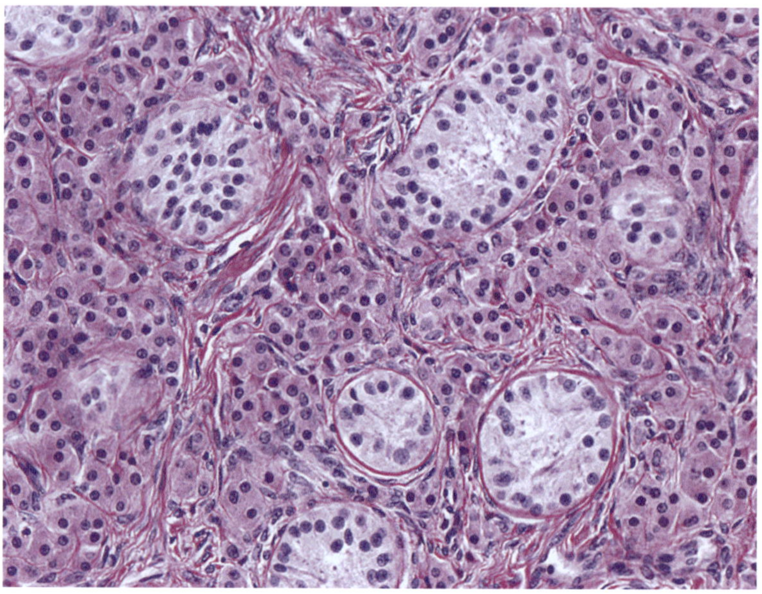

Fig. 18.23 Complete androgen insensitivity syndrome (46,XY), adult; tubulointerstitial hamartoma. Seminiferous tubules show prepubertal maturation and eosinophilic bodies in the apical pole of Sertoli cells. The interstitium shows Leydig cell hypertrophy and hyperplasia in a densely collagenized stroma (H&E stain)

Defect in 5α-Reductase

5α-Reductase defect, a rare type of male undermasculinization in males with karyotype 46,XY, is characterized by a defect in the enzyme responsible for converting testosterone to dihydrotestosterone and occurs in peripheral tissues as well as the external genitalia and prostate. Infants with this disorder have a clitoriform penis, a bifid scrotum, and a vagina that opens into the urethra or urogenital sinus. The ejaculatory ducts open into the vagina, and the testes are located in the inguinal duct or labioscrotal folds. The differential diagnosis is incomplete androgen insensitivity, because both diseases may show a similar phenotype and elevated basal testosterone levels. The presence of well-developed androgen-dependent structures

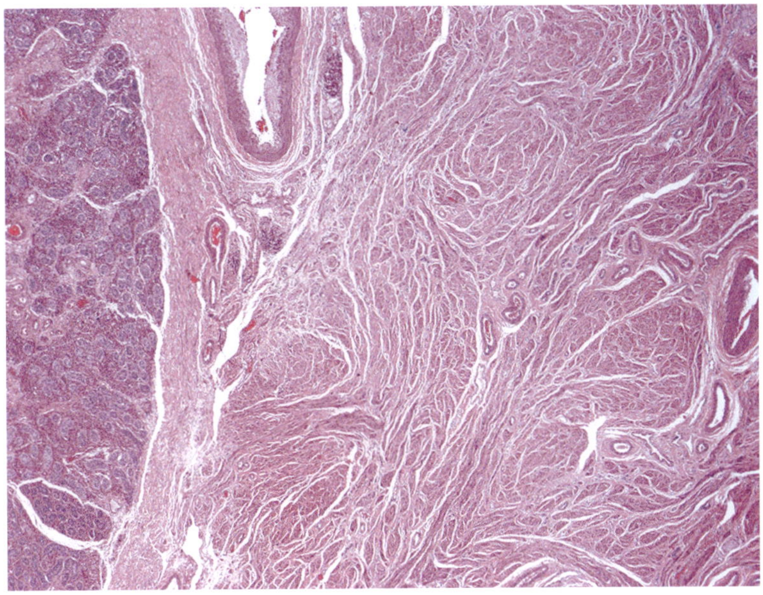

Fig. 18.24 Complete androgen insensitivity syndrome (46,XY), adult. Shown is a smooth muscle hamartoma next to the albuginea and a testis with immature seminiferous tubules and Leydig cell hyperplasia (H&E stain)

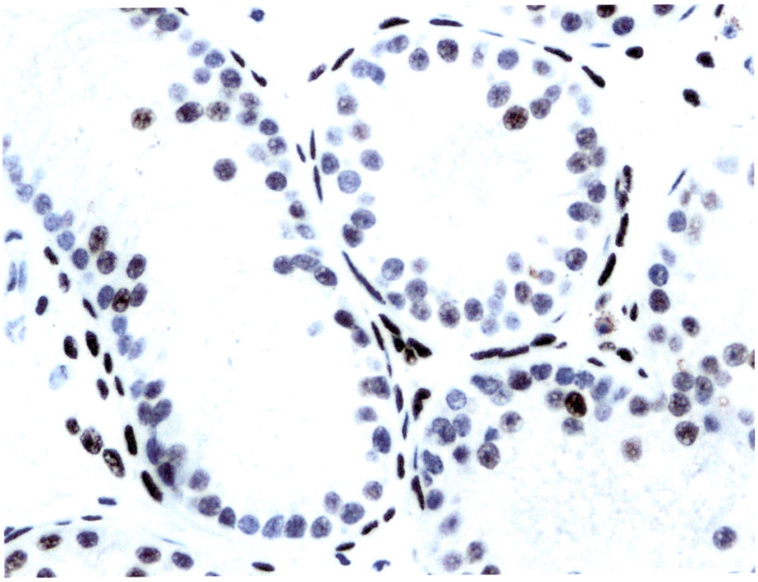

Fig. 18.25 Partial androgen insensitivity syndrome (46,XY), adult. The seminiferous tubules show prepubertal maturation (absence of triangular nuclei with prominent nucleoli in Sertoli cells). Note the positive expression in some isolated Sertoli cells, as well as peritubular cell expression in all the seminiferous tubules (H&E stain)

(epididymis, vas deferens) supports the diagnosis of 5α-reductase deficit. During the first years of life, the testes may appear normal. From puberty, however, important changes occur. The seminiferous tubules show a reduced diameter, with most lacking a lumen; the seminiferous epithelium consists of Sertoli cells, which are increased in number and show incomplete maturation. Spermatogenesis is exceptional. The testicular interstitium contains an increased number of Leydig cells, which show an apparently normal morphology.

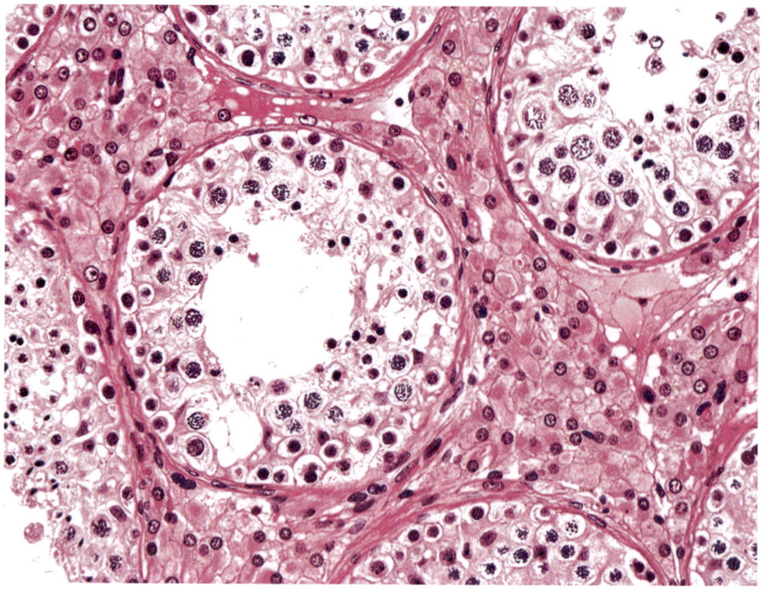

Fig. 18.26 Minimal androgen insensitivity syndrome (46,XY), adult. This image is from an infertile patient with oligoasthenoteratozoospermia. The seminiferous tubules show complete germ cell maturation, although it is quantitatively abnormal. Leydig cell hyperplasia and hypertrophy are present in the interstitium (H&E stain)

References

1. Lee PA, Houk CP, Ahmed SF, Hughes IA, International Consensus Conference on Intersex organized by the Lawson Wilkins Pediatric Endocrine Society and the European Society for Paediatric Endocrinology. Consensus statement on management of intersex disorders. International Consensus Conference on Intersex. Pediatrics. 2006;118:e488–500.
2. Damiani D, Guerra-Júnior G. New definitions and classifications of the intersexual states: in which the Chicago Consensus has contributed to the state of the art? Arq Bras Endocrinol Metabol. 2007;51:1013–7.
3. Cools M, Stoop H, Kersemaekers AM, Drop SL, Wolffenbuttel KP, Bourguignon JP, et al. Gonadoblastoma arising in undifferentiated gonadal tissue within dysgenetic gonads. J Clin Endocrinol Metab. 2006;91:2404–13.
4. Jirasek J. Principles of reproductive embryology. In: Simpson JL, editor. Disorders of sexual differentiation. New York: Academic Press; 1976. p. 51–111.
5. Krstic ZD, Smoljanic Z, Vukanic D, Varinac D, Janjic G. True hermaphroditism: 10 years' experience. Pediatr Surg Int. 2000;16:580–3.
6. Yordam N, Alikasifoglu A, Kandemir N, Caglar M, Balci S. True hermaphroditism: clinical features, genetic variants and gonadal histology. J Pediatr Endocrinol Metab. 2001;14:421–7.
7. Brinkmann AO. Molecular basis of androgen insensitivity. Mol Cell Endocrinol. 2001;179:105–9.
8. Imperato-McGinley J, Canovatchel WJ. Complete androgen insensitivity: pathophysiology, diagnosis and management. Trends Endocrinol Metab. 1992;3:75–81.

Index

D. Ježek (ed.), *Atlas on the Human Testis*,
DOI 10.1007/978-1-4471-2763-5, © Springer-Verlag London 2013

MIX
Papier aus verantwortungsvollen Quellen
Paper from responsible sources
FSC® C105338

If you have any concerns about our products,
you can contact us on
ProductSafety@springernature.com

In case Publisher is established outside the EU,
the EU authorized representative is:
Springer Nature Customer Service Center GmbH
Europaplatz 3, 69115 Heidelberg, Germany

Printed by Libri Plureos GmbH
in Hamburg, Germany